The Automotive Chassis

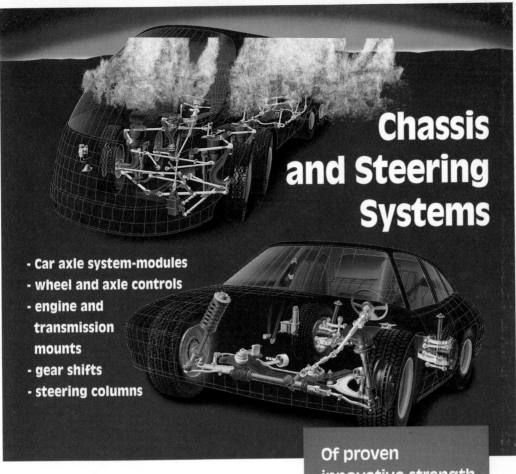

Chassis and Steering Systems

- Car axle system-modules
- wheel and axle controls
- engine and transmission mounts
- gear shifts
- steering columns

Of proven innovative strength, we are the natural value adder partner of automobile builders with close customer contact worldwide.

136 858

The Automotive Chassis: Engineering Principles

Types of drive and suspension
Tyres and wheels
Axle kinematics
Steering
Springing
Chassis and vehicle overall

Prof. Dipl.-Ing. Jörnsen Reimpell
Dipl.-Ing. Helmut Stoll

ARNOLD

A member of the Hodder Headline Group
LONDON • SYDNEY • AUCKLAND

First published in Great Britain 1996 by
Arnold, a member of the Hodder Headline Group,
338 Euston Road, London NW1 3BH

Original copyright 1986 Vogel-Buchverlag, Würzburg
Second German edition 1995

English edition © 1996 Arnold

British Library Cataloguing in Publication Data
A catalogue record for this book is available from the British Library

ISBN 0 340 61443 9 (hb)

Typeset in 10/11 Times by Wearset, Boldon, Tyne and Wear
Printed and bound in Great Britain by St Edmundsbury Press,
Bury St Edmunds, Suffolk and Hartnolls Ltd., Bodmin, Cornwall

Contents

Preface

The book series 'The Automotive Chassis' first published in the Federal Republic of Germany in 1970, covers all aspects of the chassis, beginning with a description of the design and construction of the various assemblies, deals with their function and the way they work and covers their behaviour and interaction in the vehicle as a whole.

'The Engineering Principles' volume was first published in Germany in 1986; 1995 sees its third edition. This volume gives a comprehensive outline of the current state of the art in vehicle chassis engineering. A great deal of interest in this volume from throughout Europe, America and Japan has prompted publication of this foundation volume in English.

Descriptions are clear and easy to understand, with example calculations, cross-references and the inclusion of more than 400 illustrations, which clearly show the relationship between the vehicle parts; this layout has, over the years, proved itself as the best way of imparting the necessary information. The authors' many years of experience in chassis engineering back up the practical bias of the description and will help engineers, inspectors, students and technicians in companies operating in the automotive industry and its suppliers to understand the context. The comprehensive index of key words and many cross-references make this book an invaluable work of reference.

Acknowledgement

We should like to thank Dipl.-Ing. Uwe Köttgen of Michelin Tyres for his help in preparing the introduction to international regulations and standards in Chapter 2 'Tyres and wheels'.

Jörnsen Reimpell
Cologne
Helmut Stoll
Nauheim/Rüsselsheim

Symbols and dimensions

The index of the symbols and dimensions follows the international standards:

ISO 31   Quantities
ISO 2416   Passenger cars – Mass distribution
ISO 8855   Road vehicles – Vehicle dynamics and road holding ability – Vocabulary
ISO 1000   SI – Units and recommendations
SAE J670e   Vehicle dynamics terminology

and the German standards:

DIN 1301   Einheiten
DIN 1304   Formelzeichen
DIN 70 000   Straßenfahrzeuge, Begriffe der Fahrdynamik

In a few cases, the connection between the various standards could not be achieved. In these cases priority was given to the ISO standards or specific suffixes or symbols have been selected.

1 Reference points in figures

B   body centre of gravity
C to H   reference points, in general
M   centre point
N   centre of tyre contact
O   pitchpole
P   rollpole
Q   centre of driving joint
R   roll centre
S   vehicle centre of gravity
T and U   tie rod or linkage point
U   wheel centre point

Some points can additionally have an f (front) or r (rear) or a numerical figure for further distinction.

2 Suffixes

The multitude of symbols requires the usage of suffixes for clear identification. In cases where more than one is needed, a comma is set between them. A few cases, where small letters follow capital ones, deviate from this rule. The various suffixes have the following meanings.

τ	the caster
φ	the body roll angle
0	the zero-point position, starting point
1	to the top, in jounce, in compression
2	to the bottom, in rebound
A	the drive-off condition, accelerating in general or the acceleration of both wheels or the overall vehicle
A	Ackermann steer angle
a	driven, accelerating (one wheel only)
ax	axial
B	the braking (overall vehicle)
b	braking (one wheel only)
b	the baggage, the luggage
Bo	the body
c	the inertia
co	the cornering
D	the axle drive
d or D	the damping
dr	drivable
dyn	dynamic
e	due to the elasticity or compliance
f	front
fix	fixed, idle
fr	the friction
H	the steering wheel
h	the hand
hyd	hydraulic
i	inside of curve, inner wheel
i	the excess, excessive
k	kinematic
l	left, left side
lo	the loaded condition
M	the motor, the engine
m	the mass
m or med	middle, mean, medium
max	maximum
min	minimum

O	orifice closing plate in shock absorbers
o	outside of curve, outer wheel
Ov	the overhang
p	the passenger, the person
pe	permissible, maximum authorized
Pi	the piston
pl	partial loaded
Pr	the piston rod
R	rolling (wheel)
r	rear
rad	radial
Re	residual, remaining
Ro	the body roll centre
rs	right, right side
rsl	resulting
S	the steering
S	the anti-roll bar, the stabilizer
s	side-..., lateral
sl	the slipping, sliding
Sp	the spring
stat	static
T	the tyre
T	the rod, the tie rod, the linkage
t	total
t	the depth
Th	the trailer hitch
Tr	the single-axle trailer
tr	transportable
u	the unsprung weight (axle weight)
ul	unloaded, the empty condition
V	the overall vehicle
W	the wheel
x	the longitudinal direction
y	the lateral direction
z	the vertical direction
z, W	vertical, in the centre of tyre contact

3   For lengths and distances in mm, cm or m

a to r	distances and lengths in general
Δb	tread width or track offset at rigid axle
b, B	width, breadth in general
b_D	distance of shock absorber (damper) attachment points (at solid axles)
$b_{f\,or\,r}$	tread width front or rear
b_s	distance of anti-roll bar (stabilizer) attachment points (at solid axles)

b_{Sp}	effective spring distance (at solid axles)
C_R	dynamic rolling circumference at 60 km h^{-1}
$C_{R,dyn}$	dynamic rolling circumference at top speed
d or D	diameter, in general
d_O	diameter of orifice or closing plate
D_S	track circle diameter, path radius
D_{tc}	turning circle diameter, wall to wall
$D_{tc,cb}$	turning circle diameter, kerb to kerb
e	wheel offset, rim offset
f	diagonal spring travel
$h_{Ro,f\,or\,r}$	height of roll centre at front or rear axle
h,H	height, in general
h_{Bo}	height of body centre of gravity
h_{ul}	height of the unloaded vehicle
l	wheel base
$l_{S,f\,or\,r}$	distance of vehicle centre of gravity to centre of front or rear axle
$l_{Bo,f\,or\,r}$	distance of body centre to the centre of the front or rear axle
L	distance, length in general
L_{fix}	idle length of the shock-absorber
L_t	total length of the vehicle
$n_{\tau,k}$	kinematic lateral force lever arm due to caster ($n_{\tau,k} = r_{dyn} \sin \tau$)
$n_{\tau,t}$	lateral force lever arm, in total ($n_{\tau,t} = n_{\tau,k} + r_{\tau,T} \cos \tau$)
OD_T	outer diameter of the tyre
R	radius
r	effective control arm length or force lever in general
r_Δ	static toe-in (one wheel only)
$r_{\Delta,t}$	total static toe-in (both wheels)
r_τ	caster offset at ground, in general
$r_{\tau,e}$	elastokinematic caster offset at ground
$\pm r_{\tau,k}$	kinematic caster offset at ground (positive or negative)
$r_{\tau,T}$	caster offset of tyre
$r_{\tau,t}$	overall caster offset at ground ($r_{\tau,t} = r_{\tau,k} + r_{\tau,T}$)
$\pm r_{\tau,W}$	caster offset at wheel centre (positive or negative)
r_a	force lever of longitudinal or tractive force
r_b	force lever of brake force
r_{dyn}	dynamic rolling radius (tyre)
r_n	force lever of vertical (normal) force
r_S	kingpin offset at ground (scrub-radius)
$r_{S,t}$	total kingpin offset at ground ($r_{S,t} = \pm r_S - r_T$)
r_{stat}	static loaded radius of the tyre
r_T	force offset in the centre of tyre contact
s	travel or stroke, in general
s_1	wheel travel in jounce
s_2	wheel travel in rebound
s_{Re}	residual wheel travel
s_T	static tyre deflection
s_t	total wheel travel
t	the pitch of gear wheels

4   For masses, loads and weights in kg

m	mass, load and weight in general
$m_{1,Bo,f\,or\,r}$	part of body mass on one side of the front or rear axle
$m_{1,f\,or\,r}$	weight of one side of front or rear axle
m_b	mass of luggage (baggage) related to one passenger
m_{Bo}	vehicle body weight
$m_{Bo,f\,or\,r}$	part of body mass on front or rear axle
m_p	mass of one passenger
m_t	nominal design pay mass (minimum required)
$m_{t,pe}$	permissible payload
m_{Th}	weight of the trailer hitch
m_{Tr}	trailer load
Δm_{Tr}	tongue load, trailer
m_{tr}	nominal mass of transportable goods
$m_{u,f\,or\,r}$	unsprung axle mass front or rear
Δm_V	weight of vehicle options
$m_{V,dr}$	weight of drivable vehicle (with driver)
$m_{V,f\,or\,r}$	axle load front or rear
$m_{V,max,f\,or\,r}$	maximum permissible axle load front or rear
$m_{V,t}$	gross vehicle weight (GVW)
$m_{V,t,max}$	maximum gross vehicle weight
$m_{V,ul}$	kerb weight (actual weight of unloaded vehicle)
$m_{V,ul,O}$	kerb weight as published by the car manufacturer
m_W	weight of one wheel

5   Forces in N or kN

A lower-case subscript letter after the symbol F means that the force refers only to one side of the axle; an upper-case letter refers to the whole axle. An exception is F_R, the rolling resistance of the tyre. However, this can also refer to a wheel, an axle or the whole vehicle; the subsequent further subscript enables the difference to be recognized. The forces at the reference points, or at the links B to U of the wheel suspension, are denoted by the letter of that particular point and the direction: see Fig. 3.1.

dF	change of force
F	force, in general
F_1	compressive force
F_2	rebound force
F_A	traction (accelerating) force in the centre of tyre contact of both wheels
F_a	traction (accelerating) force in the centre of tyre contact of one wheel
F_b	brake force in the centre of tyre contact of one wheel
$F_{B,Bo,f\,or\,r}$	brake reaction force to the body, front or rear
$\Delta F_{Bo,f\,or\,r}$	body lift or dive differential force (front or rear) during braking

$F_{\text{B,f or r}}$	brake force in the centre of tyre contact of both wheels front or rear
$F_{\text{B,t}}$	brake force at the centre of gravity of the vehicle
$F_{\text{B,u,f or r}}$	brake reaction force to the front or rear axle
F_{Bo}	static body weight (force), ($F_{\text{Bo}} = m_{\text{Bo}}\, g$)
F_{c}	centrifugal force, in general
$F_{\text{c,Bo}}$	centrifugal force at the body centre
F_{E}	spatial force at point E
F_{Ex}	component of F_{E} in the driving direction
F_{Ey}	component of F_{E} in the lateral direction
F_{Ez}	component of F_{E} in the vertical direction
F_{fr}	friction force in general or related to one side of the axle
F_{H}	steering wheel force
$F_{\text{i},x}$	excessive force, inertia force in the x-direction
F_{L}	aerodynamic drag
F_{ml}	force of maximum payload
F_{Pi}	piston rod extensive force
F_{R}	rolling resistance of the tyre
F_{Sp}	spring force (one side of the axle)
F_{T}	rod force (tie or pushrod)
$F_{\text{u,f or r}}$	axle weight of the front or rear axle
ΔF_{V}	axle load (force) transfer
$F_{\text{V,f or r}}$	axle load (force) front or rear
$F_{\text{V,f,dyn or r,dyn}}$	dynamic axle load (force) front or rear $(F_{\text{V,f,dyn or r,dyn}} = F_{\text{V,f or r}} \pm \Delta F_{\text{V}})$
$F_{\text{V,t}}$	gross vehicle weight (force), ($F_{\text{V,t}} = m_{\text{V,t}}\, g$)
F_x	longitudinal force
F_y	lateral force or disturbance force in lateral direction
$F_{y,\text{W}}$	lateral (side) force at the centre of tyre contact
F_z	vertical force the small x, y and z indicate that only one side of the axle is observed
$F_{z,\text{W}}$	vertical force at the centre of tyre contact
$\Delta F'_{z,\text{W}}$	change of vertical force
$F'_{z,\text{W}}$	vertical force without the axle weight of one axle side $(F'_{z,\text{W}} = F_{z,\text{W}} - m_1\, g)$

6 Spring rates in N mm^{-1} and moments in N m^{-1}

c_φ	rate of the body supporting spring at reciprocal springing related to the centre of tyre contact
$c_{\text{f or r}}$	rate of the body supporting spring at parallel springing, related to the centre of tyre contact of one axle side, front or rear
c_{S}	rate of the anti-roll bar (stabilizer) at reciprocal springing
$c_{\text{S},\varphi}$	rate of the anti-roll bar related to the centre of tyre contact
c_{Sp}	static rate of the spring
c_{T}	spring rate of the tyre

M_A	driving moment related to one axle
M_B	braking moment related to one axle
M_{fr}	friction moment
M_H	steering wheel moment
M_M	engine (motor) moment
M_R	rolling resistance moment
$M_{S,A}$	wheel self-aligning moment due to the drive torque
$M_{S,a\,or\,b}$	kingpin moment due to braking or driving torque
$M_{S,x}$	wheel aligning moment due to longitudinal force
$M_{S,y}$	wheel aligning moment due to lateral force
$M_{S,z}$	wheel aligning moment due to vertical force
$M_{T,f}$	tyre aligning moment at the front wheels
$M_{T,r}$	tyre aligning moment at the rear wheels
M_z	yawing moment
T	torsional moment, torque

7   For angles in degrees or radians

α	torsional angle of a joint or bushing
α	top view angle of the semi-trailing arm twist axis
α	angle of gradient of the road
α, α'	inclination (rear view) angle of upper control arm (double wishbone axle)
$\alpha_{f\,or\,r}$	slip angle of the front or rear wheel
β	attitude angle, sideslip angle
β	rear view angle of the semi-trailing arm twist axis
β, β'	inclination (rear view) angle of lower control arm (double wishbone or McPherson axles)
β'	driving angle of the axle
$\Delta_{f\,or\,r}$	static toe-in angle of front or rear wheels
δ_A	Ackermann steer angle
$\delta_{A,o}$	Ackermann steer angle, nominal value at outside of curve
δ_H	steering wheel angle
δ_m	mean steering angle
$\delta_{o\,or\,i}$	actual steer angle, outside or inside of curve
$d\delta$	change of steering angle of both wheels
$\Delta\delta$	differential steer angle (actual value $\Delta\delta = \delta_i - \delta_o$)
$\Delta\delta_A$	differential steer angle according to Ackermann (nominal value $\delta_A = \delta_i - \delta_{A,o}$)
$\Delta\delta_e$	part of steer angle due to compliances (elasticities)
$\Delta\delta_F$	steering flaw
$\Delta\delta_{H,e}$	part of steering wheel angle due to compliances (elasticities)
$\Delta\delta_{H,h}$	part of steering wheel angle due to manual steer by hand
$\Delta\delta_{H,Re}$	residual angle at the steering wheel
$\Delta\delta_{H,S}$	mean part of steer angle of both wheels due to steer
ε	brake reaction support angle

$\Delta\varepsilon_W$ or $d\varepsilon_W$	change of camber angle
$\Delta\varepsilon_{W,\varphi}$	roll-camber angle
ξ	top view angle between two control arms or rods
$\theta_{A \text{ or } B}$	vehicle pitch angle under accelerating or braking
κ	acceleration reaction support angle or diagonal springing angle
λ	steering arm angle
σ	kingpin inclination angle
τ or τ_f	caster angle of the (steered) front wheels
τ_r	caster angle at rear wheels (not steered)
φ_V	body (vehicle) roll angle
$d\varphi_{V,k}$	kinematic change of body roll angle

8 Characteristics and data with no dimensions

η	efficiency
$\Phi_{f \text{ or } r}$	brake force fraction front or rear ($\Phi_{f \text{ or } r} = F_{B,f \text{ or } r} / F_{B,t}$)
μ_{rsl}	resulting coefficient of friction ($\mu_{rsl} = (\mu^2_{x,W} + \mu^2_{y,W})^{\frac{1}{2}}$)
$\mu_{x,sl}$	coefficient of sliding friction in longitudinal direction
$\mu_{x,W}$	coefficient of static friction in longitudinal direction
$\mu_{y,sl}$	coefficient of sliding friction in lateral direction
$\mu_{y,W}$	lateral coefficient of friction at rolling wheel
i_φ	ratio of the wheel to the spring, shock absorber or anti-roll bar at reciprocal springing
$i_{Bo \text{ or } V}$	inertia radius of body or vehicle centre of gravity
i_D	axle differential ratio
i_D	ratio of shock absorber (damper) to the wheel
i_{dyn}	dynamic steering ratio
i_G	gearbox ratio
i_m	mass ratio
i_R	roll resistance ratio coefficient
i_S	overall kinematic steering ratio
$i_{S'}$	steering gear ratio
i_{Sp}	ratio of spring to the wheel
i_{ul}	ratio of vehicle centre of gravity to height of the empty vehicle
k_ε	anti-dive coefficient, braking
$k_{\varepsilon W,\varphi}$	roll camber coefficient ($k_{\varepsilon W,\varphi} = d\varepsilon_W / d\varphi_V$)
k_κ	anti-dive coefficient, accelerating
k_μ	friction coefficient correction factor
k_b	ratio of tread to width (breadth)
k_l	ratio of wheel base to vehicle length
k_m	load factor
k_R	rolling resistance coefficient
$k_{R,co}$	rolling resistance coefficient when cornering
k_T	factor of the increase in tyre spring rate
k_v	velocity factor
$S_{x,a \text{ or } b}$	longitudinal slip under accelerating or braking

| S_y | lateral slip | |
| z | braking factor ($z = a_x/g$) | |

9   Other symbols with dimensions

ε	ductile yield, elongation at rupture	%
ω	circular frequency	Hz
A	area, cross-section area	m^2
a_x	longitudinal acceleration or deceleration	$m\,s^{-2}$
a_y	lateral acceleration	$m\,s^{-2}$
C_H	steering elasticity value	$Nm\,rad^{-1}$
E	modulus of elasticity	$N\,mm^{-2}$
f	frequency	Hz
g	acceleration due to gravity	$m\,s^{-2}$
HRC	Rockwell hardness	–
I	area moment of inertia	cm^4
J_{Bo}	dynamic moment of inertia of body	$kg\,m^2$
J_V	dynamic moment of inertia of vehicle	$kg\,m^2$
n	revolutions per minute, vibration frequency	min^{-1}
n	number of specified seats	–
n_0	number of seats engaged	–
p_{hyd}	hydraulic pressure	$N\,cm^{-2}$
p_T	tyre pressure	bar or psi
	(1 bar = 0.1 N mm^{-2} = 0.1 MPa = 14.5 psi)	
q	climbing capability factor ($q = \tan \alpha\ 100$)	%
R_e	yield strength	$N\,mm^{-2}$
R_m	tensile strength	$N\,mm^{-2}$
$R_{p0.2}$	0.2% yield strength	$N\,mm^{-2}$
R_t	surface roughness	µm
v	velocity, vehicle speed	$m\,s^{-1}$ or $km\,h^{-1}$
$v_{D,med}$	mean piston velocity in the shock absorber	$m\,s^{-1}$
v_W	circumferential tyre velocity	$m\,s^{-1}$ or $km\,h^{-1}$

1

Types of drive and suspension

This chapter deals with the aspects directly related to drives and suspensions.

1.1 Front and rear axles – general

There is a difference between whether a suspension is used as a steerable front axle or as a rear axle, and whether or not this contains the drive. A further distinction is drawn between rigid axles and independent wheel suspensions. The latter include

- double wishbone suspensions and
- McPherson struts,

which require only a little space at the side, i.e. in the centre of the vehicle (e.g. for the engine or the axle drive, see Fig. 1.0) which makes the steering angle possible, as well as

- trailing link axles and semi-trailing link axles,

which take up little vertical space, making it possible to have a wide boot with a flat floor, but which can have considerable diagonal springing (see Section 5.4.4).

Half-way between the rigid axle and independent wheel suspension described below is the compound crank axle. This is an extremely space-saving component and, apart from a few high-roof cars, is only fitted to front-wheel drive vehicles as the rear suspension (Figs 1.1 and 1.41). On all rigid axles, the axle casing also moves over the entire spring travel. Consequently, the space that has to be provided above this reduces the boot at the rear and makes it more difficult to house the spare wheel. At the front, the axle casing would be located under the engine, and to achieve sufficient jounce travel the engine would have to be raised or moved further back. For this reason, rigid front axles are found only on commercial vehicles and four-wheel drive, general-purpose passenger cars (Figs 1.1a and 1.49).

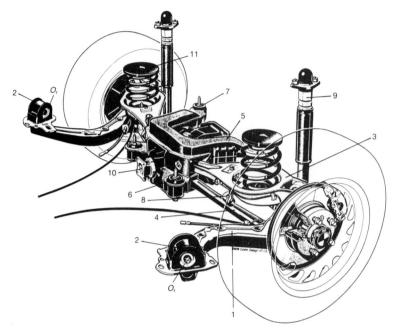

Fig. 1.0 A multi-link rear axle, a type of suspension system, which is progressively replacing the semi-trailing link axle, and consists of a trailing link on each side. This link is guided by two (or even three) transverse control arms (Figs 1.45 and 1.59). The trailing link simultaneously serves as a wheel hub carrier and (on four-wheel steering) allows the minor angle movements required to steer the rear wheels. The main advantages are, however, its good kinematic and elastokinematic characteristics.

BMW calls the design shown in the illustration and fitted in the 3-series a 'central link axle'. The trailing links 1 are made from cast iron with spherical graphite. They absorb all longitudinal forces and braking moments and transfer them via the points 2 – the centres of which also form the pitch poles O$_r$ (Figs 3.122 and 3.123) – on the body. The lateral forces generated at the centre of tyre contact are absorbed at the subframe 5, which is fastened to the body with four rubber mountings (items 6 and 7) via the transverse control arms 3 and 4. The upper arms 3 carry the minibloc springs 11 and the joints of the anti-roll bar 8. Consequently, this is the place where the majority of the vertical forces are transferred between the axle and the body.

The shock absorbers, which carry the additional polyurethane springs 9 at the top (Fig. 5.35), are fastened in a good position behind the axle centre at the ends of the trailing links. For reasons of noise, the differential 10 is attached elastically to the subframe 5 at three points. When viewed from the top and the back, the transverse control arms are positioned at an angle so that, together with the differing rubber hardness of the bearings at points 2, they achieve the desired elastokinematic characteristics. These are:

- toe-in under braking forces (Figs 3.50 and 3.64)
- lateral force compliance understeer during cornering (Figs 3.62 and 3.63)
- prevention of torque steer effects (see Section 2.10.4)
- lane change and straight running stability

For reasons of space, the front eyes 2 are pressed into parts 1 and bolted to the attachment bracket. Elongated holes are also provided in this part so toe-in can be set.

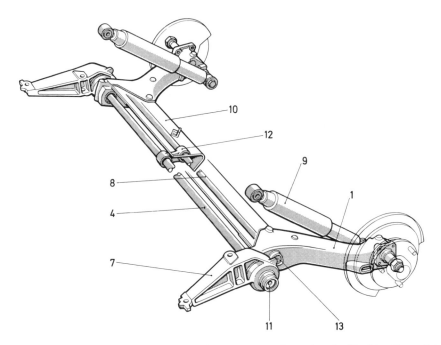

Fig. 1.1 An extremely compact four-bar compound crank axle fitted by Renault in several models, which has two torsion bar springs both for the left and right axle sides (items 4 and 8). The V-shape profile of the cross-member 10 has legs of different lengths, is resistant to bending but less torsionally stiff and absorbs all moments generated by vertical, lateral and braking forces. It also partially replaces the anti-roll bar.

At 23.4 mm, the rear bars 8 are thicker than the front ones (ϕ 20.8 mm, item 4). On the outside, parts 8 grip into the trailing links 1 with the serrated profile 13 and on the inside they grip into the connector 12. When the wheels bottom out, a pure torque is generated in part 12, which transmits it to the front bars 4, subjecting them to torsion. On the outside (as shown in Fig. 1.45a) the bars with the serrated profile 11 grip into the mounting brackets 7 to which the rotating trailing links are attached. The pivots also represent a favourably positioned pitch centre O_r (Fig. 3.123). The mounting brackets (and therefore the whole axle) are fixed to the floor pan with only four screws.

On parallel springing, all four bars work, whilst on reciprocal springing, the connector 12 remains inactive and only the thick rear bars 8 and the cross member 10 are subject to torsion. The layout of the bars means soft body springing and high roll stability can be achieved, leading to a reduction of the body roll pitch during cornering. To create a wide boot without side domes, the pressurized monotube shock absorbers 9 are inclined to the front and are able to transmit forces upwards to the side members of the floor pan.

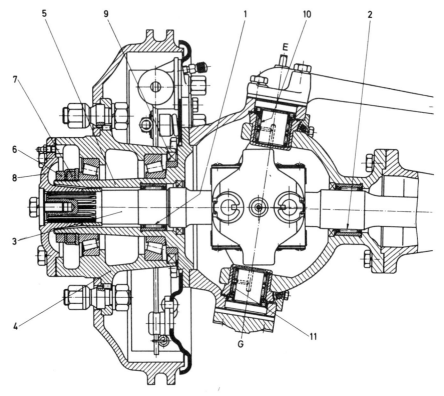

Fig. 1.1a   GKN double cardan universal joint fitted in the rigid front axle of an all-terrain, four-wheel drive lorry. Needle bearings 1 and 2 provide support for the inner and
outer drive shaft with the advantage that the outer drive shaft 3 is subjected exclusively to torsional stress. The bending moments generated by vertical, lateral and longitudinal forces pass into the wheel hub 4, which is supported on the neck 5 of the joint
casing via two taper roller bearings. The two lock nuts 6 and 7 on the outside are
secured against one another and make it possible to set the axial play once the com-
panion flange 8 has been removed.

The diameter of the shaft gasket 9, providing the seal to the brake, has been specially selected so that the gasket moves away undamaged over the stationary rollers
of the inner bearing when the hub is removed. The two short swivel kingpins turn in
the compact INA combined needle bearings 10 and 11. The bearings can absorb
both axial and radial forces. As can be seen here, the joint centre should lie in the
steering axis $\overline{EG}$.

1.2   Independent wheel suspensions – general

1.2.1 Requirements

The chassis of a passenger car must be able to handle the engine power
installed, and ever-improving acceleration, higher peak and cornering speeds

and deceleration demand safer chassis. Independent wheel suspensions follow this trend. Their main advantages are:

- small space requirement
- a kinematic or elastokinematic toe-in change tending towards understeering is possible (see Section 3.6)
- easier steerability with existing drive
- low weight
- no mutual wheel influence.

The last two characteristics are important for good road-holding, especially on bends with a bumpy road surface.

Transverse links and trailing links ensure the desired kinematic behaviour of the rebounding and compressing wheels and also transfer the forces to the body (Fig. 1.1b). Lateral forces also generate a moment which, unfortunately, reinforces the roll pitch of the body during cornering. The suspension control arms require mountings that yield under load and can also influence the springing. This effect is either hardened by twisting the rubber parts in the bearing elements, or the friction increases due to the parts shifting against one another (Fig. 1.6), and the driving comfort decreases. The wheels incline with the body (Fig. 1.2). The wheel on the outside of the bend, which has to absorb most of the lateral force, goes into a positive camber and the inner wheel into a negative camber, which reduces the lateral grip of the tyres. To avoid this, the kinematic change of camber needs to be adjusted to take account of this disadvantage (see Section 3.5.4) and the body roll in the bend should be kept as small as possible. This can be achieved with harder springs, additional anti-roll bars or a body roll centre located high up in the vehicle (Sections 3.4.3 and 5.4.3).

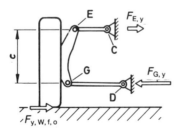

Fig. 1.1b   On front independent wheel suspensions, the lateral cornering force $F_{y,W,f,o}$ causes the reaction forces $F_{E,y}$ and $F_{G,y}$ in the links joining the axle with the body. Moments are generated on both the outside and the inside of the bend and these adversely affect the roll pitch of the body. The effective distance c between points E and G on a double wishbone suspension should be as large as possible to achieve small forces in the body and link bearings and to limit the elastic movement of the rubber elements fitted.

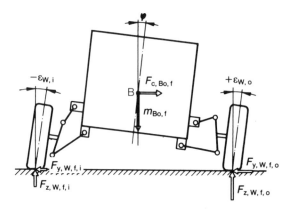

Fig. 1.2 If the body inclines by the angle φ during cornering, the outer independent-ly-suspended wheel takes on a positive camber $\epsilon_{W,o}$ and the inner wheel takes on a negative camber $\epsilon_{W,i}$. The ability of the tyres to transfer the lateral forces $F_{y,W,o}$ or $F_{y,W,i}$ decreases causing a greater slip angle (Fig. 3.44 and Equation 2.16), $m_{Bo,f}$ is the pro-portion of the weight of the body over the front axle and $F_{c,Bo,f}$ the centrifugal force act-ing at the level of the centre of gravity S. One wheel rebounds and the other compresses, i.e. this vehicle has 'reciprocal springing', that is:

$$F_{z,W,f,o} = F_{z,W,f} + \Delta F_{z,W,f}$$
$$F_{z,W,f,i} = F_{z,W,f} - \Delta F_{z,W,f}$$

1.2.2 Double wishbone suspensions

The last two characteristics above can most easily be achieved using a double wishbone suspension (Fig. 1.3). This consists of two transverse links (control arms) either side of the vehicle, which are mounted to rotate on the frame, suspension subframe or body and, in the case of the front axle, are connected on the outside to the steering knuckle or swivel heads via ball joints. The greater the effective distance c between the transverse links (Fig. 1.1b), the smaller the forces in the suspension control arms and their mountings become, i.e. component compliance is smaller and wheel control more pre-cise.

The main advantage of the double wishbone suspension is its kinematic possibilities. The positions of the suspension control arms relative to one another – in other words the size of the angles α and β (Fig. 3.19) – can deter-mine both the height of the body roll centre and the pitch pole (angles α' and β', Fig. 3.120). Moreover, the different lengths can influence the angle move-ments of the compressing and rebounding wheels, i.e. the change of camber and, irrespective of this, to a certain extent also the track width change (Figs 3.41 and 3.4). With shorter upper suspension control arms the compressing wheels go into negative camber and the rebounding wheels into positive. This counteracts the change of camber caused by the roll pitch of the body (Fig. 1.2). The vehicle pitch pole O indicated in Fig. 6.16 is located behind the

Fig. 1.3 Front axle on the VW light commercial vehicle Lt 28 to 35 with an opposed steering square. A cross-member serves as a subframe and is screwed to the frame from below. Springs, stops, shock absorbers and both pairs of control arms are supported at this force centre. Only the anti-roll bar, steering gear, idler arm and the tension rods of the lower control arms are fastened to the longitudinal members of the frame. The rods have longitudinally-elastic rubber bushings at the front that absorb the dynamic rolling hardness of the radial tyres.

wheels on the front axle and in front of the wheels on the rear axle. If O_τ can be located over the wheel centre (Fig. 3.124), it produces not only a better anti-dive mechanism, but also reduces the squat on the driven rear axles (or lift on the front axles). These are also the reasons why the double wishbone suspension is used as the rear axle on more and more passenger cars, irrespective of the type of drive, and why it is progressively replacing the semi-trailing link axle (Figs 1.0, 1.45 and 1.59).

1.2.3 McPherson struts and strut dampers

The McPherson strut is a further development of the double wishbone suspension. The upper transverse link is replaced by a pivot point on the wheel house panel, which takes the end of the piston rod and the coil spring. Forces from all directions are concentrated at this point and these cause bending stress in the piston rod. To avoid detrimental elastic camber and caster changes, the normal rod diameter of 11 mm (in the shock absorber) must be increased to at least 18 mm. With a piston diameter of usually 30 mm or 32 mm the damper works on the twin-tube system and can be non-pressurized or pressurized (see Section 5.8).

The main advantage of the McPherson strut is that all the parts providing the suspension and wheel control can be combined into one assembly. As can be seen in Fig. 1.4, this includes

- the spring seat to take the underside of the coil spring
- the auxiliary spring 11 or a compression buffer (see Fig. 5.35)
- the rebound buffer (Fig. 5.39)
- the pendulum anti-roll bar (item 7) via rod 5 and
- the steering knuckle.

The steering knuckle can be welded, soldered or bolted firmly to the outer tube (Fig. 1.40). Further advantages are:

- lower forces in the body-side mounting points E and D due to large effective distance c (Fig. 1.1b)
- short distance b between points G and N (Fig. 3.23)
- long spring travel
- three bearing positions no longer needed
- better design options on the front crumple zone
- space at the side permitting a wide engine compartment, which
- makes it easy to fit transverse engines (Fig. 1.35).

Nowadays, design measures have ensured that the advantages are not outweighed by the inevitable disadvantages on the front axle. These disadvantages are:

- less favourable kinematic characteristics (Sections 3.3 and 3.5.2)
- introduction of forces and vibrations into the inner wheel house panel and therefore into the front end of the vehicle
- it is more difficult to insulate against road noise – an upper strut mount is necessary (Fig. 1.5), which should be as decoupled as possible (item 10 in Fig. 1.4 and item 5 in Fig. 1.39)
- the friction between piston rod and guide impairs the springing effect; it can be reduced by shortening distance b (Figs 1.6 and 3.23)
- in the case of high positioned rack and pinion steering, long tie rods and, consequently, more expensive steering systems are required (Figs 1.40 and 4.0); in addition, there is the unfavourable introduction of tie-rod forces in the middle of the shock-absorbing strut (see Section 4.2.4) plus additional steering elasticity
- greater sensitivity of the front axle to tyre imbalance and radial runout deviations (see Section 2.5)

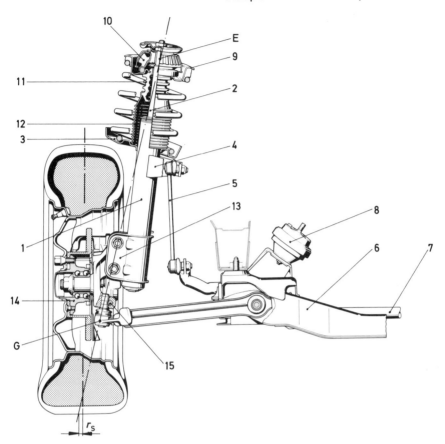

Fig. 1.4 Rear view of the left-hand side of the McPherson front axle on the Vauxhall Carlton with negative kingpin offset at ground (scrub radius) r_S and pendulum-linked anti-roll bar. The coil spring is positioned diagonally to decrease friction between piston rod 2 and the rod guide. Part 2 and the upper spring seat 9 are fixed to the inner wheel house panel via the decoupled strut mount 10. The additional elastomer spring 11 is joined to seat 9 from the inside and on the underside it carries the dust boot 12, which contacts the spring seat 3 and protects the chrome-plated piston rod 2. When the wheel bottoms out, the elastomer spring rests on the cap of the supporting tube 1.

Brackets 4 and 13 are welded to part 1, on which the upper ball joint of the anti-roll bar rod 5 is fastened from inside. Bracket 13 takes the steering knuckle in between the U-shaped side legs. The upper hole of bracket 13 has been designed as an elongated hole so that camber can be set precisely at the factory (see Fig. 3.80a). A second-generation, double-row, angular (contact) ball bearing (item 14) controls the wheel.

The ball pivot of the guiding joint G is joined to the steering knuckle by means of clamping forces. The transverse screw 15 grips into a ring groove of the joint bolt and prevents it from slipping out in the event of the screw loosening.

The subframe 6 is fixed to the body. In addition to the transverse control arms, it also takes the engine mounts 8 and the back of the anti-roll bar 7. The drop centre rim is asymmetrical to allow negative wheel offset at ground (scrub radius).

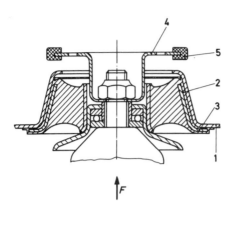

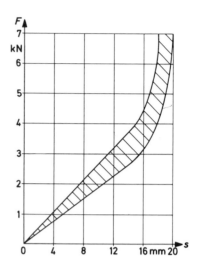

Fig. 1.5 McPherson strut mount on the VW Golf and Vento with a thrust ball bearing, which absorbs the rotary movement of the McPherson strut whereby the rubber part improves noise insulation. Initially the deflection curve remains linear and then becomes highly progressive in the main work area, which is between 3 kN and 4 kN. The chart shows the scatter. Springing and damping forces are absorbed together so the support bearing is not decoupled (as in Fig. 1.39).

In the car final assembly line the complete strut mount is pressed into a conical sheet metal insert on the wheel house inside panel 1. The rubber layer 2 on the outside of the bearing ensures a firm seat and the edge 3 gives the necessary hold in the vertical direction. The rubber ring 5 clamped on plate 4 operates when the wheel rebounds fully and so provides the necessary security (figure: Lemförder Metallwaren).

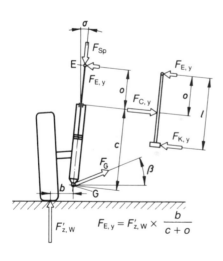

$$F_{E,y} = F'_{z,w} \times \frac{b}{c + o}$$

Fig. 1.6 If coil spring and the lower guiding joint G are not offset from the centre of the damper, the lateral force $F_{E,y}$ continually acts in the body-side fixing point E of the McPherson strut as a result of the force F_{Sp}. This generates the reaction forces $F_{C,y}$ and $F_{K,y}$ on the piston rod guide and piston. This is $F_{C,y} = F_{E,y} + F_{K,y}$, and the greater this force becomes, the further the frictional force F_{fr} increases in the piston rod guide and the greater the change in vertical force needed for it to rip away.

As the piston has a large diameter and also slides in shock-absorber fluid, lateral force $F_{K,y}$ plays only a subordinate role (see Fig. 5.39). $F_{C,y}$ can be reduced by offsetting the springs at an angle and shortening the distance b (see Figs 1.39 and 3.23 and Equation 3.4a).

- sometimes the space between the tyres and the damping element (Fig. 1.25) is very limited.

This final constraint, however, is only important on front-wheel drive vehicles as it may cause problems with fitting snow chains. On non-driven wheels, at most the lack of space prevents wider tyres being fitted. If such tyres are absolutely necessary, disc-type wheels with a smaller wheel offset e are needed and these lead to a detrimentally larger positive or smaller negative kingpin offset at ground r_s (Figs 2.5b and 3.79).

Over the last ten years, McPherson struts have become widely used as front axles, but they are also fitted as the rear suspension on front-wheel drive vehicles. The vehicle tail, which has been raised for aerodynamic reasons, allows a larger bearing span between the piston rod guide and piston. On the rear axle (Fig. 1.7)

- the upper strut mount is no longer necessary
- longer cross-members, which reach almost to the vehicle centre, can be

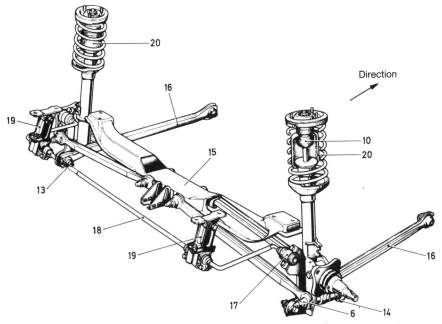

Fig. 1.7 The McPherson strut rear axle on the Lancia Delta with equal length transverse links of profiled steel fixed a long way inside the cross-member 15. As large a distance as possible is needed between points 6 and 14 on the wheel hub carrier to ensure unimpaired straight running. The fixing points 13 of the longitudinal links 16 are behind the wheel centre, exactly like mounting points 17 of the anti-roll bar 18. The back of the anti-roll bar is flexibly joined to the body via tabs 19. The additional springs 10 attached to the top of the McPherson struts are covered by the dust tube 20. The cross-member 15 helps fix the assembly to the body.

used, producing better camber and track width change (Figs 3.11 and 3.39) and a body roll centre that sinks less under load (Fig. 3.23)

- the outer points of the braces can be drawn a long way into the wheel to achieve a shorter distance *b*
- the boot can be dropped and, in the case of damper struts, also widened
- however, rubber hardness and the corresponding distance of the braces on the hub carriers (points 6 and 14 in Fig. 1.7) are needed to ensure that there is no unintentional elastic self-steer (Figs 3.62 and 3.63).

1.2.4 Rear axle trailing link suspension

This suspension – also known as a crank axle – consists of a control arm lying longitudinally in the driving direction and mounted to rotate on a suspension subframe or on the body on both sides of the vehicle (Figs 1.8 and 1.45a). The control arm has to withstand forces in all directions, and is therefore highly subject to bending and torsional stress (Fig. 1.8a). Moreover, no camber and toe-in changes are caused by vertical and lateral forces.

The trailing link axle is relatively simple and is popular on front-wheel drive vehicles. It offers the advantage that the car body floor pan can be

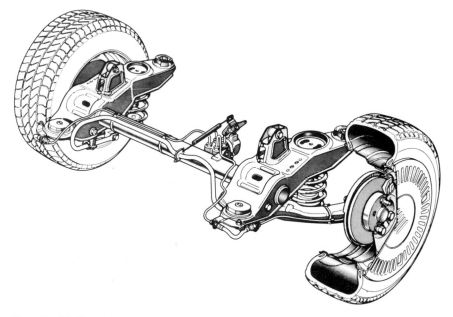

Fig. 1.8   Trailing link rear axle on the Fiat Tempra; in order not to restrict the size of the boot, the coil springs are supported low on the links. The shock absorbers have been moved forward. The highly cranked anti-roll bar is subject to torsion and bending with reciprocal springing and is fastened to the links from below.

A transverse tube and two outriggers pointing backwards form a suspension subframe to which all the components are fixed and which can be joined to the body at four points as one assembly unit.

Fig. 1.8a On rear axle longitudinal links, the vertical force $F_{z,W}$ together with the lateral forces $F_{y,W}$ cause bending and torsional stress, making a closed box profile necessary. A force from inside causes the largest torsional moment

$$T = F_{z,W} \times a + F_{y,W} \times r_{dyn}$$

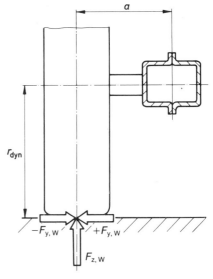

smooth and the fuel tank and/or spare wheel can be positioned between the suspension control arms. If the pivot axes lie parallel to the floor, the compressing and rebounding wheels undergo no track width, camber or toe-in change, and the wheel base simply shortens slightly. If torsion springs are applied, the length of the control arm can be used to influence the progressivity of the springing to achieve better vibration behaviour under load. The control arm pivots also provide the pitch pole O; i.e. during braking the tail end is drawn down at this point (Fig. 3.123).

The low body roll centre at floor level is a disadvantage (Fig. 3.26) as is the fact that the wheels incline more with the body when cornering than with other independent wheel suspensions (Figs 1.2 and 3.45).

1.2.5 Semi-trailing link rear axles

This is a special type of trailing link axle, which is fitted mainly in rear-wheel and four-wheel drive passenger cars, but which is also found on front-wheel drive vehicles (Figs 1.9, 1.9a, 1.28, 1.54 and 1.55). Seen from the top, the control arm axis of rotation $\overline{EG}$ is diagonally positioned at an angle $\alpha = 10°$ to 25°, and from the back an angle $\beta \leqslant 5°$ can still be achieved (Fig. 3.29). When the wheels compress and rebound they cause spatial movement, so the drive shafts need two joints per side with angular mobility and length compensation (Figs 1.9b and 1.29).

When the control arm is a certain length, the following kinematic characteristics can be positively affected by angles α and β (Fig. 3.15):

- height of the roll centre
- position of the pitch pole
- change of camber and
- toe-in change.

1.3   Rigid axles – general

Rigid axles can have a whole series of disadvantages – which are a consideration in passenger cars, but can be accepted in commercial vehicles – for example

- weight, if the differential is located in the axle casing (Fig. 1.10), produces a tendency for wheel hop to occur on bumpy roads
- mutual wheel influence (Fig. 1.11)
- the space requirement above the axle corresponds to the spring compression travel
- the wheel load changes due to drive (Fig. 1.12) and (particularly on twin tyres)
- there is a poor support basis b_{Sp} for the body, which can only be improved following costly design work (Fig. 1.26).

The effective distance b_{Sp} of the springs is generally less than the tracking width b_r, so the reciprocal spring rate c_f is lower (Fig. 1.13). As can be seen in

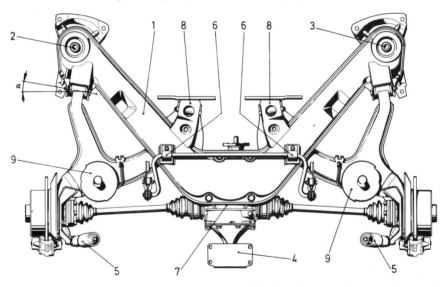

Fig. 1.9 Top view of the rear axle on the Vauxhall Omega; the final drive 7 is screwed firmly to the subframe 1, which also carries the back of the anti-roll bars (in bearings 6). The entire assembly is held on the body by the specially-designed rubber bearings 2, 3 and 4. The two outriggers 8 are used to take the inner link bearings, and the barrel-shaped coil springs sit on the spring seats 9. In order to achieve a flat boot floor, parts 9 were moved in front of the drive shafts; at 1.5, the ratio i_{Sp} (wheel to spring) is relatively high. The shock absorbers 5 are located behind the axle centre, at $i_D = 0.86$ the ratio is good. Furthermore, the top view angle also amounts to $\alpha = 10°$, the rear view angle to $\beta = -1°20'$ (Fig. 3.29), the camber with two persons in the vehicle to $\epsilon_{W,pl} = -1°40'$ (and with a permissible axle load to $\epsilon_{W,lo} = -2°45'$) and the level of the body roll centre to $h_{Ro,r} = 100$ mm.

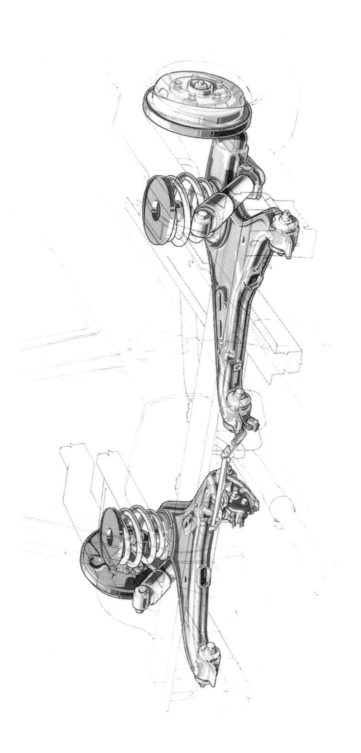

Fig. 1.9a Flat-design semi-trailing link rear axle on the VW Transporter, suited to taking the rear-wheel drive on the four-wheel drive model. Short, barrel-shaped coil springs and forward-inclined shock absorbers fitted under the underbody give a wide, low-level cargo space between the wheel housings. In addition to a small top view angle α, a negative rear view angle β was created (Fig. 3.29). This was achieved by siting the inner link mounts 15 mm above the outer ones. This permits body roll understeering of the rear axle (Fig. 3.58) and a reduction in the kinematic change of the camber (Fig. 3.40).

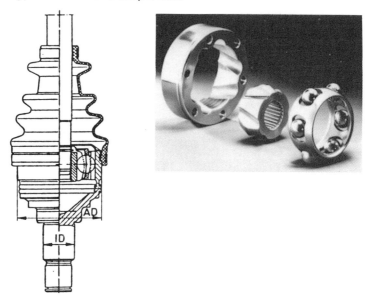

Fig. 1.9b CV slip joint VL, manufactured by GKN, allows a bending angle of $\beta \leqslant 22°$ and lateral paths up to 45 mm. Where $\beta \leqslant 10°$ it transfers torques up to 5.9 kNm. The inner shaft ID, butt welded to the joint housing, grips into the gearing in the case of front-wheel axle shafts. The slanting ball races, shown in the figure on the right-hand side, on the inner and outer joint casing, mean the six ball bearings are middle-centring if two CV slip joints are needed on rear drive shafts. The balls are housed in a cage.

Fig. 1.44, the springs, or the spring dampers, for this reason should be mounted as far out as possible (see also Section 5.3).

The centrifugal force ($F_{c,Bo}$, Fig. 1.2) acting on the body's centre of gravity during cornering increases the roll pitch where there is a rigid axle (see Section 5.4.3.5). Thanks to highly developed suspension parts and the appropriate design of the springing and damping, it has been possible to improve the behaviour of driven rigid axles. Nevertheless, they are no longer found in standard-design passenger cars, but only on four-wheel drive and special all-terrain vehicles (Figs 1.27 and 1.50).

Because of its weight, the driven rigid axle is outperformed on bumpy, uneven roads (and especially on bends) by independent wheel suspension, although the deficiency in road-holding can be partly overcome with pressurized mono-tube dampers. These are more expensive, but the compressive valve curve can be set to be harder without a perceptible loss of comfort. With this, a responsive damping force is already opposing the compressing wheels. This is the simplest and perhaps also the most economic way of overcoming the main disadvantage of rigid axles. Section 5.6.4 contains further details.

Compared with standard-design vehicles, front-wheel drive vehicles are a different story. The advantages of the rigid rear axle outweigh the disadvan-

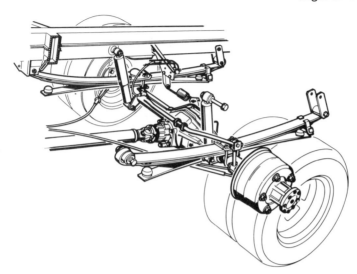

Fig. 1.10 Rear axle on the VW light commercial vehicle LT. The long, parabola-shaped rolled-out, dual leaf springs cushion the frame well and are progressive. The rubber buffers of the support springs come into play when the vehicle is laden. Spring travel is limited by the compression stops located over the spring middles, which are supported on the side-members. The spring layers are prevented from shifting against one another by the spring clips located behind them, which open downwards (see also Fig. 1.50).

The anti-roll bar is fixed outside the axle housing. The benefits of this can be seen in Fig. 1.13. The shock absorbers, however, are unfortunately located a long way to the inside and are also angled forwards so that they can be fixed to the frame side-members (Fig. 5.15).

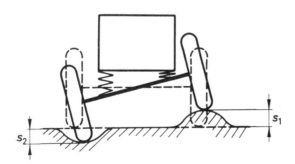

Fig. 1.11 Mutual influence of the two wheels of a rigid axle when travelling along a road with pot-holes, shown as 'reciprocal springing'. One wheel extends along the path s_2 and the other compresses along the path s_1.

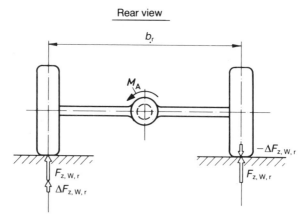

Rear view

Fig. 1.12 If the differential is located in the body of the rigid axle, the driving torque M_A coming from the engine is absorbed at the centres of tyre contact, resulting in changes to vertical force $\pm\Delta F_{z,W,r}$.

In the example, M_A would place an additional load on the left rear wheel $(F_{z,W,r} + \Delta F_{z,W,r})$ and reduce the vertical force $(F_{z,W,r} - \Delta F_{z,W,r})$ on the right one. On a right-hand bend the right wheel could spin prematurely leading to a loss in lateral force in the entire axle and the car tail suddenly breaking away (Fig. 2.38).

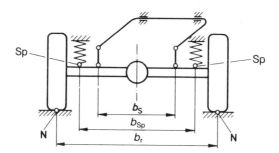

Fig. 1.13 When considering the roll pitch of the body with the rigid axle the distances b_{Sp} (of the springs F) and b_S (of the anti-roll bar linkage points) are included in the calculation of the transfer with reciprocal springing. i_φ is squared to give the rate c_φ:

$$i_\varphi = b_r/b_{Sp} \text{ and } c_\varphi = c_r i^2_\varphi$$

The greater the ratio, the less the roll reaction support of the body, i.e. springs and anti-roll bar arms should be fixed as far out as possible on the rigid axle casing (see Section 5.4.3.5 and Equations 5.20 and 5.21).

tages (Fig. 1.13a). As Section 6.1.3 explains, the rigid rear axle weighs no more than a comparable independent wheel suspension and also gives the option of raising the body roll centre (which is better for this type of drive, see Fig. 3.34). Further advantages, including those for driven axles, are:

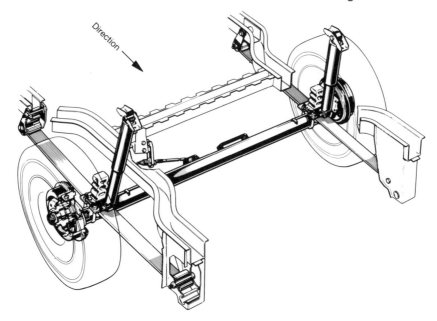

Fig. 1.13a   The rear axle on the Ford Escort Express delivery vehicle with a 635 kg payload (plus 75 kg for the driver).
Single leaf springs carry the axle and support the body well at four points. The shock absorbers (fitted vertically) are located close to the wheel to give narrow wheel housings. The additional elastomer springs sit over the axle tube and act on the side members of the body when the suspension bottoms out.

- they are simple and economical to manufacture
- there are no changes to track width, toe-in and camber when bottoming out, thus giving
- low tyre wear and sure-footed road holding
- there is no change to wheel camber when the body pitches during cornering (Figs 1.2 and 3.45), therefore there is constant lateral force transmission of tyres
- the absorbtion of lateral force moment $M_y = F_T\, h_{Ro,r}$ by a transverse link, which can be placed at almost any height (e.g. Panhard rod, Fig. 1.14), therefore
- the lateral force compliance steering can be tuned towards under- or oversteering (Figs 3.63a and 1.17).

There are many options for fitting a rigid rear axle under the body or frame. Longitudinal leaf springs are often used as a single suspension control arm, which is both supporting and springing at the same time, as these can absorb forces in all three directions as well as drive-off and braking moments (Figs 1.15 and 5.12). This economical type of suspension also has the advantage that the load area on lorries and the body of passenger cars can be supported in two places at the back: at the level of the rear seat and under the

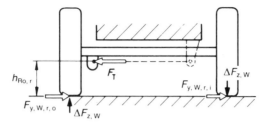

Fig. 1.14 On rigid axles the axle body absorbs the bending moments which arise as a result of lateral forces. Only the force F_T occurs between the suspension and the body, and its size corresponds to the lateral forces $F_{y,W,r,o}$ and $F_{y,W,r,i}$. On a horizontal Panhard rod, the distance $h_{Ro,r}$ is also the height of the body roll centre. The higher this is above ground, the greater the wheel force change $\pm\Delta F_{z,W}$.

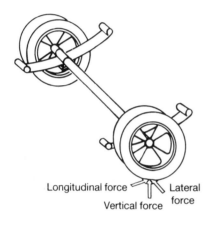

Fig. 1.15 Longitudinal leaf springs can absorb both forces in all directions and the drive-off, braking and lateral force moment.

Longitudinal force   Lateral force
Vertical force

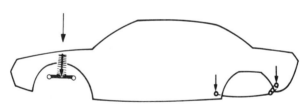

Fig. 1.15a Longitudinal rear leaf springs support the body of a car in two places – under the back seats and under the boot – with the advantage of reduced bodywork stress.

boot (Fig. 1.15a). This reduces the stress on the rear end of the car body when the boot is heavily laden, and also the stress on the lorry frame under full load (Fig. 1.10).

The longitudinal leaf springs can be fitted inclined, with the advantage that during cornering the rigid rear axle (viewed from above) is at a small angle to the vehicle longitudinal axis (Fig. 1.16). To be precise, the side of the wheel

Fig. 1.16 Angled longitudinal leaf springs fixed lower to the body at the front than at the back cause the rigid rear axle to self-steer towards understeering. Where there is body roll pitch, the wheel on the outside of the bend, which is compressing along the path s_1, is forced to accommodate a shortening of the wheel base Δl_1, whilst the wheel on the inside of the bend, which is extending by s_2, is forced to accommodate a lengthening of the wheelbase by Δl_2. The axle is dislocated at the steering angle δr (see also Fig. 3.58b).

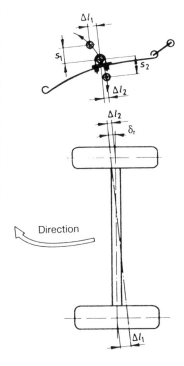

base on the outside of the bend shortens somewhat, whilst the side on the inside of the bend lengthens by the same amount. The rear axle steers into the bend and, in other words, it is forced to self-steer towards 'understeering' (Fig. 1.17). This measure can, of course, have an adverse effect when the vehicle is travelling on bad roads, but it does prevent the standard passenger car's tendency to oversteer when cornering. Even driven rigid axles exhibit – more or less irrespective of the type of suspension – a tendency towards the load alteration (torque steering) effect, but not to the same extent as semi-trailing link axles. Details can be found in Section 2.10.4. On front-wheel drive vehicles, the wheels of the non-driven axle can take on a negative camber. This perhaps improves the lateral grip somewhat, but does not promote perfect tyre wear. This is also possible on the compound crank axle (a suspension-type halfway between a rigid axle and independent wheel suspension) which, up to now, has been fitted only on front-wheel drive vehicles. This axle has only a few disadvantages (details are given in Fig. 1.1 and Section 1.6.4.1).

1.4 Standard design, front-mounted engine, rear-mounted drive

In passenger cars and estate cars, the engine is approximately in the centre of the front axle and the rear wheels are driven (Fig. 1.18). To put more weight on the rear axle and obtain a more balanced weight distribution, Porsche, for instance, has designed the gearbox and differential as a single unit on the 928 GTS and 968 (Fig. 1.19). Alfa Romeo and Volvo have also adopted this solution.

With the exception of light commercial vehicles, all lorries have the engine at the front or centrally between the front and rear axles together with rear-

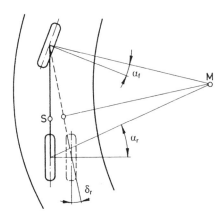

Fig. 1.17 If a rigid rear axle steers with the angle δr towards understeer, the tail moves out less in the bend and the driver has the impression of more neutral behaviour. Moreover, there is increased safety when changing lanes quickly at speed. The same occurs if the outside wheel of an independent wheel suspension goes into toe-in and the inside wheel goes into toe-out (see Fig. 3.62).

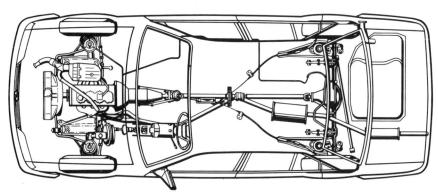

Fig. 1.18 Standard drive design concept. The engine is mounted longitudinally over the front axle with a flanged gearshift and propshaft connection to the rear axle differential. The large amount of space required by the rigid axle and the critical positioning of the fuel tank in the rear impact zone are clearly shown. The front overhang $L_{Ov,f}$ is the distance from the body front edge (bumper) to the centre of the front axle.

Fig. 1.19 Porsche design. Overhead view of the chassis with a double wishbone suspension front axle, semi-trailing link rear axle and rigid connecting tube between the engine and the rear located gearbox with differential. Known as the 'transaxle-unit', this design is used to give better loading to the driven rear-wheels. With two persons in the vehicle the axle load distribution front to back is 49%/51%.

wheel drive vehicles. Hardly any other option gives the same long load area. Articulated lorries, where a major part of the trailer weight – the trailer hitch load – is carried over the rear wheels, have the same configuration. On buses, however, the passengers are spread evenly throughout the whole interior of the vehicle, which is why there are models with front, central and rear engines.

1.4.1 Advantages and disadvantages of the standard design

The standard design has a series of advantages on passenger cars and estate cars:

- there is hardly any restriction on engine length, making it particularly suitable for more powerful vehicles (in other words with 8 to 12 cylinder engines)
- there is low load on the engine mounting, as only the maximum engine torque times the conversion of the lowest gear has to be absorbed
- insulation of engine noise is relatively easy
- under full load most of the vehicle mass is on the driven rear axle (important for estate cars and trailers (Table 1.21 and Fig. 6.22))
- a long exhaust system with good silencing is possible, with no major difficulties housing a catalytic converter
- good front crumple zone, together with the 'submarining' power plant unit, i.e. one that goes underneath the floor panel during frontal collision
- simple and varied front axle designs are possible
- more even tyre wear
- uncomplicated gear shift mechanism
- optimum gearbox efficiency in direct gear because no force-transmitting cog pair is in action (Fig. 6.19)
- sufficient space for housing the steering system in the case of a recirculating ball steering gear
- good cooling because the engine and radiator are at the front; a power-saving fan can be fitted
- effective heating due to short hot-air and water paths.

The following disadvantages mean that, in recent years, only a few saloon cars under 2 l engine displacement have been launched internationally using the standard design:

- unstable straight-running ability (Fig. 1.20), which can be fully corrected by special front axle settings, appropriate rear axle design and suitable tyres
- the driven rear axle is slightly loaded when there are only two persons in the vehicle, leading to poor traction behaviour in wet and wintry road conditions – linked to the risk of the rear wheels spinning, particularly when narrow bends are being negotiated at speed. This can be improved by setting the unladen axle load distribution at 50%/50% which, however,

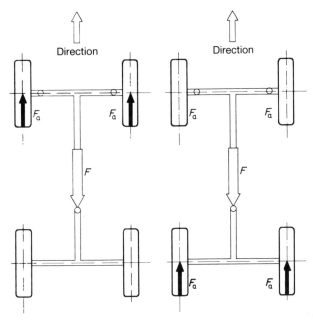

Fig. 1.20   On a front-wheel drive (left) the vehicle is pulled. The result is a more stable relationship between the driving forces F_a and the inertia force F. Conversely, in the case of driven rear wheels an unstable condition is theoretically evident; front axle settings ensure the necessary stabilization.

is not always possible (Table 1.21 and Fig. 6.22). It can be prevented by means of drive-slip control
- a tendency towards the torque steer effect (Figs 2.37 and 2.38) and therefore
- complex rear independent wheel suspension with chassis subframe, differential gear case and axle drive causing
- restrictions in boot size

Table 1.21   Average proportional axle load distribution based on drive type and loading condition. With the standard design saloon, when the vehicle is fully laden, the driven rear wheels have to carry the largest load. With the front-wheel drive, however, with only two persons in the vehicle, the front wheels bear the greater load.

	Front wheel drive		Standard drive		Rear engine	
	front	rear	front	rear	front	rear
Empty	61	39	50	50	40	60
2 passengers at the front	60	40	50	50	42	58
4 passengers	55	45	47	53	40	60
5 passengers and luggage	49	51	44	56	41	59

- the need for a propshaft between the manual gearbox and differential (Fig. 1.18), and therefore
- a tunnel in the floor pan is inevitable, plus an
- unfavourable interior to vehicle-length ratio.

1.4.2 Non-driven front axles

The standard design passenger and estate cars that have come onto the market in recent years have McPherson struts on the front axle, as well as double wishbone suspensions. The latter type of suspension is becoming more and more popular because of its low friction levels and kinematic advantages. Even some light commercial vehicles have McPherson struts or double wishbone axles (Fig. 1.3). However, like almost all medium-sized and heavy commercial vehicles, most have rigid front axles. In order to be able to situate the engine lower, the axle subframe has to be offset downwards (Fig. 1.22).

The front wheels are steerable; to control the steering knuckle 5 (Fig. 1.23) on double wishbone suspensions, there are two ball joints that allow mobility in all directions, this being the bottoming out of the wheels and the steering angle. The wishbone, which accepts the spring, must be carried on a supporting joint (item 7) in order to be able to transmit the vertical forces. A regular ball joint transferring longitudinal and lateral forces (item 8) is generally sufficient for the second suspension control arm. The greater the distance between the two joint points, the lower the forces in the components. Figure 1.23a shows a front axle with ball joints a long way apart. The base on McPherson struts is better because it is even longer. Figure 1.24 shows a standard design and Fig. 1.4 the details.

The coil spring is inclined at an angle to reduce the friction between piston rod 2 and the rod guide. The lower guiding joint (point G) performs the same function as on double wishbones, whilst point E is fixed in the shock tower, which is welded to the wheel house panel. As the wheels bottom out, piston rod 2 moves in the cylinder tube (which sits in the carrier or outer tube, see Fig. 5.39) and when there is a steering angle the rod and spring turn in an upper strut mount, which insulates noise and is located at point E (Fig. 1.5).

Wheel controlling damper struts do not require such a complex mount. The piston rod turns easily in the damping part (Fig. 1.25). Only the rod needs noise insulation. The coil spring sits separately on the lower control arm, which must be joined to the steering knuckle via a supporting joint. The damper is lighter than a shock-absorbing strut and allows a greater bearing span in the damping part, permits a wider, flatter engine compartment (which is more streamlined) and is easier to repair. However, it is likely to be more costly and, placing the spring apart from the damper (Figs 1.4 and 1.6), may cause slip-stick problems with a loss of ride comfort.

Figure 1.63 shows the additional difficulties with front-wheel drive vehicles.

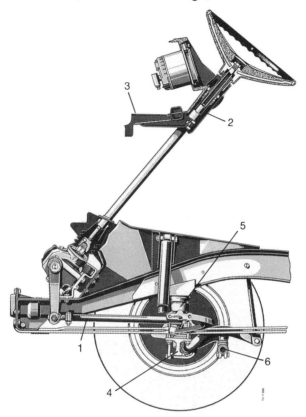

Fig. 1.22 The front rigid axle on the Mercedes-Benz light commercial vehicle of the 207 D/308 series with recirculating ball steering gear and steering rod 1 parallel to the two-layer parabola spring. This rod has to be slightly shorter than the front side of the spring, so that both parts take on the same arch shape when the axle bottoms out (see also Fig. 4.5). The brace 3 running from the steering column jacket 2 to the body bends on impact. The T-shaped axle casing 4, which is drawn downwards and to which the springs are fastened, can be seen in the section. The elastomer spring 5 sits on the longitudinal member of the frame and the two front wheels are joined by the tie rod 6. The safety steering wheel has additional padding.

1.4.3 Driven rear axles

Because of their cost advantages, robustness, ease of repair and long service life, rigid axles are fitted in practically all commercial vehicles in combination with leaf springs, coil springs or air springing (Figs 1.10 and 1.26). They are also used in all-terrain vehicles (Fig. 1.27), although they are no longer found in saloons and coupés. In spite of the advantages described in Section 1.3, the weight of the axle is noticeable on this type of vehicle and comfort reasons are more dominant.

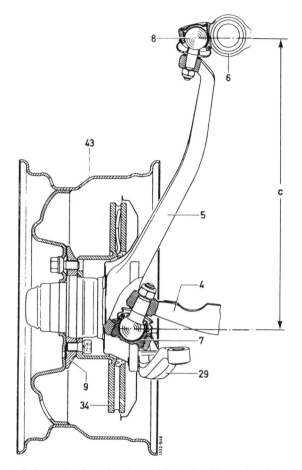

Fig. 1.23 Front hub carrier (steering knuckle) on the Mercedes-Benz S-class with a large effective distance *c* (see also Fig. 1.1b). The upper transverse control arm 6 forms the casing for the ball pivot of the guiding joint, whilst the lower supporting joint 7 is pressed into the hub carrier 5. The ventilated brake disc 34 (drawn inwards), the wheel hub 9, the double-hump rim 43 with asymmetrical drop centre and the space for the brake caliper (not included in the picture) are clearly shown.

Up until the end of the 1980s, the semi-trailing link axle, shown in Figs 1.9 and 1.28, was widely used among the various independent wheel suspensions. This suspension has a chassis subframe with which the control arms are linked, and the differential is fixed or elastically joined to give additional noise and vibration insulation. The springs sit approximately in the centre of the suspension control arms. This gives a flat, more spacious boot, but with the disadvantage that the forces in all components, particularly the control arm bushings, are higher.

Fastening spring dampers to the suspension control arms in the centre of

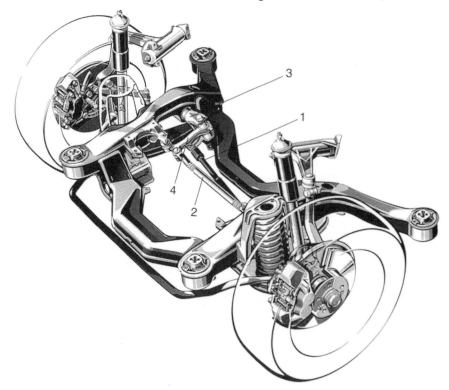

Fig. 1.23a Double wishbone suspension front axle on the Mercedes-Benz S class. The wheel hub carriers are taken through semi-trailing links at the top and through widely based transverse control arms at the bottom. Coil springs and shock absorbers are anchored on the lower transverse control arms.

Outriggers, which are fixed to the chassis subframe, support the upper sides of the springs on both sides and the link mountings are fixed on the underside of this front axle member. This part, which also absorbs the deformation energy on collisions, is joined with the car body in four places. The positioning and precisely-defined elasticity of the rubber bearings permits a decoupling between the front axle and body and thus suppresses audible and perceptible vibrations.

The subframe also carries the recirculating ball hydraulic steering gear (see Section 4.3.3) and the idler arm 4, which is joined to the pitman arm by intermediate rod 2. The steering damper 1 is connected on one side to part 2 and on the other side to the fixed bracket for the mount 3 of lever 4. The anti-roll bar in front is fastened to the body and to the lower transverse control arms.

the axle (as shown in Fig. 1.28) instead of separate coil springs, means high stresses can be avoided, but does necessitate shock towers, which restrict the boot at the side. Figure 1.30 shows a good solution on a light commercial vehicle. Because of its ride and handling advantages, more and more passenger cars have double wishbone suspension rear axles.

Figure 1.0 shows a modern design of the so-called multi-link axle and Fig. 1.59 shows a design installed in one four-wheel drive passenger car. Most

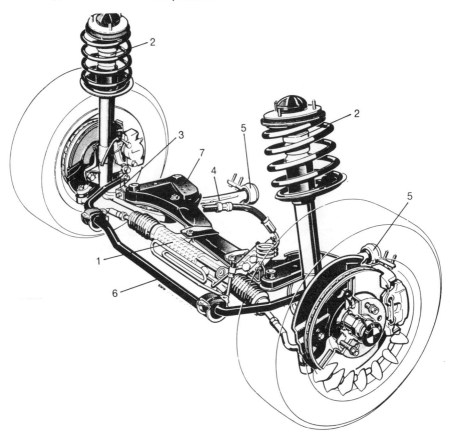

Fig. 1.24 McPherson front axle on the BMW 3 series. The polyurethane springs 2 sit inside the coil springs that are angled to reduce friction (Fig. 1.6). The anti-roll bar 6 is joined with the lower link via rods 3.

The cross-member 7, acting as a suspension subframe, carries the power-assisted rack and pinion steering gear 1 at the front and underneath the crescent-shaped transverse control arm 4, the ends of which are supported in the laterally elastic bearings 5. As shown in Fig. 3.65a, the radial tyre dynamic rolling hardness is absorbed in these rubber elements by means of defined springing. The large-diameter, internally ventilated brake discs (15" rim Fig. 2.6a) and the third-generation, two-row angular ball bearings, the outside ring of which also acts as a wheel hub, are clearly shown.

The kingpin offset at ground (scrub radius) depends on the tyre width and therefore the wheel offset (Figs 2.5b, 2.14 and 3.79). It is:

r_S = + 10 mm on 185/65 R 15 tyres and r_S = + 5 mm on 205/60 R 15 tyres

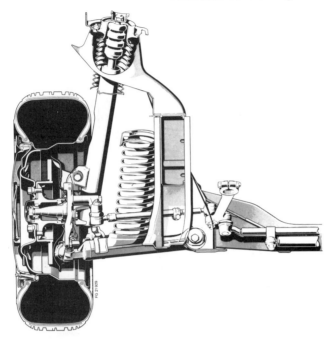

Fig. 1.25 Front axle on the Mercedes-Benz E class. The damper strut is screwed in three places to the wheel hub carrier. It is depressed next to the tyre so that the negative kingpin offset at ground ($r_S = -14$ mm) can be achieved with tyre size 195/65 R 15 90 H on the $6\frac{1}{2}$ J × 15 rim. The supplementary polyurethane spring comes into play on top of the damper tube. The wheel hub is carried by two taper roller bearings which can be adjusted via a ring nut on the outside to remove play. The brake disc is set onto the hub flange from outside. Further details are illustrated in Fig. 4.10.

independent wheel suspensions have an easy-to-assemble chassis subframe for better wheel control and noise insulation. However, all configurations (regardless of the design) require drive shafts with length compensation. This is carried out by the sliding cv joints fitted both at the wheel and the differential. Figure 1.9b shows a section through a joint of this type, whilst Fig. 1.29 shows a typical modern bearing of a driven rear wheel.

1.5   Rear and central engine drive

The rear-mounted power plant consists of the engine and the differential and manual gearbox in one assembly unit, and it drives the rear wheels. The power plant can sit behind the axle (Fig. 1.30, rear-mounted engine) or in front of it (Fig. 1.31, central engine). This configuration makes it impossible to have a rear seat as the engine occupies this space. The resulting two-seater is only suitable as a sports or rally car. The disadvantages of rear and central engine drive on passenger cars are:

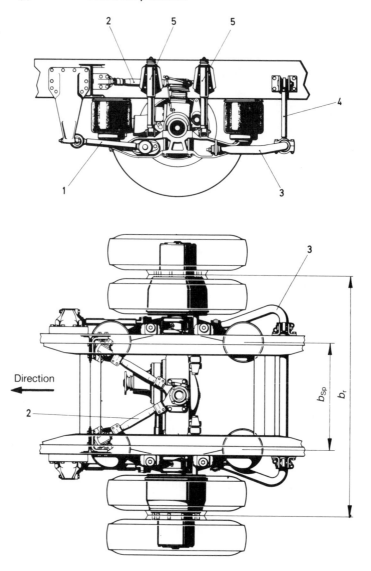

Fig. 1.26 Driven rear axle with air springs of the Mercedes-Benz lorry 1017 L to 2219 L 6 × 2. The axle is carried in the longitudinal and lateral directions by the two struts 1 and the upper wishbone type control arm 2. The four spring bellows sit under the longitudinal frame members and, because of the twin tyres, they have a relatively low effective distance b_{Sp}. The tracking width b_r divided by b_{sp} yields approximately the ratio $i_\varphi = 2.2$. As shown in Equation 5.19, with reciprocal springing the rate is $c_{\varphi,r}$ which amounts to only 21% of the rate c_r with parallel springing. To reduce body roll pitch the anti-roll bar 3 was placed behind the axle and is supported on the frame via the rod 4. The four shock absorbers 5 are almost vertical and are pushed close to the wheels to enable pitch movements of the body to fade more quickly.

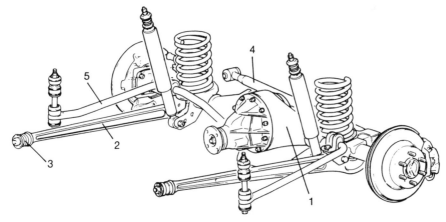

Fig. 1.27   The rear axle on the all-terrain, general-purpose passenger car, Mitsubishi Pajero. The rigid axle casing 1 is taken through the longitudinal control arms 2. These absorb the drive-off and braking forces (and the moments which arise) and transmit them to the frame. The rubber mountings 3 in the front fixing points 1, which also represent the vehicle pitch pole O$_r$ (Fig. 3.123), are designed to be longitudinally elastic to keep the road harshness due to the dynamic rolling hardness of the radial tyre away from the body. The Panhard rod 4 absorbs lateral forces. The anti-roll bar 5 is (advantageously) fastened a long way out on the frame (Fig. 1.13). The disc brakes, coil springs and almost vertical shock absorbers can be clearly seen.

- moderate straight running abilities (caster offset at ground angles of up to $\tau = 8°$ are factory-set)
- sensitivity to side winds
- indifferent cornering behaviour at the threshold level (central engine)
- extreme oversteering behaviour on bends (rear-mounted engine, see Fig. 2.29a)
- difficult to steer on ice because of low weight on the front wheels
- uneven tyre wear front to rear (high rear axle load, see Fig. 1.21)
- the engine mounting must absorb the engine moment times the total gear ratio
- the exhaust system is difficult to design because of short paths
- the engine noise suppression is problematic
- complex gear shift mechanism
- long water paths with front radiators (Fig. 1.31)
- high radiator performance requirement because of forced air cooling, the electric fan can only be used on the front radiator
- the heating system has long paths for hot water or warm air
- the fuel tank is difficult to house in safe zones, and
- the boot size is severely limited.

These are also the reasons why no saloons with rear-mounted engines have come onto the market for over 15 years.

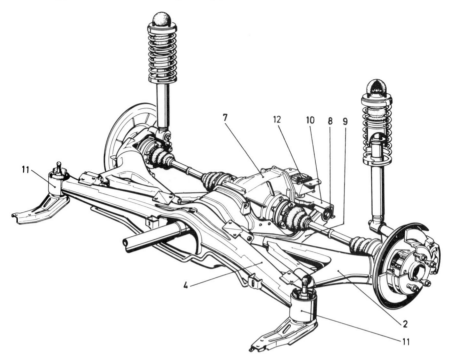

Fig. 1.28 Semi-trailing link rear axle on the BMW 5 series. The suspension subframe 4 is fixed at the front to the body with the mounts 11. The two backward pointing outriggers 9 are attached to a third connection point on the body (rubber element 12) via the transverse link 10. To insulate noise better, the final drive 7 is joined to the subframe by elastic elements – there is one rubber mounting at the front (not shown), whilst two are needed at the back (item 8).

The illustration does not show the anti-roll bar and two supplementary links, which improve the kinematic characteristics (Fig. 3.15) and which sit underneath the outer link mounts.

The few advantages, given below, may also be the reason why Porsche has kept the rear engine design in the 911, as has Renault in the Alpine V 6, whereas the VW Transporter, Fig. 1.30, has not been built in this design since 1991. The advantages are:

- very good drive-off and climbing capacity, almost irrespective of load (Fig. 6.22)
- a short power flow because the engine, gearbox and differential form one compact unit
- light steering due to low front axle load
- good braking force distribution
- simple front axle design
- easy engine dismantling (only on rear engine)
- no tunnel or only a small tunnel in the floor pan, and
- a small overhang to the front is possible.

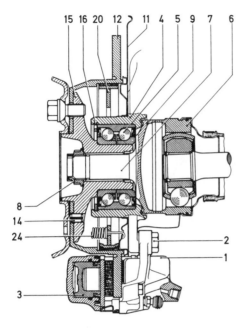

Fig. 1.29   Rear axle wheel hub carrier with wheel and brake. The drive shaft 7 is butt-welded to the CV slip joint 6. The drive shaft transmits the driving torque to the wheel hub 15 via a serrated profile. Part 15 is carried by the maintenance-free, two-row angular (contact) ball bearing 5. The one-part outer ring is held in the hub carrier 4 by the snap ring 16.

The seal rings on both sides sit in the permanently-lubricated bearing unit. The covering panel 11 (that surrounds the brake disc 12) acts as additional dirt-protection outside as does collar 9 of the CV joint on the inside. This grips into a cut-out in the wheel hub carrier 4 and creates a maze. The centrifugal effect of the bell-shaped joint housing prevents ingress of dirt and water. The brake disc 12 is pulled from outside against the flange 15 and fixed by dowel 14 until the wheel is mounted. The jaws 20 of the drum brake acting as a handbrake act on the inside of part 12.

At the lower end, the illustration shows the fixed calliper 1 of the disc brake. Two hexagonal bolts (item 2) fix it to the wheel hub carrier 4. Piston 3 and the outer brake pad are shown cut away (photograph: Mercedes-Benz).

1.6   Front-wheel drive

The engine, differential and gearbox form one unit, which can sit in front of, over, or behind the front axle. The design is very compact and, unlike the standard design, means that the vehicle can either be around 100 to 300 mm shorter, or the space for passengers and luggage can be larger. These are probably the main reasons why, worldwide, more and more car manufacturers have gone over to this design. In recent years only a few saloons of up to 2 l capacity without front-wheel drive have come onto the market. Nowadays,

Fig. 1.30 VW Transporter, a combination vehicle which could be used either as an eight-seater bus or as a light commercial vehicle for transporting goods, and which has the optimal axle load distribution of 50%/50% in almost all loading conditions. The double wishbone suspension at the front, the semi-trailing link rear axle and the rack and pinion steering, which is operated via an additional gear set in front, can be seen clearly. To achieve a flat load floor throughout, VW changed the Transporter to front-wheel drive in 1990 (Figs 1.3 and 1.32).

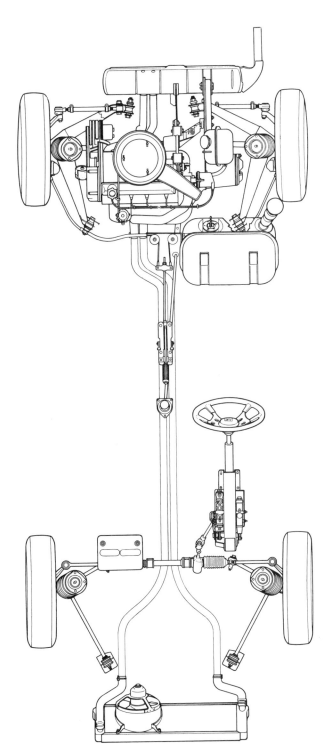

Fig. 1.31 On the Fiat X 1/9 the standard engine is transverse-mounted in front of the rear axle. This makes the two-seater coupé a centre-engine vehicle. However, the location of the fuel tank directly behind the driver's seat is less safe in case of a rear collision. The car has McPherson struts both front and rear, rack and pinion steering, with the pinion shifted to the middle to obtain longer rack travel, and larger wheel angles. An angled intermediate shaft connects it to the collapsible safety steering column.

The radiator, which is augmented by an electric fan, is located at the front, necessitating very long and costly water piping. Additional tie rods give precise rear wheel control.

front-wheel drive vehicles are manufactured with V 6 and V 8 engines and performances in excess of 150 kW.

However, this type of drive is not suitable for commercial vehicles as the rear wheels are highly loaded and the front wheels only slightly. Nevertheless, some light commercial vehicle manufacturers accept this disadvantage so they can lower the load area and offer more space or better loading conditions (Fig. 1.32). The propshafts necessary on standard passenger cars would not allow this.

1.6.1 Types of design

1.6.1.1 Engine mounted longitudinally in front of the axle In-line or V engines mounted in front of the axle – regardless of the wheelbase – give a high front axle load, whereby the vehicle centre of gravity is pushed a long way forwards (Fig. 1.33). Good handling in side winds and good traction, especially in the winter, confirm the merits of a high front axle load, whilst the heavy steering from standing (which can be equalized by power-assisted steering), distinct (and sometimes too much) understeering during cornering and poor braking force distribution would be evidence against it.

This type of design is preferred in the larger saloons when transverse mounting of relatively large engines is not possible due to space constraints. The first vehicles of this type were the Audi 80 and 100. Inclining the in-line engine and placing the radiator beside it means the front overhang length can be reduced.

Automatic gearboxes need more space because of the torque converter. This space is readily available with a longitudinally mounted engine.

1.6.1.2 Transverse engine mounted in front of the axle In spite of the advantage of the short front overhang, only limited space is available between the front wheel housings (Figs 1.34 and 1.35). This restriction means that engines larger than an in-line four cylinder or V 6 cannot be fitted in a medium-sized passenger car. Transverse, asymmetric mounting of the engine and gearbox may also cause some performance problems. The unequal length of the drive shafts affect the steering. During acceleration the vehicle

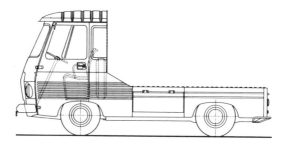

Fig. 1.32 The low cargo area on the Peugeot light commercial vehicle J 5/J 7 is achieved due to front-wheel drive and a semi-trailing link axle to the rear (similar to the one in Fig. 1.45a).

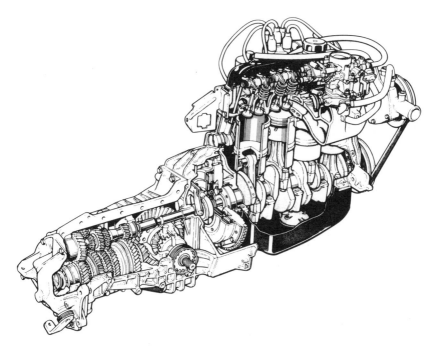

Fig. 1.33 In front-wheel drive vehicles the engine can be mounted longitudinally in front of the front axle with the manual gearbox behind. The shaft goes over the transverse differential (photograph: Renault).

rises and the drive shafts take on different angular positions, causing uneven moments around the steering axes.

The difference between these moments to the left and to the right causes unintentional steering movements resulting in a noticeable pull to one side (Fig. 3.67a); drive shafts of equal length are therefore desirable.

The large bending angle of the short axle shaft can also limit the spring travel of the wheel. To eliminate the adverse effect of unequal length shafts, passenger cars with more powerful engines have an additional bearing next to the engine and an intermediate shaft, the ends of which take one of the two sliding cv joints with angular mobility (Figs 1.36 and 1.9b). Moreover, 'bending oscillation' of the long drive shaft can occur in the main driving range. Its natural frequency can be shifted by clamping on a suppression weight (Fig. 4.0).

1.6.2 Advantages and disadvantages of front-wheel drive

Regardless of the engine position, front-wheel drive has numerous advantages:

• there is load on the steered and driven wheels

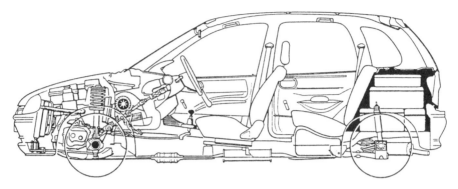

Fig. 1.34 Compact power train unit on the Vauxhall Corsa. The engine is transverse-mounted with the gearbox on the left. The McPherson front axle and safety steering column can be seen clearly.

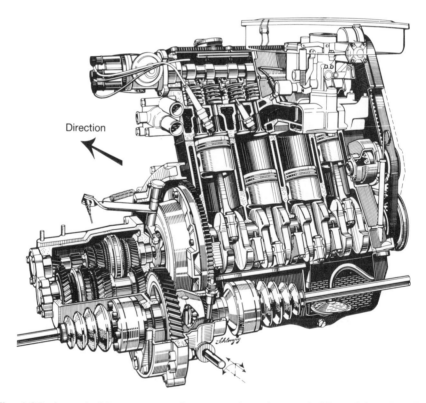

Fig. 1.35 Layout of transverse engine, manual gearbox and differential on the VW Polo. Because the arrangement is offset, the axle shaft leading to the left front wheel is shorter than that leading to the right one. The shifter shaft between the two can be seen clearly. The total mechanical efficiency should be around $\eta \approx 0.9$.

Fig. 1.36 Gearbox unit on the Lancia Thema, located beside the transverse engine and between the front axle McPherson struts. Owing to the high engine performance, the design features two equal-length axle shafts joined by an intermediate shaft. There are also internally ventilated disc brakes.

- good road-holding, especially on wet roads and in wintry conditions; the car is pulled and not pushed (Fig. 1.20)
- good drive-off and sufficient climbing capacity with only few people in the vehicle (Fig. 6.22)
- understeering in cornering, therefore self-braking (Fig. 2.29)
- insensitive to side wind (Fig. 3.96)
- the front axle is loaded due to the weight of the drive unit, although the steering is not necessarily heavier (in comparison with standard cars) during driving
- simple rear axle design (a rigid axle is not necessarily a disadvantage)
- long wheel base making high ride comfort possible
- short power flow because the engine, gearbox and differential form a compact unit
- good engine cooling (radiator in front), and an electric fan can be fitted
- effective heating due to short paths
- smooth car floor pan
- exhaust system with long path (important on cars with catalytic converters)
- a large boot with a favourable crumple zone for rear end crash.

The disadvantages are:
- under full load, poorer drive-off capacity on wet and icy roads and on inclines (Figs 1.21 and 6.22)
- with powerful engines, increasing torque and vibrational influence on steering, also the engine length is limited

- with high front axle load, high steering ratio or power steering is necessary (see Section 4.2)
- with high located, dash-panel mounted rack and pinion steering, centre take-off tie rods become necessary (Figs 1.40 and 4.24) or significant kinematic toe-in change practically inevitable (Fig. 3.52a)
- the power plant mounting has to absorb the engine moment times the total gear ratio (Figs 3.85, 6.20 and Equation 6.36)
- it is difficult to design the power plant mounting; booming noises, resonant frequencies in conjunction with the suspension, tip in and let off torque effects etc., need to be suppressed
- with soft mountings, wavy road surfaces excite the power plant to natural frequency oscillation (so-called 'front end shake', see Section 5.1.3)
- there is bending stress on the exhaust system from the power plant movements during drive-off and braking (with the engine)
- there is a complex front axle, so inner drive shafts need a sliding cv joint (Figs 1.36a and 1.29)
- the turning and track circle is restricted due to the limited bending angle (up to 50°) of the drive joints (see Section 3.7.2)
- there is high sensitivity in the case of tyre imbalance and non-uniformity on the front wheels
- there is higher tyre wear in front, because the highly loaded front wheels are both steered and driven
- poor braking force distribution (about 75% to the front and 25% to the rear)
- there is a complex gear shift mechanism which can also be influenced by power plant movements.

The disadvantage of the decreased climbing performance on wet roads and those with packed snow can be compensated with a drive slip control (ASR) or by shifting the weight to the front axle. On the XM models, Citroën has moved the rear axle a long way to the rear resulting in an axle load distribution of about 65% to the front and 35% to the back. The greater the load on

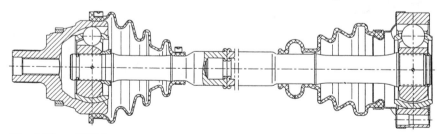

Fig. 1.36a GKN front-wheel drive shaft with a CV slip joint on the gearbox side and a fixed CV joint on the outside. The maximum joint angles are 22° and 50° with shift travel up to 48 mm possible. To take the wheel hub, the outer joint casing has a spigot with a short involute profile and a fine thread (inside) into which a shank bolt is inserted and tightened to a certain torque to ensure that the hub sits securely on the profile. To avoid vibration and save weight, the annealed steel shaft parts are hollow-drilled and joined together by friction welding.

the front wheels, the more the car tends to understeer, causing adverse steering angles and heavy steering, which makes power steering mandatory (see Section 4.2.5).

1.6.3 Driven front axles

The following are fitted as front axles on passenger cars, estate cars as well as light commercial vehicles:

- double wishbone suspensions
- McPherson struts, and (only in very few cases)
- damper struts (Fig. 1.63).

On double wishbone suspensions the drive shafts require free passage in those places where the coil springs are normally located on the lower suspension control arms. This means that the springs must be placed higher up (Fig. 1.37) with the disadvantage that (as on McPherson struts) vertical forces are introduced a long way up on the wheel house panel. It is better to leave the springs on the lower suspension control arms and to attach these to the stiffer

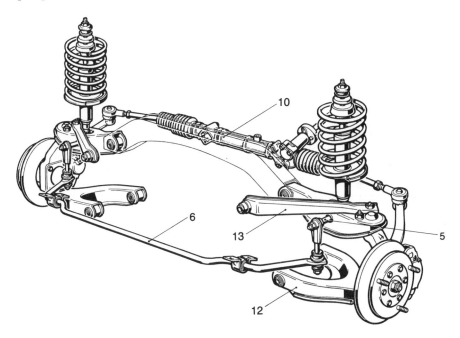

Fig. 1.37   Double wishbone front axle fitted by Renault in several models. To make space for the drive shafts, the spring shock absorbers were placed on the upper control arm. The rack and pinion steering 10 is fixed to one body side cross member. The screws 5 act as a force centre; joining both the spring shock absorbers and the rod forces leading to the anti-roll bar 6. The supporting ball joint, which is located up in the wheel hub carriers, absorbs all the vertical forces. Low-friction guiding joints can therefore be fitted in the lower links.

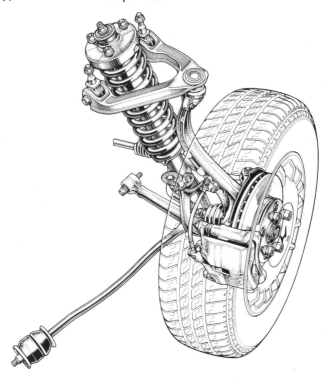

Fig. 1.38 Double wishbone front suspension on the Honda models Prelude and Accord with short upper wishbones with widely spaced bearings, lower transverse control arms and longitudinal rods whose front mounts absorb the dynamic rolling hardness of the radial tyres. The spring shock absorbers are supported via fork-shaped struts on the transverse control arms and are fixed within the upper link mounts. This point is a good force centre. Despite the fact that the upper wheel carrier joint is located high, which gives favourable wheel kinematics, the suspension is compact and the bonnet can be low to give aerodynamic advantages. The large effective distance c between the upper and lower wheel hub carrier joints, which can be seen in Fig. 1.1b, results in lower forces in all mounts and therefore less elestic deflection and better wheel control.

body area where the upper control arms are fixed. Shock absorbers and springs can be positioned behind the drive shafts or sit on split braces, which grip round the shafts and are jointed to the lower suspension control arms. The axle is flatter and the front end (hood) can be drawn further down. Figure 1.38 shows this type of design. The upper suspension control arms are relatively short and have mountings that are wide apart. This increases the width of the engine compartment and the spring shock absorber unit can also be taken through the suspension control arms, however, sufficient clearance to the axle shaft is prerequisite. Due to the slight track width change, the change of camber becomes favourable. Furthermore, the inclination of the control arms provides an advantageous pitch pole position and anti-dive when braking.

Most front-wheel drives coming onto the market today have McPherson struts. It was a long time after their use in standard design cars that McPherson struts were used at the front axle on front-wheel drive vehicles. The drive shaft requires passage under the damping part (Fig. 1.39). This can lead to a shortening of the effective distance 1–o, which is important for the axle (Fig. 1.6), with the result that larger transverse forces F_{Cy} and F_{Ky} occur on the piston and rod guide and therefore increase friction.

On front-wheel drive vehicles there is little space available to fit rack and pinion steering. If the vehicle has spring dampers or damper struts, and if the steering gear is housed with short outer take-off tie rods, a toe-in change is almost inevitable (Figs 4.3 and 3.52a). A high steering system can readily be attached to the dash panel (Fig. 1.40), but a centre take-off is then necessary and the steering system becomes more expensive (Figs 4.0, 4.9 and 4.24). Moreover, the steering force applied to the strut is approximately halfway between mountings E and G (Fig. 1.6).

The inevitable, greater yield in the transverse direction can increase the steering loss angle and make the steering less responsive.

1.6.4 Non-driven rear axles

1.6.4.1 Compound crank axle   The compound crank axle could be described as the new rear axle design of the 1970s (Figs 1.41 and 1.1) and it is still used in today's small and medium-sized front-wheel drive vehicles. It consists of two trailing links that are welded to a twistable cross member. This member absorbs all vertical and lateral force moments and, because of its offset to the wheel centre, must be less torsionally stiff and function simultaneously as an anti-roll bar. The axle has numerous advantages and is therefore found on a number of passenger cars which have come onto the market.

From an installation point of view:

- the whole axle is easy to assemble and dismantle
- it needs little space
- a spring damper unit or the shock absorber and springs are easy to fit
- no need for any control arms and rods, and thus
- only few components to handle.

From a suspension point of view:

- there is a favourable wheel to spring damper ratio
- there are only two bearing points O_l and O_{rs}, which hardly affect the springing (Fig. 1.42)
- low weight of the unsprung masses (see Section 6.1.3) and
- the cross-member can also function as an anti-roll bar.

From a kinematic point of view:

- there is negligible toe-in and track width change on reciprocal and parallel springing

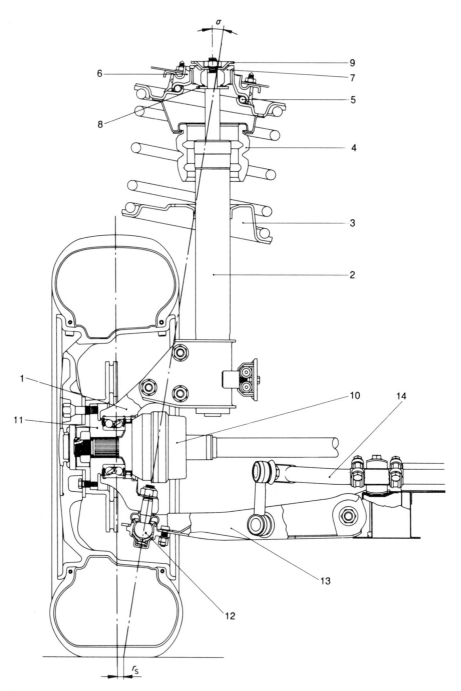

Fig. 1.39 Lancia front axle. The McPherson strut consists of the wheel hub carrier 1 and the damping part 2; the two are connected by three screws. The lower spring seat 3 sits firmly on the outer tube and also acts as a buffer for the supplementary spring 4. This surrounds the outer tube 2 giving a longer bearing span (path l – o,

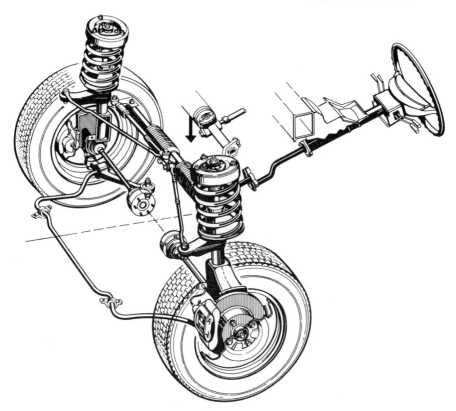

Fig. 1.40 Driven McPherson front axle on different Audi models. The dynamic rolling hardness of the radial tyre is absorbed by the rubber bearings shown in Fig. 3.66, which sits in the lower transverse control arms. These bearings take the arms of the anti-roll bar, which act as a trailing link. To avoid greater toe-in changes when the wheels are bottoming out, centre take-off tie rods are used on the highly located rack and pinion steering (Fig. 4.24). The steering damper is fastened to these rods and to the end of the steering rack. The engine is mounted longitudinally, which means the drive shafts are of equal length (see Section 3.6.5.3).

Fig. 1.6). The bearing 5 is positioned diagonally and matches the offset position of the coil spring. The rubber bearing 6 absorbs the spring forces, and the softer part 7 absorbs the forces generated by the damping. Disc 8 acts as a compression buffer and plate 9 acts as a rebound buffer for this elastic bearing. Both parts come into play if the damping forces exceed certain values. This strut mount design is called 'decoupled', i.e. springing and damping forces are absorbed separately (see Fig. 1.5).

The centre of the CV joint 10 lies in the steering axis and the wheel hub 11 fits onto a two-row angular (contact) ball bearing. Guiding joint 12 sits in a cone of the wheel hub carrier 1 and is bolted to the lower transverse control arm 13. Inelastic ball joints provide the connection to the anti-roll bar 14. The steering axis inclination σ between the centre point of the upper strut mount and guiding joint 12 and the (here slightly positive) kingpin offset at ground (scrub radius) r_S are included.

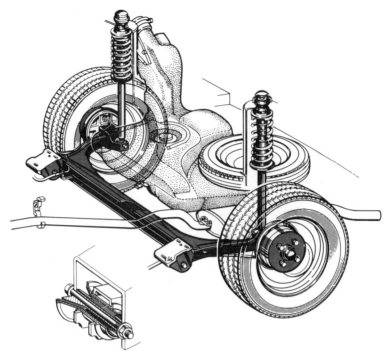

Fig. 1.41 Compound crank rear axle on the VW Golf. The toe-aligning bearing, which connects the axle to the body, is shown enlarged bottom left.

It has the advantage of barely yielding under longitudinal forces ($\pm F_{O,x}$, Fig. 1.42), if actual cornering forces $F_{y,W,r,o}$ and $F_{y,W,r,i}$ occur in the lateral direction. The spring shock absorbers, positioned vertically in the centre of the axle (see Fig. 5.36), the down-sized special spare wheel and the position of the plastic fuel tank are clearly shown.

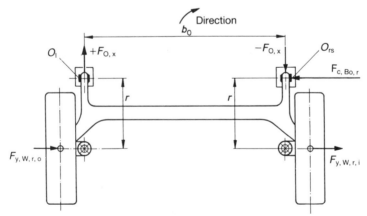

Fig. 1.42 The lateral forces $F_{y,W,r,o}$ and $F_{y,W,r,i}$ occurring at the centres of tyre contact during cornering are absorbed at the bearing points O_l and O_{rs}. This results in a moment $M_y = (F_{y,W,r,o} + F_{y,W,r,i})\, r = F_{O,x}\, b_O$ which (depending on the elasticity of the rubber bearing) can cause 'lateral force oversteering'. The longer the control arms (distance r) and the closer points O_l and O_{rs} (distance b_O), the greater the longitudinal forces $\pm F_{Ox}$.

- there is a low change of camber under lateral forces (Figs 3.45 and 3.46b)
- there is low load-dependent body roll understeering of the whole axle (Fig. 3.61) and
- good pitch pole positions O_l and O_{rs} (Fig. 1.42), which reduce tail-lift during braking.

There are only a few disadvantages:

- a tendency to compliance oversteer due to lateral force (Fig. 3.57)
- torsion and shear stress in the cross-member
- high stress in the weld seams, which means
- the permissible rear axle load is limited in terms of strength.

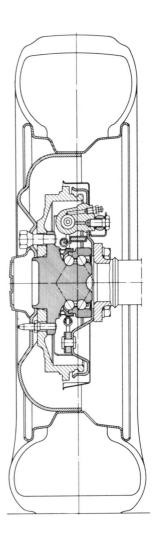

Fig. 1.43 Rear wheel bearing on the Fiat Panda with a third generation, two-row angular (contact) ball bearing. The wheel hub and inner ball bearing ring are made of one part, and the square outer ring is fixed to the rigid axle casing with four bolts (picture: SKF).

1.6.4.2 Rigid axle Non-driven rigid axles can be lighter than comparable independent wheel suspensions. Their advantages outweigh the disadvantages because of the almost non-variable track and camber values during drive. Figure 1.13a illustrates an inexpensive yet effective design:

- axle casing in steel tubing
- suspension on single leaf springs.

The lateral and longitudinal wheel control characteristics are sufficient for passenger cars in the medium to small vehicle range and delivery vehicles. The resultant hard springing is acceptable and may even be necessary because of the load to be moved. The wheel bearing can be simple on such axles (Fig. 1.43). Faster, more comfortable vehicles, on the other hand, require coil springs and, for precise axle control, trailing links and a good central guide (Fig. 1.43a) or Panhard rod. This is generally positioned behind the axle (Fig. 1.44).

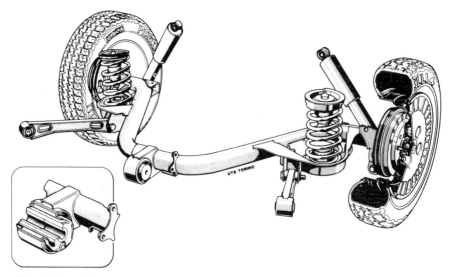

Fig. 1.43a 'Omega' rear wheel suspension on the Lancia Y 10 and Fiat Panda, a drawbar axle with a U-shaped tube, drum brakes, inclined shock absorbers and additional elastomer springs seated inside the low positioned coil springs. The rubber element in the shaft axle bearing point, shown separately, has cut-outs similar to the front bearings of the two longitudinal trailing links and achieves the longitudinal elasticity necessary for radial tyre dynamic rolling hardness. The middle bearing point is also the body pitch pole. The body roll centre is located in the centre of the axle but is determined by the level of the three mounting points on the body. The lateral forces are absorbed here. The angled position of the longitudinal trailing links is chosen to reduce the lateral force oversteering that would otherwise occur (shown in Figs 1.42 and 3.62). The coil springs are located in front of the axle centre and so have to be harder, with the advantage that the body is better supported on bends.

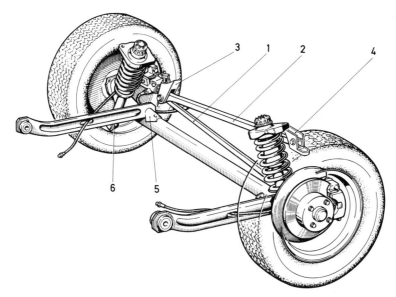

Fig. 1.44 Torsion crank axle on the Audi 100 with spring dampers fixed a long way out at points 6 and which largely suppress body roll vibrations. The longitudinal control arms therefore had to be welded further in to the U profile acting as a cross-member and reinforced by shoe 5. The U-profile is also raised at the side to achieve higher torsional resistance. The anti-roll bar is located inside the U-profile.

Brace 2 distributes the lateral forces coming from the Panhard rod 1 to the two body-side fixing points 3 and 4. Bar 1 is located behind the axle, and the lateral force understeering thus caused and shown in Fig. 3.63a, could be largely suppressed by the length of the longitudinal control arms. Furthermore, it was possible to increase the comfort and to house an 80 l fuel tank as well as the main muffler in front of the axle. The only disadvantage is that the link fixing points, and therefore the body pitch pole O_r, moves further forward and, as a result, reduces the 'anti-dive' described in Fig. 3.123.

1.6.4.3 Independent wheel suspension

An independent wheel suspension is not necessarily better than a rigid axle in terms of handling properties. The wheels incline with the body and the lateral grip characteristics of the tyres decrease (Figs 3.44 and 3.45), and there are hardly any advantages in terms of weight (see Section 6.1.3). This suspension usually needs just as much space as a compound crank axle.

Among the various types, McPherson struts (Fig. 1.7), semi-trailing or trailing link axles (Figs 1.45a, 1.8 and 1.34) and double wishbone suspensions, which have grown in popularity for some years now, and different forms of so-called multi-link axles (Figs 1.45 and 1.0) are all used. The latter are currently the best solution due to:

- kinematic characteristics
- elastokinematic behaviour

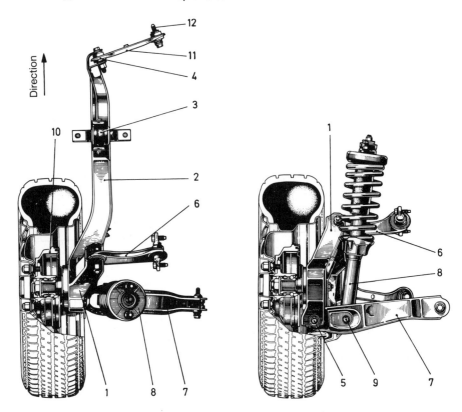

Fig. 1.45 Top view and rear view of the double wishbone rear axle on the Honda Civic. The trailing arm 2, which is stiff under flexure and torsion, and the wheel hub carrier 1 form a unit and, along with the two widely spaced lower transverse control arms 7 and 11, ensure precise wheel control and prevent unintentional toe-in changes. The rubber bearing in point 3, which represents the so-called 'vehicle pitch pole' O_r, provides the real longitudinal wheel control of the axle (Fig. 3.123). The lateral control of part 1 is performed by the short upper transverse control arm 6 and the longer lower one 7, which accepts the spring shock absorber 8 in point 9. The length difference in the control arms gives favourable camber and track width change (Figs 3.4 and 3.41).

During braking, bearing 3 yields in the longitudinal direction and, due to the angled position of the links 11 when viewed from the top, the front point 4 moves inwards (Fig. 3.64) and the wheel goes into toe-in. Behaviour during cornering is similar: the axle understeers due to lateral force and body roll (see Sections 3.6.3 and 3.6.4 and Figs 1.0 and 1.59). The wheel is carried by 'third generation' angular (contact) ball bearings on which the outside ring is also designed as a wheel hub. In models with smaller engines, brake drums (item 10) are used, which are fixed to the wheel hub.

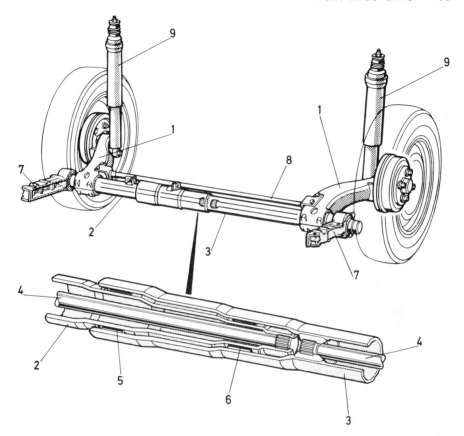

Fig. 1.45a Compact trailing link rear axle, fitted by Renault to less powerful medium-sized vehicles. The short torsion bar springs grip into the guide tubes 2 and 3 in the centre of the vehicle. Parts 2, 3 and 4 are jointly subjected to torsional stresses and so the torsional stiffness of the transverse tubes contributes to the spring rate. On the outside, the cast trailing arms 1 are welded to the transverse tubes, which (pushed into each other) support each other on the torsionally-elastic bearings 5 and 6. This creates a sufficiently long bearing basis, which largely prevents camber and toe-in changes when forces are generated.

The entire assembly is fixed by the brackets 7 in which both the guide tubes 2 and 3 are mounted and rotate as well as the outer sides of the two torsion bars 4. The two arms thus transfer all vertical forces plus the entire springing moment to the body. The anti-roll bar 8 is connected to the two trailing arms via two U-shaped tabs. The two rubber bearings 5 and 6 located between the tubes 2 and 3 also contribute to the stabilizing effect. The compression and rebound bumpers are fitted into the shock absorber 9 (see Section 5.6.8). As shown in Fig. 1.1, on the newer models the dampers would be inclined so that they can be fixed to the side members of the floor pan which also leads to more space between the wheel housings.

- space requirements
- axle weight
- the possibility of being able to retrofit the differential on four-wheel drive (Figs 1.59 and 1.0). See also Section 1.4.3.

1.7 Four-wheel drive

All wheels of a passenger car or commercial vehicle are continuously – in other words permanently – driven, or one of the two axles is always linked to the engine and the other can be selected manually or automatically. This is made possible by what is known as the 'centre differential lock'.

This section deals with the most current four-wheel drive designs. In spite of the advantages of four-wheel drive – as shown in Fig. 1.46 – suitable tyres should be fitted in winter.

1.7.1 Advantages and disadvantages

In summary, the advantages of passenger cars with permanent four-wheel drive over those with only one driven axle are:

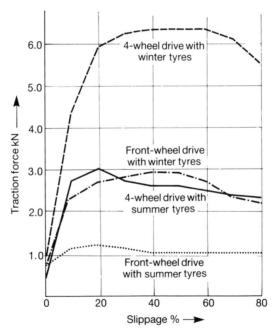

Fig. 1.46 With a loaded Vauxhall Cavalier on compacted snow ($\mu_{x,W} \approx 0.2$) driving forces are measured on the flat as a function of the slip (Fig. 2.21). The illustration shows the advantage of four-wheel drive, and the necessity, even with this type of drive, of fitting correct tyres. Regardless of the type of drive, winter tyres also give shorter braking (stopping) distances on these road surface conditions.

- better traction on smooth surfaces in all road conditions, especially in wet and wintry weather (Figs 1.46 and 1.47)
- an increase in the drive-off and climbing capacity regardless of load
- better acceleration in low gear, especially with high engine performance
- reduced sensitivity to side wind
- stability reserves when driving on slush and compacted snow tracks
- better aquaplaning behaviour
- particularly suitable for towing trailers
- more balanced axle load distribution
- minor torque steer effect (Figs 2.38 and 3.57)
- even tyre wear.

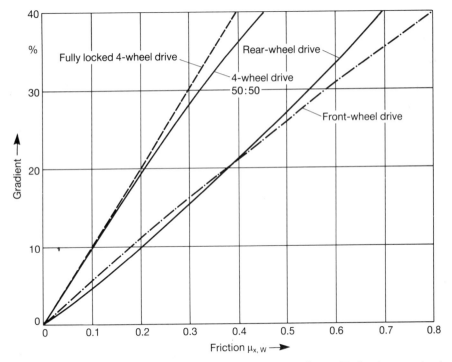

Fig. 1.47 Hill-climbing capacity on a homogeneous surface with front, rear-wheel and four-wheel drive, and with locked centre differential and a driving force distribution of 50%/50% on four-wheel drive. Of the cars studied, the front/rear axle load distribution was:

front-wheel drive 57%/43%
rear-wheel drive 51%/49%
four-wheel drive 52%/48%
(see Fig. 1.21).

However, the disadvantages given below should not be ignored:

- acquisition costs
- around 6% to 10% higher kerb weight of the vehicle
- generally somewhat lower maximum speed
- 5% to 10% increased fuel consumption
- lack of ABS-compatibility of some systems
- not always clear cornering behaviour
- smaller boot compared with front-wheel drive vehicles.

The available engine moment is transmitted to all four wheels, so there must be an equalization mechanism or a 'centre differential lock' in the prop-shaft between the front and rear axles (Figs 1.54 and 1.55), and a front and rear differential. However, on roads with different coefficients of friction on the left and right wheels, known as 'μ-split', and with traditional differentials, each driven axle can, at most, transmit double the propulsion force of the wheels running on the side with the lower coefficient of friction (μ-low). Higher driving forces can only be achieved with an 'axle differential lock'. This can only be 100% effective on the rear axle (Fig. 1.48) as, at the front, there would no longer be problem-free steering control. The lock partially or completely stops equalization of the number of revolutions between the left and right wheel of the respective axle and prevents wheelspin on the μ-low side.

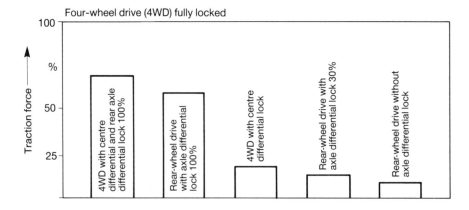

Fig. 1.48 Influence of the type of drive and differential lock on the propulsion force with 'μ split', in other words a slippery road surface with $\mu_{x,W} = 0.1/0.8$ on one side only. 100% locking of the rear axle differential gives most benefits. Some car manufac-turers offer this option as ASR (or EDL) or using a hydraulic manual selection clutch (ASD at Mercedes-Benz). However, only 25% to 40% locking is provided on the multi-disc, limited slip differentials which have usually been fitted on vehicles to date.

In passenger cars, automatic locking differentials are used between front and rear axles. These can operate mechanically – such as multi-disc limited slip differentials, the Torsen differential (Fig. 1.53), or based on fluid friction (visco lock, Fig. 1.56) – and produce a locking degree of usually 25% to 40%. Higher values severely impede cornering due to the tensions in the power train (Fig. 1.51) and are therefore only accepted for straight ahead driving or in motor sport, which takes up to 80% of these differentials.

Automatic locking differentials (ASD, see Section 1.7.5) deliver a locking effect that is optimally matched to every operating situation; the locking effect is produced by an electronically controlled and hydraulically operated multi-disc lock, which can be varied by adjusting the oil pressure. On pure front or rear-wheel drive vehicles, drive-slip control systems (ASR) are a substitute for four-wheel drive to a certain extent. With these systems, the spinning wheel can be influenced both by engine action via the throttle valve, ignition point or injection (or both) or via braking action. The combination of engine and brake control makes these systems superior to automatic locking differentials and they are preferable if the vehicle is fitted with an anti-lock braking system (ABS) and the wheel revolution speed signals are already available.

This might be the reason why an economical ASR combined with ABS is increasingly in demand on saloons, estate cars and coupés.

The all-terrain, general purpose passenger car (Fig. 1.49) is a different story, and here four-wheel drive is essential. However, ABS can only be installed provided the vehicle is designed to be compatible with the system. Reference is made to this characteristic in the following sections. It can, for

Fig. 1.49 The Mercedes G all-terrain vehicle, a so-called 'all-purpose passenger car', has high ground clearance and short overhangs both front and rear. This makes it particularly suitable for off-road driving.

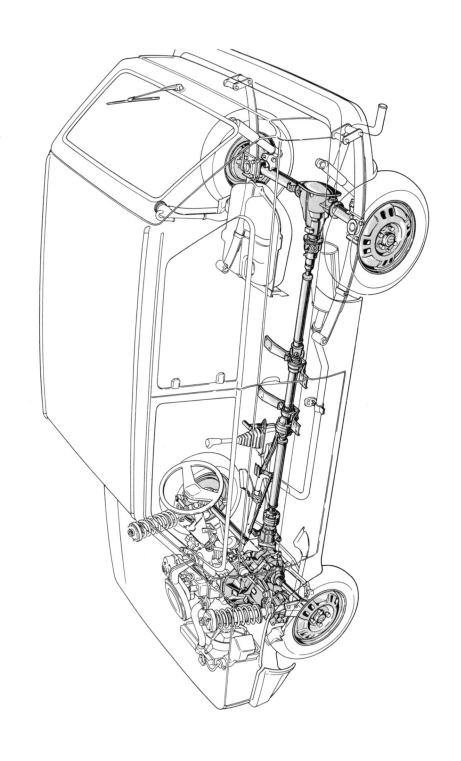

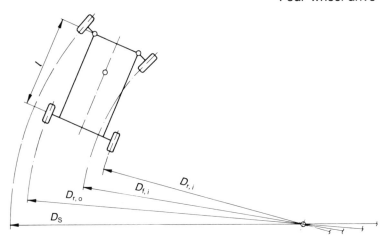

Fig. 1.51 The front wheel on the outside of the bend draws the largest arc during slow cornering, the track circle diameter D_S, whilst the inner wheel draws the considerably smaller arc $D_{f,i}$. This is the reason for the differential in the driven front axle of the front-wheel drive. The bend diameters $D_{r,o}$ and $D_{r,i}$ to the rear are even smaller, so the rolling distance of the two wheels of this axle decreases further and there can be tensions in the drive train if both axles are rigidly connected, a bend is being negotiated, or when a dry road surface makes wheel slip more difficult because of high coefficients of friction.

instance, be achieved with a Torsen central differential, an overrunning clutch or by disengaging the rear wheel drive during braking. Unfortunately, these measures increase both the cost and the weight.

1.7.2 Manual selection four-wheel drive on passenger cars and estate cars

The manual selection four-wheel drive (without centre differential) only distributes the engine moment to all four wheels when needed. This design is popular on smaller and low-priced vehicles, which do not have ABS anyway. The design complexity, and therefore the costs, are lower than on permanent drive. Usually there is no rear axle differential lock, a feature that is important where the road is extremely slippery, but would result in price and weight disadvantages.

Fig. 1.50 The Fiat Panda Treking 4 × 4, a passenger car based on front-wheel drive with transverse engine. The vehicle has McPherson struts at the front and a rigid axle with longitudinal leaf springs at the back. The propshaft leading to it is divided into three to be able to take the rotational movements of the rigid axle around the transverse (y) axis during drive-off and braking and to absorb movements of the drive unit. The Fiat Panda is an estate car with the ratio

$$k_1 = \frac{2159 \text{ mm}}{3689 \text{ mm}} = 0.59 \text{ (see Equation 3.1a)}$$

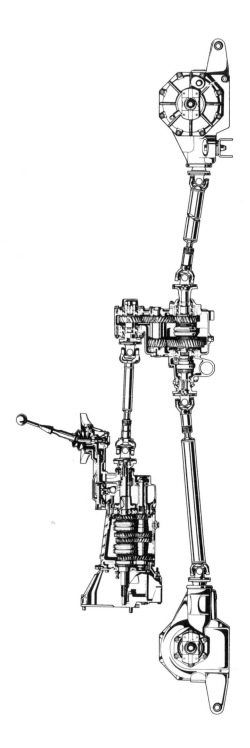

Fig. 1.52 Complex power distribution on the Fiat Campagnolo, a four-wheel drive, all-purpose passenger car. The drive moment is transferred from the manual gearbox via a centrally located two-gear power take off gear to the differentials of the front and rear axles. Efficiency is not likely to be especially good.

Front-wheel drive is suitable as a basic version and the longitudinal engine certainly has advantages here (Fig. 1.33). However, as it requires a lot of space, it is no longer found in the vehicle classes that are dealt with here. With the transverse engine, the force from the manual gearbox is transmitted via a bevel gear (item 3 in Fig. 1.55) and a divided propshaft, to the rear axle with a differential (Fig. 1.50). There is relatively little additional complexity compared with the front-wheel drive design, even if, on the Fiat Panda (Trecking 4 x 4), there is a weight increase of about 11% (90 kg), not least because of the heavy, driven, rigid axle. It is possible to select rear-wheel drive during a journey using a shift lever that is attached to the cardan tunnel.

Manual selection on the Subaru Justy operates pneumatically at the touch of a button (even whilst travelling). This vehicle has independent rear-wheel suspension and weighs only 6% more than the basic vehicle with front-wheel drive. Traction is always improved considerably if the driver recognizes the need in time and switches the engine force onto all four wheels. In critical situations, this usually happens too late, and the abrupt change in drive behaviour becomes an additional disadvantage. Conversely, if the driver forgets to switch to single axle drive on a dry road, tensions occur in the power train during cornering, as the front wheels travel larger arcs than the back ones (Figs 1.51 and 3.70). The tighter the bend, the greater the stress on the power train and the greater the force to detrimental tyre slip.

A further problem is the braking stability of these vehicles. If the front axle locks on a wet or wintry road during unplanned braking, the rear one is taken with it due to the rigid power train. All four wheels lock simultaneously and the car goes into an uncontrollable skid.

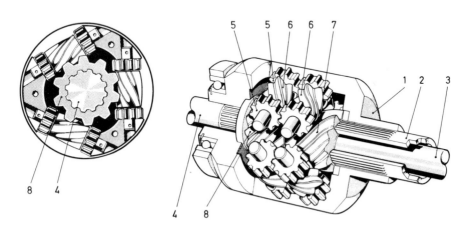

Fig. 1.53 Torsen central differential fitted in all Quattro models by Audi. It consists of two worm gears, which are joined by spur gears and, depending on the traction requirement, can distribute the driving torque up to 75% to the front or rear axle. Under normal driving conditions 50% goes to both axles.

Fig. 1.54 Vauxhall Cavalier and Calibra 4 × 4. For reasons of noise and vibration and because of the movement of the drive unit during drive-off and braking (with the engine) around the transverse (y)-axis, the propshaft is divided in two places. The propshaft connects the power take-off gear with the rear axle drive. The drive casing is fixed to the suspension subframe with elastic bearings. McPherson struts are provided as the front axle and semi-trailing links are fitted to the rear (Fig. 1.9).

1.7.3 Manual selection four-wheel drive on commercial and all-terrain vehicles

The basis for this type of vehicle is the standard design which, because of the larger ground clearance necessary in off-road vehicles (Fig. 1.49), has more space available between the engine and front axle differential and between the cargo area and the rear axle. Figure 1.52 shows the design details:

- a central power take-off gear with manual selection for the front axle, plus a larger ratio off-road gear, which can be engaged if desired
- three propshafts
- complex accommodation of the drive joints if there is a rigid front axle (Fig. 1.1a).

1.7.4 Permanent four-wheel drive; basic passenger car with front-wheel drive

All four wheels are constantly driven; this can be achieved between the front and rear axle with different design principles:

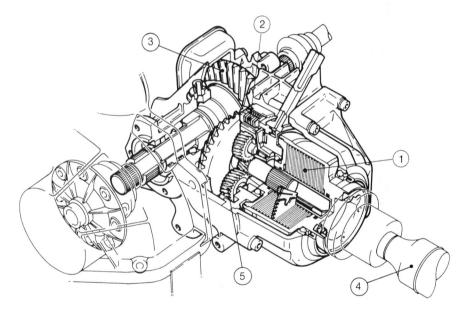

Fig. 1.55 Power take-off gear on the Vauxhall Cavalier and Calibra 4 × 4. The transverse engine makes the bevel gear 3 for power distribution to the longitudinal propshaft 4 necessary. The planet gear centre differential 5, the visco clutch 1 and the clutch 2 sit between the two. Whilst the clutch 1 automatically regulates the drive moment distribution to the rear axle, hydraulic multi-disc clutch 2 switches off the drive to the rear axle within fractions of a second each time braking occurs. The vehicles are thus fully ABS-compatible.

- a bevel centre differential with or without manual lock selection
- a Torsen centre differential with moment distribution, based on the traction requirement (Fig. 1.53)
- a planet gear central differential with fixed moment distribution and additional visco clutch, which automatically takes over the locking function when a difference in the number of revolutions occurs (Figs 1.54 and 1.55) or a magnet clutch (which is electronically controlled, Fig. 1.61) or

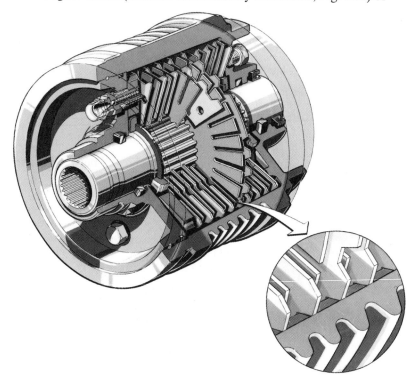

Fig. 1.56 Visco clutch on the VW Golf syncro with slip-dependent drive moment distribution. Two different packages sit in the closed drum-shaped housing: radially slit steel discs, which are moved by the serrated profile of the hollow shaft, and perforated discs which grip (as can be seen below) into housing keys. The shaft is joined with the differential and the casing with the propshaft going to the rear axle.

The discs are arranged in the casing so that a perforated disc alternates with a slit one. The individual parts have no definite spacing but can be slid against one another axially. The whole assembly is filled with viscous silicone fluid and the torque behaviour (therefore the locking effect) can be adjusted via the filling level.

If slip occurs between the front and rear axle, the sets of discs in the clutch rotate relative to one another and shearing forces are transferred via the silicone fluid. These increase with increasing slip and ensure a moment increase in the rear axle. The power consumed in the visco clutch leads to warming and thus to growing inner pressure. This causes an increase in the transferable torque which, under conditions of extreme torque requirement, ultimately leads to an almost slip-free torque transfer (rigid drive).

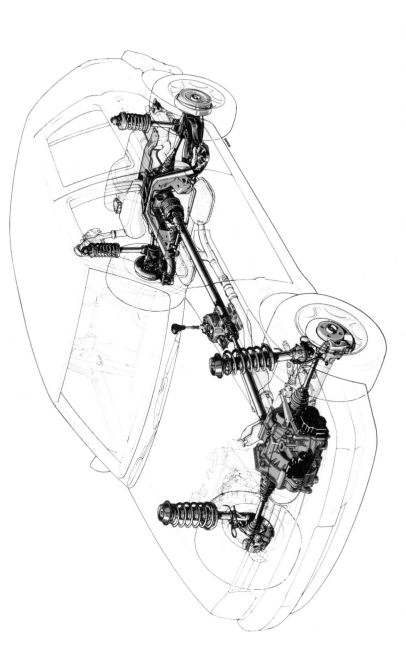

Fig. 1.57 VW Golf syncro with bevel gear behind the transverse mounted drive unit, McPherson struts on the front and semi-trailing links with spring dampers on the rear. The capacity of the fuel tank located in the rear remains at 55 l, while this model weighs around 90 kg more than the basic design. Boot capacity was reduced as the rear axle needed more space. The three-piece propshaft has to absorb the movement of the drive unit during drive-off and braking (with the engine). The visco clutch is located in front of the rear axle differential.

The VW Golf is an estate car with the ratio $\quad k_1 = \dfrac{2475\ \text{mm}}{4020\ \text{mm}} = 0.62$ (see Equation 3.1a).

- a visco-clutch in the propshaft power train, which selects the initially undriven axle depending on the tyre slip (Fig. 1.56).

Here too, the front-wheel drive passenger car is suitable as a basic vehicle. In 1979, Audi was the first company to bring out a car with permanent four-wheel drive, the Quattro, and today vehicles with this type of drive are available throughout the entire Audi range. On a longitudinally-mounted engine, a Torsen centre differential distributes the moment according to the traction requirement (Fig. 1.53). The four-wheel drive increases the weight by around 100 kg.

VW used a 'visco-clutch' in the power train (without centre differential) for the first time on the Transporter (Fig. 1.30) and then subsequently used it in the Golf syncro (Figs 1.56 and 1.57). The clutch has the advantage of the engine moment distribution being dependent on the tyre slip. If the slip on the front wheels, which are otherwise driven at the higher moment, increases on a wet or frozen surface or off-road, more drive is applied to the rear

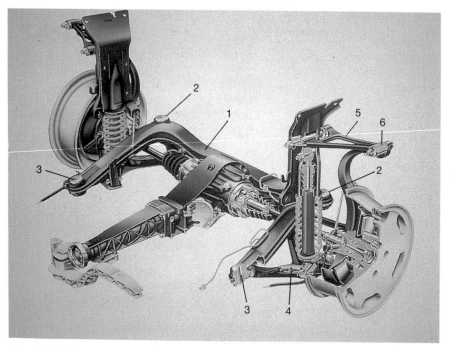

Fig. 1.58 Double wishbone rear axle on the Audi 80 Quattro. The suspension subframe 1 is fixed to the body with four widely spaced rubber mountings (items 2 and 3) and houses the differential casing and transverse control arms (items 4 and 5). The springs and shock absorbers are mounted next to the fixings for the upper control arms. The location 6 of the wheel hub carrier was raised (long base c, Fig. 1.1b) and drawn outwards. The lower transverse control arm 4 is fixed to part 1 with widely spaced mountings. These measures ensure a wide boot and low forces, making it easier to attain the desired kinematic characteristics.

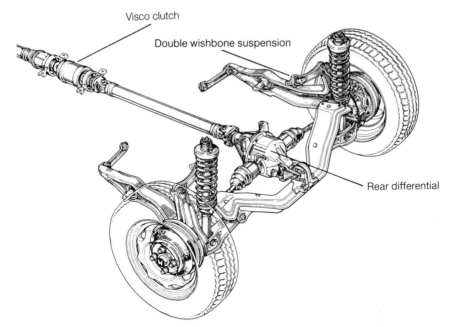

Visco clutch

Double wishbone suspension

Rear differential

Fig. 1.59 Double wishbone rear axle of the Honda Civic Shuttle 4 WD. The visco clutch sits (held by two shaft bearings) in the centre of the divided propshaft. The rear axle differential has been moved forwards and is mounted to the rear on the body via a cross-member. Apart from the different type of wheel bearings and the lower transverse control arm positioned somewhat further back (to make it possible to bring the drive shafts through in front of the spring dampers), the axle corresponds to Fig. 1.45 and resembles the suspension shown in Fig. 1.0.

wheels. No action on the part of the driver is either necessary or possible. The transverse engine makes a bevel gear in front of the split propshaft necessary. The visco clutch sits in the rear differential casing and there is also an overrunning clutch, which ensures that the rear wheels are automatically disengaged from the drive when pushing to guarantee proper braking behaviour. This type of drive is fully ABS-compatible. When reverse is engaged, a sliding sleeve is moved, which bridges the overrunning clutch to make it possible to drive backwards.

When selecting their rear axle design, manufacturers have gone in different directions. Audi fits a double wishbone suspension in the 80 and 100 Quattro (Fig. 1.58), Honda uses the requisite centre differential on the double wishbone standard suspension in the Civic Shuttle 4WD (Figs 1.59 and 1.45) and, in Fig. 1.57, we see the semi-trailing link axle that replaces the standard compound crank axle in the VW Golf syncro (Fig. 1.41).

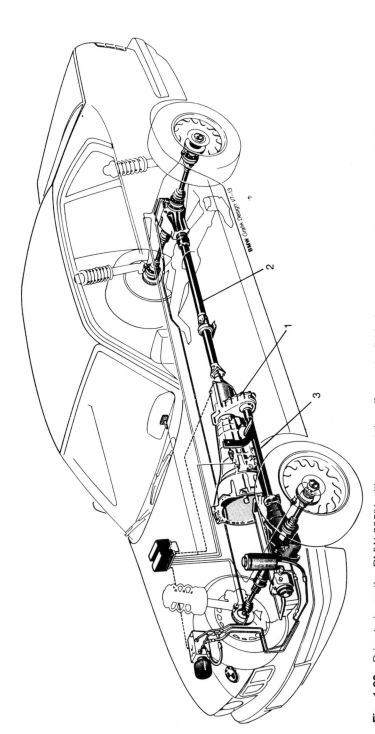

Fig. 1.60 Drive train on the BMW 525iX with power take-off gear 1, divided rear propshaft (item 3). The semi-trailing arm rear axle is the same as the standard design (Fig. 1.28) and the front suspension of the 3 series (Fig. 1.24). Only the McPherson struts were modified on this model to provide space for the drive shaft to pass. The central and rear axle differentials are up to 100% lockable. The BMW 5 vehicle is a notchback saloon with the ratio

$$k_1 = \frac{2761 \text{ mm}}{4720 \text{ mm}} = 0.58 \text{ (see Equation 3.1a)}.$$

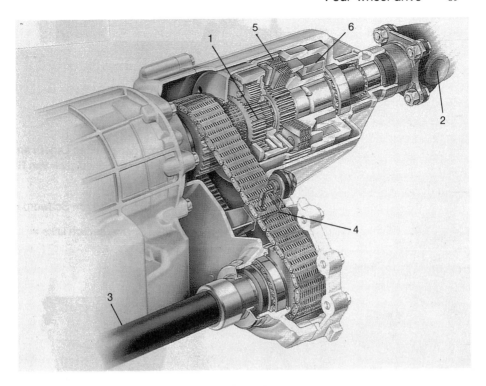

Fig. 1.61 Power take-off gear A 100 produced by ZF and fitted in the BMW 525iX. The torque from the engine is divided by the planet gear centre differential 1 into one torque to the rear propshaft 2 (64%) and one to the front 3 (36%). The serrated chain 4 bridges the offset to this shaft. The multi-disc clutch 5, which is controlled by the solenoid 6, adjusts the driving force distribution.

1.7.5 Permanent four-wheel drive, basic standard design passenger car

Giving a standard design car four-wheel drive requires larger modifications, greater design complexity and makes the drive less efficient (Fig. 1.60). A power take-off gear is required, from which a short propshaft transmits the engine moment to the front differential. A toothed chain bridges the lateral offset (Fig. 1.61). The ground clearance must not be affected and so changes in the engine oil pan are indispensable if the axle drive is to be accommodated (Fig. 1.62). The power take-off gear (Fig. 1.61) contains a planet gear centre differential which facilitates a variable force distribution (based on the internal ratio); 36% of the drive moment normally goes to the front and 64% to the rear axle. A multi-plate clutch is fitted, which can lock the differential electro-magnetically up to 100%, depending on the torque requirement (front to rear axle). Moreover, there is a further electro-hydraulically controlled lock in the rear axle differential which is also up to 100% effective.

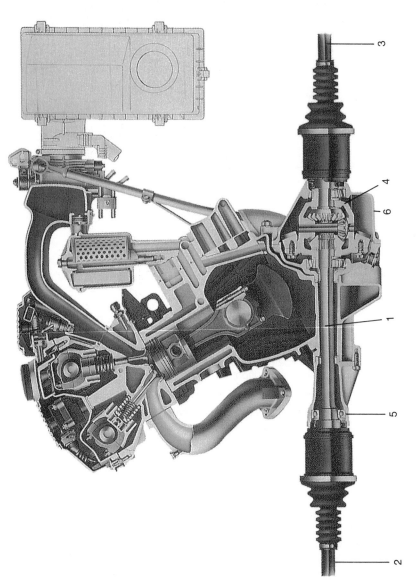

Fig. 1.62 Front view of the engine of the 525iX with a four-valve cylinder head. The basic vehicle has rear-wheel drive and, in order to also be able to drive the front wheels, the front axle power take-off 4 had to be moved into the space of the oil pan. The intermediate shaft 1 bridges the distance to the right inner CV joint and thus ensures drive shafts of equal length to both wheels (items 2 and 3 and Fig. 1.36). Part 1 is mounted on one side in the non-lockable differential 4 and on the other side in the outrigger 5. This, and the casing 6, are screwed to the oil pan.

Fig. 1.63 Driven McPherson front axle of the Mercedes E 300 with 4MATIC. The basic vehicle has the axle shown in Fig. 1.25. To keep the fundamental design, the coil springs were given a widely differing winding, with the drive shaft passing through them.

The two differentials with variable degrees of lock offer decisive advantages:

- to reach optimal driving stability, they distribute the engine moments during pushing and pulling according to the wheel slip on the drive axles, and
- they allow maximum traction without loss of driving stability (Fig. 1.48).

The locks are open during normal driving. By including the front axle differential, they make it possible to equalize the number of revolutions between all wheels, so narrow bends can be negotiated without stress in the power train and parking presents no problems. If the car is moved with locked differentials and the driver is forced to apply the brakes, the locks are released in a fraction of a second. The system is therefore fully ABS-compatible. Mercedes-Benz uses a similar power take-off gear on the 4MATIC models, as can be seen in Fig. 1.61. The system has four shift positions in this instance:

- pure rear-wheel drive
- unlocked four-wheel drive
- locked four-wheel drive (centre differential lock) and
- additional axle differential lock on the rear axle.

The shift positions are automatically activated by means of an intelligent electronic system according to the traction requirement. The first position is engaged when the driver brakes and so the system is thus fully ABS-compatible. Design complexity – and with it the cost – is considerable. Figure 1.63 shows the driven front axle, on which the shafts go through the springs in a simple manner.

2

Tyres and wheels

2.1 Tyre requirements

2.1.1 Interchangeability

All tyres and rims are standardized to guarantee interchangeability, i.e. guarantee the possibility of using tyres from different manufacturers but with the same designation on one vehicle and to restrict the variety of tyre types worldwide.

Within Europe, standardization is carried out by the European Tyre and Rim Technical Organization or ETRTO, which specifies the following:

- tyre and rim dimensions
- the code for tyre type and size
- the load index and speed symbol.

Passenger car tyres are governed by UNO regulation ECE-R 30, commercial vehicles by R 54, spare wheels by R 64, and type approval of tyres on the vehicle by EC directive 92/23/EC.

In the USA the Department of Transportation (or DOT, see item 9 in Fig. 2.11) is responsible for the safety standards. The standards relevant here are:

Standard 109   Passenger cars
Standard 119   Motor vehicles other than passenger cars.

The Tire and Rim Association, or TRA for short, is responsible for standardization.

In Australia, binding information is published by the Federal Office of Road Safety, Australian Motor Vehicle Certification Board.

ARD 23   Australian Design Rule 23/01:
             Passenger car tyres

is the applicable standard.

In Germany the DIN Standards (Deutsches Institut für Normung) and the WdK Guidelines (Wirtschaftsverband der Deutschen Kautschukindustrie, based in Frankfurt am Main) are responsible for specifying tyre data. All bodies recognize the publications of these two organizations.

At the international level, the ISO (International Organization for Standardization) also works in the field of tyre standardization and ISO Standards are translated into many languages.

2.1.2 Passenger car requirements

The requirements for tyres on passenger cars and light commercial vehicles can be subdivided into the following groups:

- driving safety
- service life
- economy
- comfort
- handling
- load capacity.

To ensure driving safety it is essential that the tyre sits firmly on the rim. This is achieved by a special tyre bead design (tyre foot) and the safety rim, which is the only type of rim in use today (Figs 2.4 and 2.12). Not only is as great a degree of tyre-on-rim retention as possible required, but the tyre must also be hermetically sealed; on the tubeless tyre this is the function of the inner lining. Its job is to prevént air escaping from the tyre, i.e. it stops the tyre from losing pressure. However, this pressure reduces by around 25 to 30% per year, which shows how important it is to check the tyres. Service life is dependent on durability and high-speed strength, both of which are tested on the road and on drum-type rotary test rigs.

Economy is determined principally by the purchase price, the mileage achievable, the wear pattern (Fig. 3.38) and the rolling resistance. Tyre pressure is an important factor here too.

Comfort includes:

- good springing and damping properties
- tyre uniformity
- low tyre noise
- low steering effort during parking and driving.

Handling includes:

- rapid delay-free response to steering inputs
- even build-up of lateral forces
- good cornering properties.

2.1.3 Commercial vehicle requirements

In principle, the same requirements apply for commercial vehicles as for passenger cars, although the priority of the individual groups changes. After safety, economy is the main consideration for commercial vehicle tyres. The following properties are desirable:

- high mileage and even wear pattern
- low rolling resistance
- good traction
- low tyre weight
- ability to take chains
- remoulding/retreading possibilities.

Compared with passenger car tyres, the rolling resistance of commercial vehicle tyres has a greater influence on fuel consumption (20 to 30%) and is therefore an important point.

2.2   Tyre designs

2.2.1 Diagonal ply tyres

In industrialized countries, diagonal ply tyres are no longer used on passenger cars, either as the original tyres or as replacement tyres, unlike areas with very poor roads where the undamageable sidewall has certain advantages. The same is true of commercial vehicles and vehicles that tow trailers, and here too radial tyres have swept the board because of their many advantages. Nowadays, diagonal ply tyres are used only for:

- temporary use (emergency) spare tyres for passenger cars (due to the low durability requirements at speeds up to 80 or 100 km h^{-1})
- motor cycles (due to the inclination of the wheels against the lateral force)
- racing cars (due to the lower moment of inertia)
- agricultural vehicles (which do not reach high speeds).

Diagonal ply tyres consist of the substructure (also known as the tyre carcass, Fig. 2.1) which, as the 'supporting framework' has at least two layers of rubberized cord fibres, which have a Zenith angle ξ of between 20° and 40° to the centre plane of the tyre (Fig. 2.2). Rayon (an artificial silk cord), nylon or even steel cord may be used, depending on the strength requirements. At the tyre feet the ends of the layers are wrapped around the core of the tyre bead on both sides; two wire rings, together with the folded ends of the plies, form the bead. This represents the frictional connection to the rim. The bead must thus provide the permanent seat and transfer drive-off and braking moments to the tyre. On tubeless tyres it must also provide the airtight seal.

The running tread (also known as the protector), which is applied to the

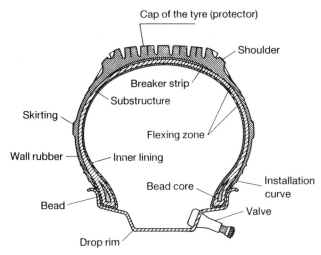

Fig. 2.1 Design of a diagonal ply tubeless car tyre with a normal drop rim and pressed-in inflating valve (see also Fig 2.5).

outer diameter of the substructure, provides the contact to the road and is profiled. Some tyres also have an intermediate structure over the carcass as reinforcement.

At the side, the running tread blends into the shoulder, which connects to the side rubber (also known as the sidewall), and is a layer which protects the substructure. This layer and the shoulders consist of different rubber blends from the running tread because they are barely subjected to wear; they are simply deformed when the tyre rolls. This is known as flexing. Protective mouldings on the sides are designed to prevent the tyre from being damaged through contact with kerbstones. There are also GG grooves, which make it possible to see that the tyre is seated properly on the rim flange.

Diagonal ply design and maximum authorized speed are indicated in the tyre marking by a dash (or a letter, Fig. 2.8) between the letters for width and rim diameter (both in inches) and a PR suffix. This ply rating refers to the carcass strength and simply indicates the possible number of plies (Fig. 2.1).

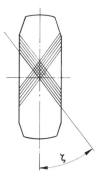

Fig. 2.2 The diagonal ply tyre has crossed layers; the zenith angle ξ was 30° to 40° for passenger cars. The 4 PR design should have two layers in each direction. Smaller angles ξ can be found in racing cars.

The marking is

5.60-15/4 PR   (VW rear-engine passenger car, tyres authorized up to 150 km h^{-1})
7.00-14/8 PR   (VW Transporter, tyres authorized up to 150 km h^{-1})
9.00-20/14 PR   (reinforced design for a commercial vehicle)

and on the temporary use spare wheel of the VW Golf, which requires a tyre pressure of $p_T = 4.2$ bar and may only be driven at speeds up to 80 km h^{-1} (F symbol)

   T 105/70 D 14 38 F

2.2.2 Radial ply tyres

The radial ply tyre consists of two bead cores joined together radially via the carcass (Fig. 2.3) – hence the name radial tyres. A belt of cords provides the necessary stiffness (Fig. 2.3a), whilst the external part of the tyre consists of the tread and sidewall and the interior of the inner lining, which ensures the tyre is hermetically sealed (Figs 2.1 and 2.4). In passenger car tyres, the carcass is made of rayon or nylon, the belt of steel cord or a combination of steel, rayon or nylon cord, and the core exclusively of steel. Due to the predominance of steel as the material for the belt, the tyres are also known as

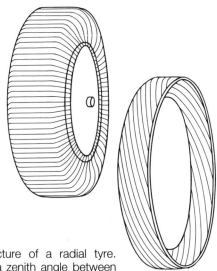

Fig. 2.3 Substructure of a radial tyre. The threads have a zenith angle between 88° and 90°.

Fig. 2.3a The belt of the radial tyre sits on the substructure. The threads are at angles of between 15° and 25° to the plane of the tyre centre.

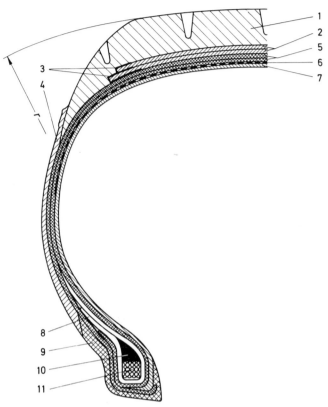

Fig. 2.4 Radial design passenger car tyres in speed category T (Fig. 2.8); the number of layers and the materials are indicated – as shown in Fig. 2.11 – on the sidewall. The components are: 1 running tread; 2 steel belt; 3 edge protection for the belt, made of rayon or nylon; 4 sidewall; 5 substructure with two layers; 6 cap; 7 inner lining; 8 flipper; 9 bead profile; 10 core profile; 11 bead core.

'steel radial tyres'. The materials used are indicated on the sidewall (Fig. 2.11, points 7 and 8). In commercial vehicle designs this is particularly important and the carcass may also consist of steel.

The stiff belt causes longitudinal oscillation, which has to be kept away from the body by axle or suspension mountings with a precisely defined longitudinal elasticity, otherwise this would cause an unpleasant droning noise in the body, when on cobbles and poor road surfaces at speeds of less than 80 km h^{-1} (see Sections 3.6.5.2 and 5.1.2). The only other disadvantage is the greater susceptibility of the thinner sidewalls of the tyres to damage compared with diagonal ply tyres. The advantages over diagonal ply tyres, which are especially important for today's passenger cars and commercial vehicles, are:

- significantly higher mileage
- greater load capacity at lower component weight
- lower rolling resistance
- better aquaplaning properties
- better wet-braking behaviour
- transferable, greater lateral forces at the same tyre pressure
- greater ride comfort when travelling at high speeds on motorways and trunk roads.

2.2.3 Tubeless or tubed

In passenger cars, the tubeless tyre has almost completely ousted the tubed tyre. The main reasons for this are that the tubeless tyre is

- easier and faster to fit
- the inner lining is able to self-seal small incisions in the tyre.

In tubeless tyres the inner lining performs the function of the tube, i.e. it prevents air escaping from the tyre (Fig. 2.4). As it forms a unit with the carcass and, unlike the tube, is not under tensional stress, if the tyre is damaged the incision does not increase in size, rapidly causing loss of pressure and failure of the tyre. The use of tubeless tyres is linked to two conditions:

- safety contour on the rim (Fig. 2.12)
- its air-tightness.

Because this is not yet guaranteed worldwide, tubed tyres continue to be fitted in some countries. When choosing the tube, attention should be paid to ensuring the correct type for the tyre. If the tube is too big it will crease, and if it is too small it will be overstretched, both of which reduce durability. In order to avoid mix-ups, the tyres carry the following marking on the sidewall:

tubeless (Fig. 2.11, point 6)
tubed or tube type.

Valves are needed for inflating the tyre and maintaining the required pressure. Various designs are available for tubeless and tubed tyres (Figs 2.5 and 2.5a). The most widely used valve is the so-called 'snap-in valve'. It comprises a metal foot valve body vulcanized into a rubber sheath, which provides the seal in the rim hole (Fig. 2.11b). The functionality is achieved by a valve insert, whilst a cap closes the valve and protects it against ingress of dirt.

At high speeds, the valve can be subjected to bending stress and loss of air can occur. Hub caps and support areas on alloy wheels can help to alleviate this (see Fig. 2.15).

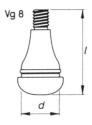

DIN	l	diameter d
43 GS 11,5	43	15,2
43 GS 16	43	19,5

Fig. 2.5 Snap-in rubber valve for tubeless tyres, can be used on rims with the standard valve holes φ 11.5 mm and φ 16 mm. The numerical value 43 gives the total length in mm (dimension l). There is also the longer 49 GS 11.5 design.

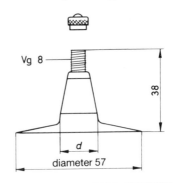

Valve specification	d
38/11,5	11,7
38/16	16,5

Fig. 2.5a Rubber valve vulcanized onto tubes. Designations are 38/11.5 or 38/16.

2.2.4 Height–width ratio

The height–width ratio H/W – also known as the cross-section ratio – influences the tyre properties and affects how much space the wheel requires (Fig. 2.5b). As shown in Fig. 2.6, the narrower tyres with a H/W ratio = 0.70 have a reduced tread and therefore good aquaplaning behaviour (Fig. 2.23). Wide designs make it possible to have a larger diameter rim and bigger brake discs (Table 2.6a) and can also transmit higher lateral and longitudinal forces.

W is the cross-sectional width of the new tyre (Fig. 2.7); the height H can easily be calculated from the rim diameter given in inches and the outside diameter of the tyre OD_T. The values OD_T and W are to be taken from the new tyre mounted onto a measuring rim at a measuring tyre pressure of 1.8 bar (or 2.3 bar on VR tyres, Table 2.10):

$$H = 0.5 \, (OD_T - d) \tag{2.1}$$
$$1'' = 1 \, \text{in} = 25.4 \, \text{mm} \tag{2.1a}$$

The 175/65 R 14 82 H tyre (Table 2.10) mounted on the measuring rim 5J x 14 can be taken as an example:

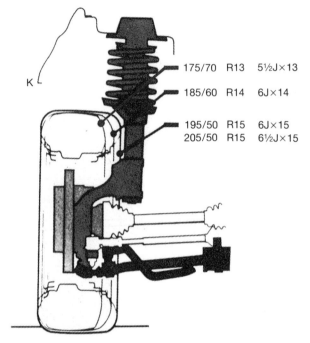

K

175/70   R13   5½J×13

185/60   R14   6J×14

195/50   R15   6J×15
205/50   R15   6½J×15

Fig. 2.5b   Tyre sizes and associated rims used on the VW Golf. All tyres fit flush up to the outer edge of the wing (wheel house outer panel) K. To achieve this, differing wheel offsets (depth of impression) *e* are used on disc type wheels (Fig. 2.14) with the advantage of a more negative rolling radius r_S on wider tyres (Fig. 3.79). A disadvantage then is that snow chains can no longer be fitted.

$OD_\mathrm{T} = 584$ mm, $d = 14 \times 25.4 = 356$ mm and $W = 177$ mm
$H/W = [0.5\,(OD_\mathrm{T} \times d)]/W = 114/177 = 0.644$

The cross-section ratio is rounded to two digits and given as a percentage. We talk of 'series', and here the ratio is 65% as shown in the tyre marking – in other words it is a 65 series tyre. A wider rim, e.g. 6J x 14 would give a smaller percentage.

2.2.5 Tyre dimensions and markings

2.2.5.1 Designations for passenger cars up to 270 km h^{-1}   The ETRTO standards manual of the 'European Tire and Rim Technical Organization' includes all tyres for passenger cars and delivery vehicles up to 270 km h^{-1} and specifies the following data:

- tyre width in mm
- height–width ratio as a percentage
- code for tyre design

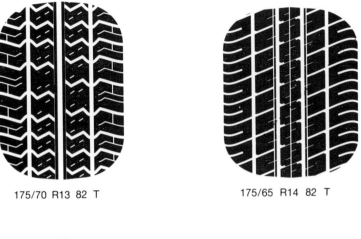

175/70 R13 82 T   175/65 R14 82 T

185/60 R14 82 H   195/50 R15 82 H

Fig. 2.6 If they have the same outside diameter and load capacity the four tyre sizes used on medium-sized passenger cars are interchangeable. The series 65 and 60 wide tyres allow a 1" larger rim (and therefore larger brake discs) and the 50 series a 2" larger rim. The different widths and lengths of the area of tyre contact, known as 'tyre print' are clearly shown (Fig. 3.91a), as are the different designs of the standard road profile. The 70 and 65 series are intended for passenger cars and the 60 and 50 series for more sporty cars. (Photograph: Kleber factory, see also Figs 2.11 and 2.11a.)

- rim diameter in inches or mm
- operational identification, comprising load index 'LI' (carrying capacity index) and speed symbol.

Taking the tyre used in the above section as an example:

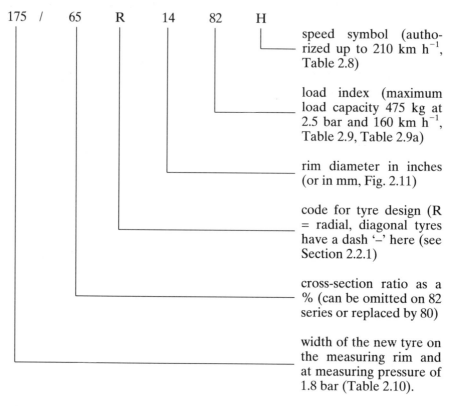

175 / 65 R 14 82 H

        speed symbol (authorized up to 210 km h^{-1}, Table 2.8)

        load index (maximum load capacity 475 kg at 2.5 bar and 160 km h^{-1}, Table 2.9, Table 2.9a)

        rim diameter in inches (or in mm, Fig. 2.11)

        code for tyre design (R = radial, diagonal tyres have a dash '–' here (see Section 2.2.1)

        cross-section ratio as a % (can be omitted on 82 series or replaced by 80)

        width of the new tyre on the measuring rim and at measuring pressure of 1.8 bar (Table 2.10).

The old markings can still be found on some tyres, e.g.:

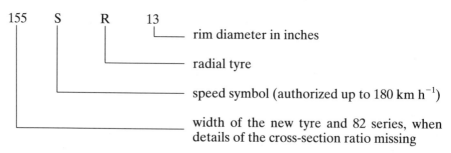

155 S R 13

        rim diameter in inches

        radial tyre

        speed symbol (authorized up to 180 km h^{-1})

        width of the new tyre and 82 series, when details of the cross-section ratio missing

2.2.5.2 Designations of discontinued sizes for passenger cars and US tyres Tyres manufactured in the USA and other non-European countries may also bear a 'P' for passenger car (see Fig. 2.10b) and a reference to the cross-section ratio:

P 155/80 R 13 79 S

Table. 2.6a The flatter the tyre, i.e. the larger the rim diameter d (Fig. 2.7) in comparison with the outside diameter OD_T, the larger the brake discs or drums that can be accommodated, with the advantage of a better braking capacity and less tendency to fade. An asymmetric drop base rim is favourable (Figs 1.4 and 2.7).

Wheel rim diameter in inches	12	13	14	15	16	17
Brake disc outer diameter in mm	221	256	278	308	330	360
Brake drum inner diameter in mm	200	230	250	280	300	325

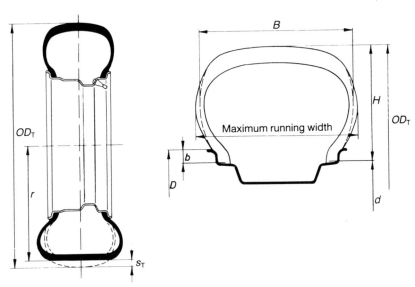

Fig. 2.7 Tyre dimensions specified in standards and directives. B is the cross-section width of the new tyre; the protective moulding (as can be seen in Fig. 2.1) is not included in the dimension. For clearances, the maximum running width with the respective rim must be taken into consideration, as should the snow chain contour for driven axles. The tyre radius, dependent on the speed, is designated r (see Section 2.2.8). Pictured on the left is an asymmetrical drop base rim, which creates more space for the brake caliper and allows a larger brake disc (Fig. 2.6a).

The old system applied up until 1992 for tyres which were authorized for over $v = 210$ km h^{-1} (or 240 km h^{-1}, Table 2.8); the size used by Porsche on the 928 S can be used as an example:

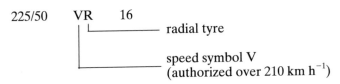

225/50   VR   16

└── radial tyre

speed symbol V
(authorized over 210 km h^{-1})

Table. 2.8 Standardized speed categories for radial tyres, expressed by means of a speed symbol and – in the case of discontinued sizes – by means of the former speed marking. Sizes marked VR or ZR may be used up to maximum speeds specified by the tyre manufacturer (see also Fig. 2.10b). The symbols F and M are intended for emergency (temporary use) spare wheels.

v_{max} in km/h	Speed symbol	Identification
80	F	
130	M	
150	P	
160	Q	
170	R	
180	S	
190	T	
210	H	
240	V	
270	W	
over 210	—	VR (old system)
over 240	—	ZR

The following should be noted for VR tyres:

- over 210 km h^{-1} and up to 220 km h^{-1} inclusive, the load may only be 90% of the otherwise authorized value
- over 220 km h^{-1} the carrying capacity reduces by at least 5% per 10 km h^{-1} speed increment.

2.2.5.3 Designation of light commercial vehicle tyres Tyres for light commercial vehicles have a reinforced substructure in comparison with those for passenger cars (Fig. 2.4) and so can take higher pressures, which means they have a higher load capacity. Previously, the suffix 'C' followed by information on the carcass strength (6, 8 or 10 PR) was used to indicate suitability for use on light commercial vehicles, or the word 'reinforced' simply appeared at the end of the marking. The new marking (as for passenger cars) retains the speed symbol as well as the load index which, behind the slash, gives the reduced load capacity on twin tyres (Fig. 3.1a). In comparison with the previous marking, the new system is as follows.

Previous	New
–	205/65 R 15 98 S (Table 2.10)
185 SR 14	185 R 14 90 S
185 SR 14 reinforced	185 R 14 94 R
185 R 14 C 6 PR	185 R 14 99/97 M
185 R 14 C 8 PR	185 R 14 102/100 M

The 185 R 14 tyre is a passenger car size which is also fitted to light commercial vehicles.

Table. 2.9 Load capacity/air pressure category specified in the directives. The load capacity on the left – also known as 'load index' LI – apply for all passenger cars up to the speed symbol W; they relate to the minimum load capacity values up to 160 km h⁻ at tyre pressure 2.5 bar (see Section 2.2.6). Further criteria, such as maximum speed, handling etc, are important for the tyre pressures to be used on the vehicle. For LI values above 100, further load increases are in 25 kg increments, i.e.

LI = 101 corresponds to 825 kg,
LI = 102 corresponds to 850 kg etc to
LI = 108 corresponds to 1000 kg.

| Load index | Wheel load capacity in kg with tyre pressure measured in bars | | | | | | | | | | |
	1.5	1.6	1.7	1.8	1.9	2.0	2.1	2.2	2.3	2.4	2.5
69	215	225	240	250	260	270	285	295	305	315	325
70	225	235	245	260	270	280	290	300	315	325	335
71	230	240	255	265	275	290	300	310	325	335	345
72	235	250	260	275	285	295	310	320	330	345	355
73	245	255	270	280	295	305	315	330	340	355	365
74	250	260	275	290	300	315	325	340	350	365	375
75	255	270	285	300	310	325	335	350	360	375	387
76	265	280	295	310	320	335	350	360	375	385	400
77	275	290	305	315	330	345	360	370	385	400	412
78	280	295	310	325	340	355	370	385	400	410	425
79	290	305	320	335	350	365	380	395	410	425	437
80	300	315	330	345	360	375	390	405	420	435	450
81	305	325	340	355	370	385	400	415	430	445	462
82	315	330	350	365	380	395	415	430	445	460	475
83	325	340	360	375	390	405	425	440	455	470	487
84	330	350	365	385	400	420	435	450	470	485	500
85	340	360	380	395	415	430	450	465	480	500	515
86	350	370	390	410	425	445	460	480	495	515	530
87	360	380	400	420	440	455	475	490	510	525	545
88	370	390	410	430	450	470	485	505	525	540	560
89	385	405	425	445	465	485	505	525	545	560	580
90	400	420	440	460	480	500	520	540	560	580	600
91	410	430	450	475	495	515	535	555	575	595	615
92	420	440	465	485	505	525	550	570	590	610	630
93	430	455	475	500	520	545	565	585	610	630	650

Table. 2.9 Continued

Load index	Wheel load capacity in kg with tyre pressure measured in bars										
	1.5	1.6	1.7	1.8	1.9	2.0	2.1	2.2	2.3	2.4	2.5
94	445	470	490	515	540	560	585	605	625	650	670
95	460	485	505	530	555	575	600	625	645	670	690
96	470	495	520	545	570	595	620	640	665	685	710
97	485	510	535	560	585	610	635	660	685	705	730
98	500	525	550	575	600	625	650	675	700	725	750
99	515	540	570	595	620	650	675	700	725	750	775
100	530	560	590	615	640	670	695	720	750	775	800

2.2.5.4 Tyre dimensions   Table 2.10 shows the important data for determining tyre size:

- size marking
- authorized rims and measuring rim
- tyre dimensions: width and outside diameter new and maximum during running
- static rolling radius (Fig. 2.7)
- rolling circumference (at 60 km h^{-1}, Table 2.10a, see also Section 2.2.8)
- load capacity coefficient (load index LI, Table 2.9)
- tyre load capacity at 2.5 bar and up to 160 km h^{-1} (see Section 2.2.6).

2.2.6 Tyre load capacities and inflation pressures

The authorized axle loads $m_{V,max,f}$ and $m_{V,max,r}$ (see Section 5.3.5), and the maximum speed v_{max} of the vehicle, determine the minimum tyre pressure. However, the required tyre pressure may be higher to achieve optimum vehicle handling (see also Section 2.10.3.5 and Fig. 2.31).

2.2.6.1 Tyre load capacity designation   The load capacities indicated in the load index (Table 2.9) and specified in the tables in the ETRTO standards manual are the maximum loads per tyre permitted for vehicle speeds up to and including 210 km h^{-1}, which cover all tyres up to the speed symbol 'H'. For tyres with an authorized speed of more than 210 km h^{-1}, load capacity has to be reduced, as shown in Table 2.9a.

Consequently, for tyres with speed symbol 'V', at a maximum speed of 240 km h^{-1} the load capacity is only 91% of the limit value. Tyres designated 'W' on the sidewall are only authorized up to 85% at 270 km h^{-1}. In both cases the load capacity values between 210 km h^{-1} ('V' tyre) and 240 km h^{-1} ('W' tyre) and the maximum speed must be determined by linear interpolation.

Table. 2.9a The tyre load capacity shown in the ETRTO standards manual in the form of the load index LI is the maximum load per tyre permitted for vehicle speeds up to 210 km h^{-1}. For higher speeds and V, W or ZR tyres, the load capacity is reduced and should not exceed the percentages shown.

		Tyre load capacity [%]	
Top speed of car		Speed symbol	
(km/h)	V	W	ZR Tyres
210	100	100	100
220	97	100	100
230	94	100	100
240	91	100	100
250	–	95	95
260	–	90	90
270	–	85	85
over 270	–	–	*

*to be determined by car manufacturer

For 'ZR' tyres the interpolation applies to the 240 km h^{-1} to 270 km h^{-1} speed range. At higher speeds, the load capacity as well as the inflating pressure will be agreed between the car and tyre manufacturers. However, this approval does not necessarily apply to tyres which are specially produced for the US market and which bear the additional marking 'P' (Fig. 2.10b and Section 2.2.5.2).

As the load capacity is also influenced by the wheel camber, the above procedures and values restrict wheel camber angles up to $\epsilon_w \leqslant 2°$. Greater angles necessitate a reduction in load capacity or an increase in tyre pressure (see Section 2.2.6.3).

2.2.6.2 Tyre pressure determination Tyre pressures listed in the tyre tables (e.g. in Table 2.10) are basic values and general minimum pressures which apply to vehicle speeds up to 160 km h^{-1}. Special operating conditions, the design of the vehicle or wheel suspension and expected handling properties can all be reasons for giving higher values in the pressure data in the vehicle specification. It is also essential to increase the pressure when the vehicle speed exceeds 160 km h^{-1}.

For tyres with speed symbols 'R' to 'V', the basic pressure has to be increased in a linear manner by up to 0.3 bar for the 160 km h^{-1} to 210 km h^{-1} speed range (see the end of Section 2.8.4).

For 'V' tyres from 210 km h^{-1} up to 240 km h^{-1} the basic pressure has to be increased by 0.3 bar and the actual tyre load must also be adjusted as per Table 2.9a.

For tyres with the speed symbol 'W', the pressures in Table 2.9 apply up to 190 km h^{-1}. After this it has to be increased by 0.1 bar for every 10 km h^{-1} up

Table. 2.10 Radial 65 series tyres, sizes, new and running dimensions, authorized rims and load capacity values (related to maximum 160 km h^{-1} and 2.5 bar); the necessary increase in pressures at higher speeds can be taken from Section 2.2.6. The tyre dimensions apply to tyres of a normal and increased load capacity design (see Section 2.2.5.3) and to all speed symbols and the speed marking ZR.

Tyre size	Measuring rim	Width of cross-section	Outer diameter	Permissible rims according to DIN 7817 and DIN 7824	Max. width	Max. outer diameter[4]	Static radius ±2.0%	Circumference +1.5% −2.5%	Load index (LI)	Wheel load capacity[5]
155/65 R 13	4.50 B × 13	157	532	4.00 B × 13[1]	158	540	244	1625	73	365
				4.50 B × 13[1]	164					
				5.00 B × 13[1]	169					
				5.50 B × 13[1]	174					
155/65 R 14	4½ J × 14	157	558	4 J × 14[2]	158	566	257	1700	74	375
				4½ J × 14[2]	164					
				5 J × 14[2]	169					
				5½ J × 14[2]	174					
165/65 R 13	5.00 B × 13	170	544	4.50 B × 13[1]	171	533	248	1660	76	400
				5.00 B × 13[1]	176					
				5.50 B × 13[1]	182					
				6.00 B × 13[1,3]	187					
165/65 R 14	5 J × 14	170	570	4½ J × 14[2]	171	579	261	1740	78	425
				5 J × 14[2]	176					
				5½ J × 14[2]	182					
				6 J × 14	187					
175/65 R 13	5.00 B × 13	177	558	5.00 B × 13[1]	184	567	254	1700	80	450
				5.50 B × 13[1]	189					
				6.00 B × 13[1,3]	194					
175/65 R 14	5 J × 13	177	584	5 J × 14[2]	184	593	267	1780	82	475
				5½ J × 14[2]	189					
				6 J × 14	194					
175/65 R 15	5 J × 15	177	609	5 J × 15[2]	184	618	279	1855	83	487
				5½ J × 15[2]	189					
				6 J × 15	194					
185/65 R 13	5.50 B × 14	189	570	5.50 B × 13[1]	191	580	259	1740	84	500
				5.50 B × 13[1]	197					
				6.00 B × 13[1,3]	202					
				6½ J × 13	207					
185/65 R 14	5½ J × 14	189	596	5 J × 14	191	606	272	1820	86	530
				5½ J × 14	197					
				6 J × 14	202					
				6½ J × 14	207					
185/65 R 15	5½ J × 15	189	621	5 J × 15	191	631	284	1895	88	560
				5½ J × 15	197					
				6 J × 15	202					
				6½ J × 15	207					
195/65 R 14	6 J × 14	201	610	5½ J × 14	204	620	277	1860	89	580
				6 J × 14	209					
				6½ J × 14	215					
				7 J × 14	220					
195/65 R 15	6 J × 15	201	635	5½ J × 15	204	645	290	1935	91	615
				6 J × 15	209					
				6½ J × 15	215					
				7 J × 15	220					
205/65 R 14	6 J × 14	209	622	5½ J × 14	212	633	282	1895	91	615
				6 J × 14	217					
				6½ J × 14	222					
				7 J × 14	227					
				7½ J × 14	233					

205/65 R 15	6 J × 15	209	647	5½ J × 15	212	658	294	1975	94[6]	670
				6 J × 15	217					
				6½ J × 15	222					
				7 J × 15	227					
				7½ J × 15	233					
215/65 R 15	6½ J × 15	221	661	6 J × 15	225	672	300	2015	96[7]	710
				6½ J × 15	230					
				7 J × 15	235					
				7½ J × 15	240					
215/65 R 16	6½ J × 16	221	686	6 J × 16	225	697	312	2090	98	750
				6½ J × 16	230					
				7 J × 16	235					
				7½ J × 16	240					
225/65 R 15	6½ J × 15	228	673	6 J × 15	232	685	304	2055	99	775
				6½ J × 15	237					
				7 J × 15	242					
				7½ J × 15	248					
				8 J × 15	253					

[1] Instead of wheel rims with the identification letter B, same-sized rims with the identification letter J may be used. For example 5½ J × 13 instead of 5.50 B × 13. (Please refer to Section 2.3.2.)

[2] Instead of wheel rims with the identification letter J, same-sized rims with the identification letter B may be used. For example 4.50 B × 14 instead of 4½ J × 14.

[3] The wheel rims without identification letters mentioned in the table are expected to be identified with DIN 7824 Part 1.

[4] The outer diameter of wheels with M + S − tread (Fig. 2.11a) can be up to 1% bigger than the standard tread.

[5] Maximum in kg at 2.5 bar.

[6] Reinforced model, 750 kg at 3.0 bar (LI 98).

[7] Reinforced model, 800 kg at 3.0 bar (LI 100).

to 240 km h^{-1}. In the case of speeds above this, the load capacity must be reduced (see Section 2.2.6.1).

On vehicles, pressure should be tested on cold tyres, i.e. these must be adjusted to the ambient temperature. If the tyre pressure is set in a warm area in winter there will be an excessive pressure drop when the vehicle is taken outside. On M & S winter tyres it has long been recommended that inflation pressures be increased by 0.2 bar over summer tyres for all speed categories, and these supplementary increases should be included in the values described above. Newer brands of tyre no longer require this adjustment and the service instructions of the car manufacturer should be double-checked in all cases to prevent any errors.

2.2.6.3 Influence of wheel camber   Wheel camber angles considerably influence tyre performance and service life. The camber angle should therefore not exceed 4° even in full wheel jounce condition. For angles above ±2° the loadability of the tyres reduces at

$2° < \epsilon_w < 3°$ to 95%
$3° < \epsilon_w < 4°$ to 90%

Intermediate values have to be interpolated. Compensation can be achieved by increasing the inflation pressure. The values are as follows:

Camber angle	2°20'	2°40'	3°	3°20'	3°40'	4°
Pressure increase	2.1%	4.3%	6.6%	9.0%	11.5%	14.1%

Taking all the influences into account, such as top speed, wheel camber and axle load, the minimum tyre pressure required can be calculated for each tyre category (size and speed symbol). Formulas are shown in the 'WdK 99' guidelines from the 'Wirtschaftsverband der Deutschen Kautschukindustrie'.

Tyre pressure limit values should, however, be adhered to. These values are

Q and T tyres	3.2 bar
H to W and ZR tyres	3.5 bar
M & S tyres (Q and T tyres)	3.5 bar

2.2.7 Tyre sidewall markings

Apart from VR and ZR tyres, all tyres used in Europe should be marked in accordance with the 'ETRTO' standards and EC regulation 30 (see Section 2.1.1). In the USA, Japan and Australia, additional markings are required to indicate the design of the tyre and its characteristics. The characters must also bear the import sizes – the reason why these can be found on all tyres manufactured in Europe (Fig. 2.11).

2.2.8 Rolling circumference and driving speed

The driving speed is:

$$v = 0.006(1 - S_{x,a}) \frac{C_{R,dyn} n_M}{i_D i_G} \text{ [km/h]} \tag{2.1b}$$

This includes:

$S_{x,a}$	the absolute traction slip (Equation 2.4e)
$C_{R,dyn}$	the dynamic rolling circumference in m (Equation 2.1d)
n_M	the engine speed in rpm
i_D	the ratio in the axle drive (differential)
i_G	the ratio of the gear engaged (Equation 6.36)

The following can be assumed for slip $S_{x,a}$:

1st gear	0.08
2nd gear	0.065
3rd gear	0.05
4th gear	0.035
5th gear	0.02

The rolling circumference C_R given in the tyre tables relates to 60 km h^{-1} and operating pressure of 1.8 bar. At lower speeds it goes down to U_{stat}:

$$C_{R,stat} = r_{stat} 2\pi \tag{2.1c}$$

The values for r_{stat} are also given in the tables. At higher speeds, C_R increases due to the increasing centrifugal force. The dynamic rolling circumference

$C_{R,dyn}$ at speeds over 60 km h^{-1} can be determined using the speed factor k_v. Table 2.10a shows the details for k_v as a percentage increasing by increments of 30 km h^{-1}. Intermediate values must be interpolated. The circumference would then be:

$$C_{R,dyn} = C_R (1 + 0.01 \, k_v) \, [mm] \tag{2.1d}$$

The dynamic rolling radius, which is sometimes required, can be calculated from $C_{R,dyn}$:

$$r_{dyn} = C_R/2\pi$$

or, at speeds of more than 60 km h^{-1}

$$r_{dyn} = C_{R,dyn}/2\pi \tag{2.2}$$

Taking as an example the tyre

175/65 R 14 82 H and $v = 200$ km h^{-1}

(Table 2.10) it is

$k_{v180} = 0.7\%$ and $k_{v210} = 1.1\%$

and interpolation gives:

$k_{v200} = 0.007 + 0.0027 = 0.0097$
$k_{v200} = 0.97\%$

The rolling circumference C_R taken from Table 2.10, according to Equation 2.1d, gives:

$$C_{R,dyn200} = 1780 \, (1 + 0.0097) = 1797 \text{ mm}$$

and thus the dynamic radius in accordance with Equation 2.2:

$r_{dyn60} = 283$ mm and $r_{dyn200} = 286$ mm

Table. 2.10a Factor k_v, which expresses the speed-dependence of the rolling circumference of passenger vehicle radial tyres above 60 km h^{-1} as a percentage. The permissible tolerances Δk_v have to be added (see Section 2.2.8), all taken from the German WDK Guideline 107, page 1.

V in km^{-1}	60	90	120	150	180	210	240
Factor k_v in %	0	+0.1	+0.2	+0.4	+0.7	+1.1	+1.6
Deviation Δk_v in %	—	±0.1	±0.2	±0.4	±0.7	±1.1	±1.6

Fig. 2.10b ZR tyres manufactured specially for the American market and marked with a 'P' do not meet the European standard and are therefore not authorized here (photograph: Dunlop factory).

The outside diameter is

$OD_T = 584$ mm and thus $OD_T/2 = 292$ mm

a value which shows the extent to which the tyre becomes upright when the vehicle is being driven: r_{dyn} is only 9 mm or 6 mm less than $OD_T/2$.

2.2.9 Influence of the tyre on the speedometer

The speedometer is designed to show slightly more than, and under no circumstances less than, the actual speed. Tyres influence the degree of advance, whereby the following play a role:

- the degree of wear

- the tolerances of the rolling circumference
- the profile design, and its associated
- slip.

The EC Council directive 75/443, in force since 1991, specifies an almost linear advance Δv,

$$+ \Delta v \leqslant 0.1\, v + 4\, [\text{km h}^{-1}] \qquad\qquad (2.2a)$$

On vehicles registered from this year onwards the values displayed may only be as follows:

Actual speed in km h^{-1}	30	60	120	180	240
Max displayed value in km h^{-1}	37	70	136	202	268

As Table 2.10 indicates, at 60 km h^{-1} the rolling circumference C_R has a tolerance range of $\Delta C_R = +1.5\%$ to -2.5%, and in accordance with Table 2.10a with a speed factor of k_v, deviations of up to $\Delta k_v = \pm 1.6\%$ are possible. When related to the dynamic rolling circumference $C_{R,dyn}$ (Equation 2.1d), the following tolerance limits (rounded to the nearest figure) may prevail and result in the displayed values when only the minus tolerances are considered, and if the speedometer has the maximum authorized advance:

Actual speed in km h^{-1}	60	120	180	240
Possible overall tolerance in %	+1.5 −2.5	+1.7 −2.7	+2.2 −3.2	+3.1 −4.1
Max display value in km h^{-1} at minus tolerance	140	208	279	

The slip should be added directly to this, which in direct gear amounts to around 2%, in other words

$S_{x,a} = 0.02$ (see Equations 2.1b and 2.4f)

If the manufacturer fully utilizes the advance specified in Equation 2.2a, it is possible that although the speedometer indicates 140 km h^{-1}, the vehicle is only moving at 120 km h^{-1}. This occurs, in particular, when the tyres are worn:

3 mm wear gives an advance of around 1%

Tyres with an M & S winter profile can, however, have a 1% larger outside

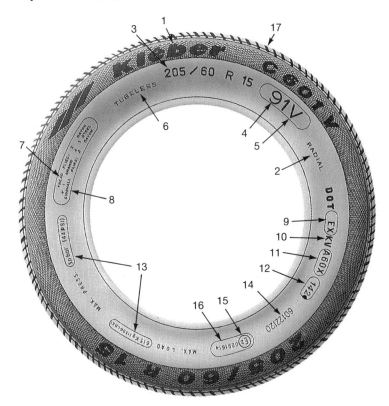

Fig. 2.11 Explanation of the marking on the sidewall of a tyre manufactured by Pneumatiques Kléber SA:

1 Manufacturer and tyre type
2 Tyre design: radial
3 Size marking (Section 2.2.5.1)
4 Load capacity index 91 (Fig. 2.9)
5 Speed symbol V (Fig. 2.8)
6 Tubeless (Section 2.2.3)
7 Tread plies 2 rayon + 1 aramide + 1 steel. The belt (Fig. 2.4) consists of two layers of rayon (an artificial silk cord), a layer of aramide and a further layer of steel
8 Sidewall plies 2 rayon: the sidewall consists of two layers of artificial silk cord
9 DOT: The tyre meets the US Department of Transportation requirements. EX: Code for the country of manufacture (France)
10 KV = size code
11 A60X = manufacture code
12 142Δ = coded date of manufacture: 14 = calender week, 2 = year two and Δ = 9th decade (in other words 1992)
13 Maximum load capacity and max tyre pressure in accordance with US standard
14 Factory own fabrication number
15 E2: certificate of European type approval in accordance with ECE 30. The number indicates the country in which the approval was carried out (2 is France)
16 0291614 is the number of the type approval
17 TWI = tread wear indicator. There are points in the main drainage channels which protrude at 1.6 mm over the minimum profile.

Fig. 2.11a Kléber tyre designs: standard road profile (top) and M & S winter profile (bottom) (see also Figs 2.6 and 2.11). The design pictured on the left is used in speed category Q (up to 160 km h^{-1}), and the one on the right in categories T and H (up to 190 or 210 km h^{-1}).

diameter so that the profile can be deeper (Table 2.10, note 5 and Fig. 2.11a). They would therefore reduce the degree by which the speedometer is advanced if the tyres are not yet worn. The same applies where the positive tolerances given in the above table are used. In this instance it is also possible that even a very precise speedometer could display too low a speed.

2.2.10 Tyre profiles

As can be seen from Fig. 2.6, the profile design depends on the height–width ratio on passenger car radial tyres. Wide tyres have a greater negative proportion (light areas) to improve aquaplaning behaviour. For the same reason, the profile is opened to the side as this provides good water drainage; Fig. 2.11a shows further profile designs.

2.3 Wheels

2.3.1 Concepts

The tyres are differentiated according to the loads to be carried, the possible maximum speed of the vehicle and whether a tubed or tubeless tyre is driven. In the case of a tubeless tyre, the air-tightness of the rim is extremely important. The wheel also plays a role as a 'styling element'. It must permit good brake ventilation and a secure connection to the hub flange. Figure 2.11b shows a passenger car rim fitted with a tubeless tyre.

2.3.2 Rims for passenger cars, light commercial vehicles and trailers

Only drop base rims are provided for these types of vehicle. The dimensions of the smallest size, at 12" and 13" diameter and rim width up to 5.0", are

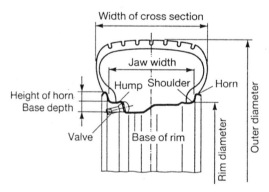

Fig. 2.11b Series 55 wide tyre designs, mounted on a double hump rim with the inflating valve shown in Fig. 2.5. The actual rim consists of the
- rim horns, which form the lateral seat for the tyre bead (the distance between the two rims is the jaw width)
- rim shoulders, the seat of the beads, generally inclined at 5° ± 1° to the centre where the force transfer occurs around the circumference (Fig. 2.4)
- well base (also known as the inner base), designed as a drop rim to allow tyre fitting, and shifted mainly to the outside (photograph: Lemmerz factory).

contained in the international standards (see Section 2.1.1). The designation for a normal rim, suitable for the 145 R 13 tyre (Fig. 2.1) for example is:

drop base rim 4.00 B x 13

This type of rim used on passenger cars up to around 66 kW (90 PS) has only a 14 mm high rim flange and is identified with the letter B.

In order to make it possible to fit bigger brakes (Table 2.6a), more powerful vehicles have larger diameter rims as follows:

- series production passenger cars: 14" to 16" rims
- sports cars: 16" to 18".

The J rim flange applied here is used on rims from 13" upwards and is 17.3 mm high. The rim base can (as shown in Fig. 2.1) be arranged symmetrically or shifted outwards. The rim diameter, which is larger on the inside, creates more space for the brake (Figs 1.4, 1.39, 2.7, 2.11b and Table 2.6a). The standards specify the rim width from $3\frac{1}{2}$" to $8\frac{1}{2}$". The definition of a normal asymmetrical rim with a 5" width, J rim flange and 14" diameter is:

drop base rim – 5 J x 14

The symmetrical design is identified by the suffix 'S'. The standards also contain precise details on the design and position of the valve hole (see also Figs 2.11b and 2.15).

C tyres for light commercial vehicles require a broader shoulder (22 mm instead of 19.8 mm), which can be referred to by adding the letters LT (light truck) at the end of the marking:

drop base rim – $5\frac{1}{2}$ J x 15 – LT

There is a preference worldwide for using tubeless radial tyres on passenger cars and light commercial vehicles. Where these tyres are used, it is essential to have a 'safety contour' at least on the outer rim shoulder. This stops air suddenly escaping if the vehicle is cornering at reduced tyre pressure.

The three different contours mainly used are (Fig. 2.12):

Hump   (H, previously H1)
Flat-hump   (FH, previously FHA)
Contre Pente   (CP)

The standards specify the dimensions of the first two designs. The 'hump' runs around the rim, which is rounded in H designs, whilst a flat hump rim is simply given a small radius towards the tyre foot. The fact that the bead sits firmly between the hump and rim flange is advantageous on both contours. An arrangement on both the outside and inside also prevents the tyre feet sliding into the drop bases in the event of all the air escaping from the tyre when travelling at low speeds, which could otherwise cause the vehicle to

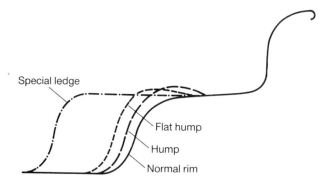

Special ledge

Flat hump

Hump

Normal rim

Fig. 2.12 Normal rim and contours of the safety shoulders which can be used on passenger cars and light commercial vehicles.

swerve. The disadvantage of hump rims is that changing the tyre is difficult and requires special tools.

A French design, intended only for passenger car rims, is the 'Contre Pente' rim, known as the CP for short. This has an inclined shoulder towards the rim base, which for rim widths between 4" and 6" is provided on one or both sides.

For years, the rims of most passenger cars have had safety shoulders on both sides, either a double hump (Figs 2.11b and 2.15) or the sharp-edged flat-hump on the outside, and the rounder design on the inside (Fig. 2.14). The desired contour must be specified in the rim designation. Table 2.13 gives the possible combinations and abbreviations which must appear after the rim diameter data. A complete designation for an asymmetrical rim would then be as follows:

Table. 2.13 Marking of the various safety shoulders when used only on the outside of the rim or on both the inside and outside. Normal means there is no safety contour (Fig. 2.1). Further details are contained in the standards and directives.

Denomination	Nature of safety shoulder		Identification letters
	Outside of rim	Inside of rim	
One-sided hump	Hump	Normal	H
Double hump	Hump	Hump	H2
One-sided flat hump	Flat hump	Normal	FH[1]
Double-sided flat hump	Flat hump	Flat hump	FH2[1]
Combination hump	Flat hump	Hump	CH[2]

[1] In place of the identification letters FH the old identification letters FHA are still permitted.
[2] In place of the identification letters CH the old identification letters FH1-H are still permitted.

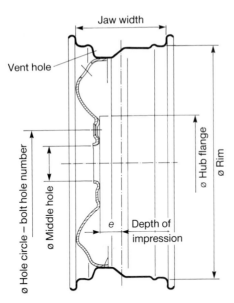

Jaw width

Vent hole

ø Hole circle – bolt hole number

ø Middle hole

e Depth of impression

ø Hub flange

ø Rim

Fig. 2.14 The sheet metal disc-type wheel used in series production vehicles consists of a rim and dish. To avoid fatigue fractures, the wheel hub flange diameter should be greater than the dish contact surface. Wheel offset e (depth of impression) and kingpin offset at ground r_S are directly correlated. A change in e can lead to an increase or a reduction in r_S. The dome-shaped dish leading to the negative kingpin offset at ground is clearly shown (figure: Lemmerz factory).

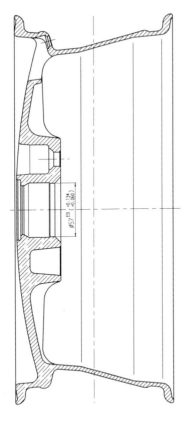

Fig. 2.15 Lemmerz alloy wheel for the Audi 80, made of the aluminium alloy GK-Al Si 7 Mg wa. The wheel has a double-hump rim (H2) and middle centring and is fixed with four spherical collar bolts. The different wall thicknesses, which are important for the strength, the shape of the bolt hole, the different shape of the drop-rim and the position of the valve hole are clearly shown. At high speeds the snap-fit valve (Fig. 2.5) is pressed outwards by the centrifugal force and supported below the rim base.

Drop base rim DIN 7817 – 5   J   ×   13   H2

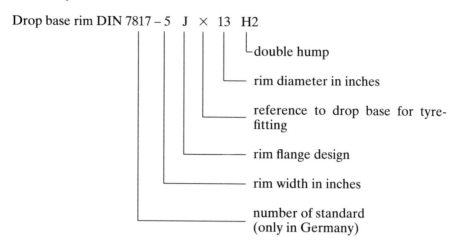

double hump

rim diameter in inches

reference to drop base for tyre-fitting

rim flange design

rim width in inches

number of standard (only in Germany)

2.3.3 Wheels for passenger cars, light commercial vehicles and trailers

Most passenger cars and light commercial vehicles are fitted with sheet metal disc wheels, because these are economic, have high stress limits and can be readily serviced. They consist of a rim and a welded-on wheel disc (also known as an attachment face, Fig. 2.14). Cold-formable sheet metal, or band steel with a high elongation, can be used (e.g. RSt37-2 to European standard 20) depending on the wheel load, in thicknesses from 1.8 to 4.0 mm for the rim and 3.0 to 6.5 mm for the attachment faces. There is a direct correlation between wheel offset e and 'kingpin offset at ground' r_S; the more positive r_S, the smaller can be the depth dimension e. A negative kingpin offset $-r_S$ on the other hand, especially on front-wheel drive, results in a significant depth e and severe bowing of the attachment faces (as can be seen in Figs 2.5b, 2.14, 2.16 and 3.79). The wheel disc can be perforated to save weight and achieve better brake cooling.

Despite the fact that they cost almost four times as much as sheet metal designs, alloy wheels are becoming increasingly popular (Figs 2.15 and 1.39). Their advantages are:

- more extensive styling options, and therefore
- better appearance
- processing allows precise centring and limitation of the radial and lateral runout (see Section 2.5)
- good heat transfer for brake-cooling.

Often incorrectly called aluminium rims, the alloy wheels are mainly manufactured using low-pressure chill casting and generally consist of aluminium alloys with a silicon content (which are sometimes heat hardenable), e.g. GK-Al Si 11 Mg, GK-Al Si 7 Mg T (T = tempered after casting) etc. Regardless of the material, the wheels must be stamped with a marking containing the most important data (Fig. 2.16).

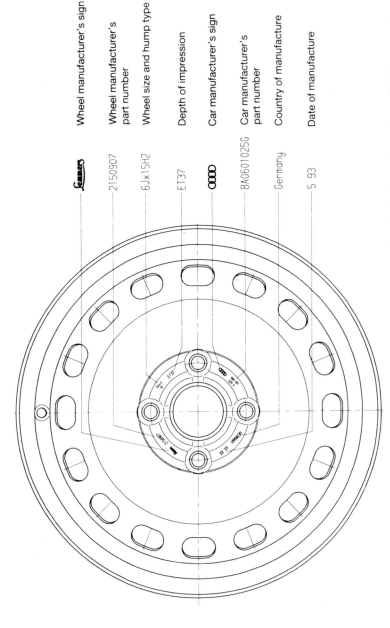

Wheel manufacturer's sign

Wheel manufacturer's part number

Wheel size and hump type

Depth of impression

Car manufacturer's sign

Car manufacturer's part number

Country of manufacture

Date of manufacture

Fig. 2.16 Double-hump sheet metal disc-type wheel with openings for cooling the brakes. Also pictured is the stamp in accordance with the German standard DIN 7829, indicating manufacturer code, rim type and date of manufacture (week or month and year).

Also specified is the wheel offset (ET 37) and, in the case of special wheels with their own ABE (General operating approval), the allocation number of the KBA, the German Federal Vehicle Licensing Office. If there is not much space the stamp may be found on the inside of the dish. (figure: Lemmerz factory).

2.3.4 Wheel mountings

Many strength requirements are placed on the wheel disc sitting in the rim (or the wheel spider on alloy wheels); it has to absorb vertical, lateral and longitudinal forces coming from the road and transfer them to the wheel hub via the fixing bolts. The important thing here is that the contact area of the attachment faces, known as the 'mirror', should sit evenly and, for passenger cars, that the hub flange should have a slightly larger diameter (Fig. 2.14), otherwise it is possible that the outer edge of the hub will dig into the contact area, with a loss of torque on the bolts. The notch effect can also cause a fatigue fracture leading to an accident.

The number of holes and their circle diameter are important in this context. This should be as large as possible to introduce less force into the flange and fixing bolts. If the brake discs are placed onto the wheel hub from the outside – which is easier from a fitting point of view – it is difficult to create a hole larger than 100 mm on 13" wheels, and using a 14" or 15" wheel should make for the best compromise (Figs 1.4, 1.25, 1.29 and Table 2.6a). ISO 898 and the German standard DIN 74361 contain further details.

The brake disc can also be fixed to the wheel hub from the inside (Fig. 1.23). However, the disadvantage of this is that the hub has to be removed before the disc can be changed. This is easy on the non-driven axle, but time-consuming on the driven axle. This brief look shows that even the brakes play a role in the problems of fixing wheels.

Nowadays, wheels are almost always fixed with four or five metric M12 × 1.5 or M14 × 1.5 spherical collar bolts. The high friction between the spherical collar and the stud hole prevents the bolts from coming loose whilst the vehicle is in motion. For this reason, some car manufacturers keep the contact surface free of paint. On sheet metal disc wheels with attachment faces up to 6.5 mm thick, the spring action of the hole surround (Fig. 2.17) is an additional safety feature, which also reduces the stress on the wheel bolts as a result of its design elasticity. Sheet metal rings are often inserted in the alloy wheels to withstand high stresses underneath the bolt head.

Generally, the spherical collar nuts also do the job of centring the wheels on the hub. Middle centring has become popular on fast passenger cars because of a possible middle or radial run-out and the associated steering vibrations. A toleranced collar placed on the hub fits into the dimensioned hole which can be seen in Fig. 2.15.

2.4 Springing behaviour

The static tyre spring rate c_T – frequently also known as spring stiffness or (in the case of a linear curve) spring constant – is the quotient of the change in vertical force $\Delta F_{z,w}$ in Newtons and the resultant change Δs_3 – the compression in mm within a load capacity range corresponding to the tyre pressure p_T (Fig. 2.18, see also Section 2.2.5.4):

$$c_T = \Delta F_{z,w}/\Delta s_3 \ [\text{N/mm}] \tag{2.3}$$

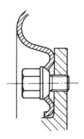

Fig. 2.17 Depression design with special springing characteristics on a passenger car sheet metal disc-type wheel. The wheel can be centred using the fixing bolts or by fitting into the toleranced hole (Fig. 2.15).

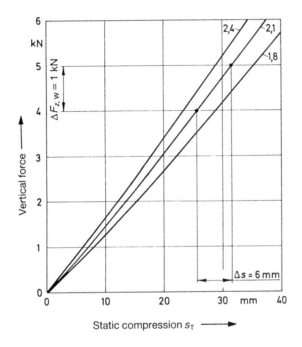

Fig. 2.18 The static tyre springing rate c_T is the quotient of the force and the deflection travel shown on the radial tyre 175/70 R 13 80 S at p_T = 1.8 bar, 2.1 bar and 2.4 bar; the example shown gives:

$$c_T = \frac{\Delta F_{z,W}}{\Delta s_T} = \frac{1000\ N}{6\ mm} = 167\ N\ mm^{-1}$$

c_T forms part of the vibration and damping calculation and has a critical influence on the wheel load impact factor (see Section 5.2). The harder the tyre, the higher the damping must be set and the greater the stress experienced by the chassis components. The following parameters influence the spring rate:

- vertical force
- tyre pressure

- driving speed
- slip angle
- camber angle
- rim width
- height–width ratio
- wheel load frequency.

As can be seen in Fig. 2.18, apart from in the low load range, the spring rate is independent of the load. A linear increase can be seen as the speed increases (Table 2.10a and Fig. 2.18a, see also Equation 5.5a), which persists even when the tyre pressure changes.

During cornering, the force $F_{y,w}$ (Fig. 3.91a) shifts the belt in a lateral direction, and so it tips relative to the wheel plane. This leads to a highly asymmetrical distribution of pressure and (as can be seen from Fig. 2.18a) to a reduction in the spring rate as the slip angles increase.

2.5 Non-uniformity

The tyre consists of a number of individual parts, e.g. carcass layers, belt layers, running tread, sidewall stock and inner lining, which – put together on a

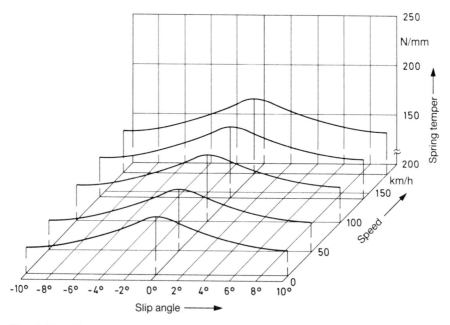

Fig. 2.18a Tyre springing rate as a function of slip angle and driving speed, measured on a radial tyre 185/70 R 13 86 S at $p_T = 2.1$ bar. Speed increases the rate as the belt stands up due to the centrifugal force; the slip angle, however, makes it softer because the belt is pushed away to the side and the shoulders take over part of the springing effect.

tyre rolling machine – give the tyre blank (Fig. 2.4). In the area where it is put together, variations in thickness and stiffness occur, which can lead to non-uniformity.

Due to the irregularities caused during manufacture, the following occur around the circumference of the tyre:

- thickness variations
- mass variations
- stiffness variations

which cause various effects when the tyre rolls:

- imbalance
- radial tyre runout
- lateral tyre runout
- variation in vertical and/or radial force
- lateral force variations
- longitudinal force variation.

Imbalance occurs when an uneven distribution of mass and the resulting centrifugal forces is not equalized. Because the uneven distribution occurs not only around the circumference, but also laterally, we have to differentiate between static and dynamic imbalance (Fig. 2.19). This is calculated in size and direction on balancing machines and eliminated with balancing weights on the rim bead outside and inside the wheel.

Radial and lateral runout are the geometrical variations in the running tread and the sidewalls. They are measured with distance sensors on a tyre-uniformity machine. The German WdK Guideline 109 contains full details. The most important of the three force variations is the radial force variation. For greater clarity, it is shown on the model spare in Fig. 2.19a, where the tyre consists of different springs whose rates fluctuate between c_1 and c_8. The resulting phenomena should be indicated on the 175 R 14 88 S steel radial tyre, loaded at $F_{z,w} = 4.5$ kN and pressurized to $p_T = 1.9$ bar. Assuming this had a mean spring rate $c_T = 186$ N m, which fluctuates by $\pm 5\%$, the upper limit would be $c_{T,max} = 195$ N mm^{-1} and the lower limit would be $c_{T,min} = 177$ N mm^{-1}. Under vertical force $F_{z,w} = 4.5$ kN = 4500 N the tyre would, according to Equation 2.3, have as its smallest jounce travel:

$$s_{T,min} = \frac{F_{z,w}}{c_{T,max}} = \frac{4500}{195}; \qquad s_{T,min} = 23.1 \text{ mm} \tag{2.3a}$$

and

$$s_{T,\,max} = 25.4 \text{ mm}$$

as the greatest travel. The difference is:

$$\Delta s_T = s_{T,\,max} - s_T,\, min = 2.3 \text{ mm}$$

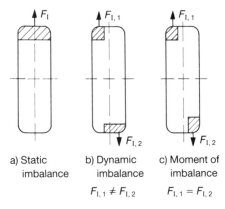

a) Static
imbalance

b) Dynamic
imbalance

c) Moment of
imbalance

$F_{1,1} \neq F_{1,2}$

$F_{1,1} = F_{1,2}$

Fig. 2.19 Different forms of imbalance, static on the left and dynamic in the centre. The imbalance is equalized in the illustration on the right.

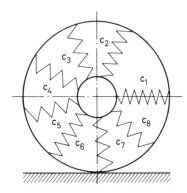

Fig. 2.19a The tyre spring rate can fluctuate depending on the manufacturing process, shown as c_1 to c_8.

This difference in the dynamic rolling radius of $\Delta s_T = 2.3$ mm would cause variations in vertical force $\Delta F_{z,W}$, which nevertheless is still smaller than the friction in the wheel suspension bearings. At a speed of perhaps 120 km h^{-1} and travelling on a completely smooth road surface, this would nevertheless lead to vibration that would be particularly noticeable on the front axle. The vehicle used as an example should have a body spring rate of $c_f = 15$ N mm^{-1} per front axle side. The travel Δs_T would then give a vertical force difference, in accordance with Equation 5.0a, of:

$$\Delta F_{z,W,f} = c_f \, \Delta s_T = 15 \times 2.3; \; \Delta F_{z,W,f} = 34.5 \text{ N}$$

The friction per front axle side is, however, rarely much below

$$F_{fr} = 100 \text{ N (Fig. 5.2b)}$$

so it can only be overcome if greater variations in vertical force occur as a result of non-uniformity in the road surface. The more softly sprung the vehicle, the more the variations in radial force in the tyre make themselves felt (see Section 5.1.2).

The lateral force variations of the tyre influence the straight-running ability of the vehicle. Even with a tyre that is running straight, i.e. where the slip angle is zero, lateral forces occur, which also depend on the direction of travel. The variations in longitudinal force that occur must be absorbed on the chassis side by the rubber bearings described in Section 3.6.5.2.

2.6   Rolling resistance

2.6.1 Rolling resistance in straight-line driving

Rolling resistance is a result of energy loss in the tyre, which can be traced back to the deformation of the area of tyre contact and the damping properties of the rubber. These lead to the transformation of mechanical into thermal energy, contributing to warming of the tyre.

Sixty to 70% of the rolling resistance is generated in the running tread (Fig. 2.4) and its level is mainly dependent on the rubber mixture. Low damping running tread mixtures improve the rolling resistance, but at the same time reduce the coefficient of friction on a wet road surface. It can be said that the ratio is approximately 1:1, which means a 10% reduction in the rolling resistance leads to a 10% longer braking distance on a wet road surface. Rolling resistance is either expressed as a rolling resistance force F_R or as the rolling resistance factor k_R – also known as the coefficient of rolling resistance:

$$F_R = k_R F_{z,W} \text{ [N]} \tag{2.4}$$

The factor k_R is important for calculating the driving performance diagram and depends on the vertical force $F_{z,W}$ and the tyre pressure p_T. Figure 2.20 shows the theoretical k_R curve of tyres of different speed classes as a function of the speed. Whilst the coefficient of rolling friction of the S tyre increases disproportionally from around 120 km h^{-1}, this increase does not occur in H and V tyres until 150 to 170 km h^{-1}. The reason for this behaviour is the

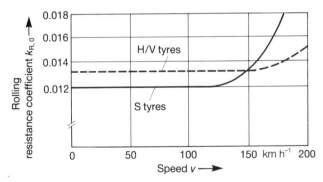

Fig. 2.20 Rolling resistance coefficients $k_{R,0}$, average values of radial tyres as a function of the speed, measured on a drum test rig. S tyres authorized up to 180 km h^{-1} have a lower rolling resistance below 140 km h^{-1} (than the H and V designs) whilst the value rises sharply above this speed. Asphalted roads cause $k_{R,0}$ to increase by around 20% as k_R and rough concrete to at least 30%. The ratios i_R are then 1.2 or 1.3 to 1.4 and the actual value of k_R is:

$$k_R = i_R k_{R,0} \tag{2.3a}$$

shape of the rolling hump that occurs at different speeds depending on the speed class, and is dependent on the stiffness of the belt, in other words on its design. The lower k_R values for the S tyres result from the usually poorer wet skidding behaviour of this speed class.

The difference is due to the different design emphases during development of the tyres. The design priorities for H and V tyres are high-speed road holding and good wet skidding and aquaplaning behaviour, whilst S tyres are designed more for economy, i.e. lower rolling resistance (which plays an important role at lower speeds and influences urban driving fuel consumption, Fig. 2.20a) and long service life.

2.6.2 Rolling resistance during cornering

Rolling resistance can change dramatically during cornering; its value depends on the speed and the rolling radius R, in other words on $\mu_{y,W}$ (see Equations 2.9 and 2.11 and Fig. 2.30) and $\alpha_{f \text{ or } r}$. The rolling resistance $k_{R,co}$, which is included in some calculations (see Equation 3.35), comprises the coefficient k_R for straight running and the increase Δk_R:

$$k_{R,co} = k_R + \Delta k_R \tag{2.4a}$$

$$\Delta k_R \approx \mu_{y,W} \sin \alpha \tag{2.4b}$$

The following data can provide an example:

Front axle force $F_{V,f} = 7\,\text{kN}$; $\mu_{y,W} = 0.7$ (asphalted road)
Tyres 155 R 13 78 S $p_T = 1.8\,\text{bar}$, $v \leqslant 120\,\text{km h}^{-1}$

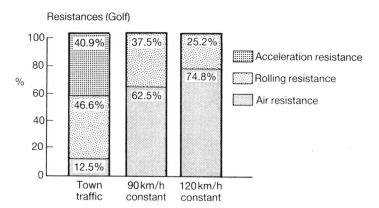

Fig. 2.20a In town and when the vehicle is travelling at low speeds on rural roads, fuel consumption is determined up to 40% by the rolling resistance, whilst at higher speeds the air drag is the determining factor. The figure shows a study carried out by VW on the Golf.

In accordance with Equation 2.11 related to one wheel:

$$F_{y,W,f} = \mu_{y,W} \, F_{z,W,f} = \mu_{y,W} \, F_{V,f}/2 = 0.7 \times 3.5 \text{ kN } F_{y,W,f} = 2.45 \text{ kN}$$

The slip angle read off at $F_{y,W,f}$ in Fig. 2.31 is $4°$ and corresponds to the values in Fig. 2.30. However the dynamic wheel load transfer seen in Fig. 1.2 plays a role during cornering, leading to a greater slip angle on the outside of the curve, and thus also on the inner wheel, than resulted from test rig measurements. On '82' series tyres, approximately $\alpha = 5°$ in accordance with Fig. 2.30:

$$\alpha \approx 7 \, \mu_{y,W} \tag{2.4c}$$

With $\sin 5°$ in accordance with Equation 2.4b there is an increase of:

$$\Delta k_R \approx 0.7 \times 0.087 = 0.061$$

The value that can be read off in Fig. 2.20 $k_{R,O} = 0.012$, in accordance with Equation 2.3a is, on asphalted road

$$k_R = i_R \, k_{R,O} = 1.2 \times 0.012 = 0.0144$$

and therefore the rolling resistance during cornering

$$k_{R,co} = 0.0144 + 0.061 \approx 0.075$$

In the case of the understeering vehicles (Fig. 2.29) $k_{R,co}$ increases as a result of the additional steering input and – if the wheels are driven – μ_{rsl} should be inserted for $\mu_{y,W}$ (see Equation 2.18); the slip angle increases further. '65 Series' tyres, on the other hand, require a smaller steering input and thus make the vehicle easier to handle:

$$\alpha = 3 \, \mu_{y,W} \tag{2.4d}$$

2.6.3 Other influencing variables

The rolling resistance increases

- in the case of a large negative or positive camber (the influence can be ignored up to $\pm 2°$)
- due to a change to track width (Fig. 3.3)
- in the case of deviations in zero toe-in around 1% per $\delta = 10'$ or $v = 1$ mm
- on uneven ground.

In general it can be said that the ratio i_R (see Fig. 2.20) will be

- around 1.5 on cobbles
- around 3 on potholed roads

- around 4 on compacted sand
- up to 20 on loose sand.

2.7 Starting and sliding friction in the longitudinal direction

2.7.1 Slip

If a tyre transfers drive or braking forces, a relative movement occurs between the road and tyre, i.e. the rolling speed of the wheel is greater or less than the vehicle speed (see Equation 2.1b). The ratio of the two speeds goes almost to ∞ when the wheel is spinning, and is 0 when it locks. Slip is usually given as a percentage. The following equation applies during braking:

$$S_{x,b} = \frac{\text{vehicle speed} - \text{circumferential speed of wheel}}{\text{vehicle speed}}$$

$$S_{x,b} = \frac{v - v_W}{v} \, 100 \, [\%] \qquad (2.4e)$$

Drive slip is governed by:

$$S_{x,a} = \frac{v_W - v}{v_W} \, 100 \, [\%] \qquad (2.4f)$$

The different expressions have the advantage that, in both cases where the wheel is spinning or locked, the value is 100% and is positive.

Further details can be found in Section 2.2.8.

2.7.2 Coefficients of friction

The higher the braking force or traction to be transmitted, the greater the slip becomes. Depending on the road condition, the transferable longitudinal force reaches its highest value between 10% and 30% slip and then reduces until the wheel locks (100% slip). The quotient from longitudinal force F_x and vertical force $F_{z,W}$ is the coefficient of friction

$$\mu_{x,W} = F_x/F_{z,W} \qquad (2.5)$$

when it relates to the maximum value, and the coefficient of sliding friction

$$\mu_{x,sl} = F_x/F_{z,W} \qquad (2.5a)$$

when it is the minimal value (100% slip) (Fig. 2.21). F_x is designated F_b during braking and F_a during traction.

In all cases $\mu_{x,W}$ is greater than $\mu_{x,sl}$; in general it can be said that on dry road $\mu_{x,\mathrm{W}} \approx 1.2\, \mu_{x,\mathrm{sl}}$ (2.6)

wet road $\mu_{x,\mathrm{W}} \approx 1.3\, \mu_{x,\mathrm{sl}}$ (2.6a)

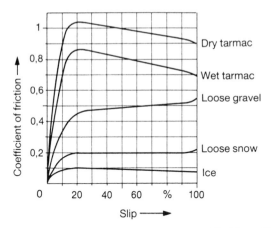

Fig. 2.21   Coefficient of friction $\mu_{x,\mathrm{W}}$ of a summer tyre with 80 to 90% deep profile, measured at around 60 km h^{-1} and shown in relation to the slip on road surfaces in different conditions. Wide tyres in the 65 series and below have the greatest friction at around 10% slip, which is important for the ABS function.

2.7.3 Road influences

2.7.3.1 Dry and wet roads   On a dry road, the coefficient of friction is relatively independent of the speed (Fig. 2.22), but a slight increase can be determined below 20 km h^{-1}. The reason lies in the transition from dynamic to

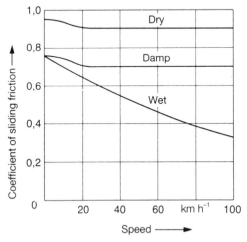

Fig. 2.22   Relationship of coefficient of sliding friction $\mu_{x,\mathrm{sl}}$ to speed on different road conditions.

static rolling radius (see the example in Section 2.2.5.4) and is therefore linked to an increasing area of tyre contact. At speeds a little over zero, on a rough surface, a toothing effect can occur, which causes a further increase in the coefficient of friction, then:

$$\mu_{x,w} \geqslant 1.3 \tag{2.6b}$$

When the road is wet, the coefficient of friction reduces but is still independent of the speed. This situation changes as the amount of water increases and also with shallower profile depth. The water can no longer be moved out of the profile grooves and the μ value falls as speed increases.

2.7.3.2 Aquaplaning The higher the water level, the greater the risk of aquaplaning. Three principle factors influence when this occurs:

- road
- tyres
- speed.

With regard to the road, the water level is the critical factor (Fig. 2.23). As the level rises, there is a disproportionate increase in the tendency towards aquaplaning. When the level is low, the road surface continues to play a role because the coarseness of the surface absorbs a large part of the volume of water and carries it to the edge of the road. Following rainfall, the water levels on roads are generally up to 2 mm; greater depths can also be found where it has been raining for a long time, during storms or in puddles.

On the tyre, the tread depth has the greatest influence (Fig. 2.34). There can be up to a 25 km h^{-1} difference in speed between a full tread and the legal minimum tread depth of 1.4 mm. High tyre pressure and low running surface radius r (Fig. 2.4) lead to the area of contact becoming narrower, giving the advantage of improved aquaplaning behaviour as the distribution of ground pressure becomes more even (Fig. 2.6). Lower tyre pressure and contours with larger radii make aquaplaning more likely; this also applies to wider tyres (Fig. 2.11b) particularly when tread depths are low.

However, the greatest influence by far is the speed, especially when the water level increases and tread depths are low. This is why reducing speed is the best way to lessen the risk of aquaplaning, and is a decision drivers can make for themselves.

2.7.3.3 Snow and ice Similar to aquaplaning, low coefficients of friction occur on icy roads, although these are highly dependent on the temperature of the ice. At close to 0°C, special conditions occur; compression of the surface can lead to the formation of water which has a lubricating effect and reduces the coefficient of friction to $\mu_{x,w} \leqslant 0.08$ (Fig. 2.24). At −25°C, a temperature that is by no means rare in the Nordic countries, values of around $\mu_{x,w} = 0.6$ can be reached. At low temperatures, coefficients of friction and sliding friction are further apart:

$$\mu_{x,w} \sim 2\,\mu_{x,sl} \tag{2.7}$$

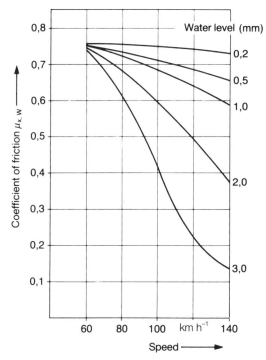

Fig. 2.23 Coefficients of friction $\mu_{x,W}$ of a summer tyre with an 8 mm deep profile dependent on speed at different water levels. Hardly any influence can be detected under 60 km h^{-1}; at higher speeds and 3 mm water depth, the curve shows $\mu_{x,W}$ which indicates the aquaplaning effect.

2.8 Lateral friction

2.8.1 Lateral forces, slip angle and coefficient of friction

A rolling tyre is only able to transfer lateral forces if it rolls at an angle to the direction of travel. If a lateral interference force acts at the centre of gravity of the vehicle (e.g. a wind, or hill tractive force F_y), as shown in Fig. 2.25, the lateral forces $F_{y,W,f,o}$, $F_{y,W,f,i}$, $F_{y,W,r,o}$ and $F_{y,W,r,i}$ occur at the centre of contact. When subject to the influence of F_y, the vehicle changes its direction of travel, i.e. it deviates from this by the angle α. The greater the interference force, the greater the slip angle α, because only where the angle has increased can the tyres transfer the higher lateral forces; force and angle are therefore dependent upon each other (Fig. 2.26). When cornering, the interference force should be equal to the centrifugal force F_c, which results from the speed v in m s^{-1} and the radius of the bend R in m, on which the vehicle centre of gravity S (Fig. 2.29a) moves. With the total weight $m_{v,t}$ of the vehicle the equation is:

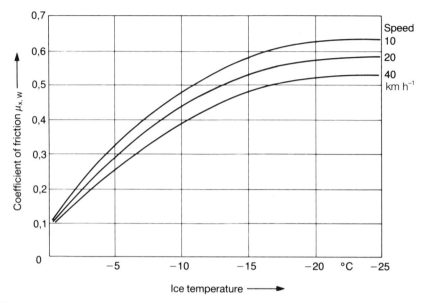

Fig. 2.24 Influence of ice temperature and car speed on the coefficient of friction $\mu_{x,w}$ of an 82 series winter tyre; the extremely low values at 0°C can be seen clearly.

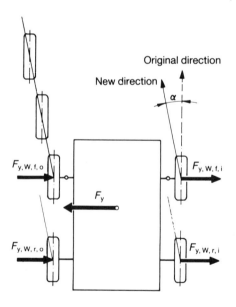

Fig. 2.25 Tyres are only able to transfer a lateral force F_y acting on the vehicle if they are rolling at an angle to the vehicle. Regardless of whether these are F_y or the centrifugal force F_c during cornering, the lateral forces $F_{y,W}$ should be regarded as being perpendicular to the wheel centre plane.

$$F_c = m_{v,t}\, v^2/R = m_{v,t}\, a_y = F_y\ [\text{N}] \tag{2.8}$$

The centrifugal or disturbance force is just as large as the lateral forces on the wheels (Fig. 2.25):

Fig. 2.26 The higher the lateral force F_y, the greater the tyre slip angle α.

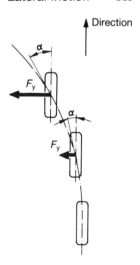

$$F_v = F_{y,W,f,o} + F_{y,W,f,i} + F_{y,W,r,o} + F_{y,W,r,i} = \Sigma F_{y,W} \tag{2.8a}$$

and

$$\Sigma F_{y,W} = \mu_{y,W} \, \Sigma F_{z,W} = \mu_{y,W} \, F_{V,t}$$

Together the two equations give

$$\mu_{y,W} \, F_{V,t} = \mu_{y,W} \, m_{V,t} \, g = m_{V,t} \, a_y$$

and

$$\mu_{y,W} = a_y / g \tag{2.9}$$

The coefficient of friction $\mu_{y,W}$ is not dependent on the radius of the curve and driving speed and is therefore more suitable for calculating cornering behaviour (see also Equation 6.13a).

The faster the vehicle negotiates a bend, the higher the coefficient of friction used and the greater the slip angles (Fig. 2.27).

2.8.2 Cornering behaviour and attitude angle

A difference should be drawn between the angles α_f and α_r at the centres of tyre contact of the front and rear wheels, respectively. The difference between the two can be regarded as the measurement or definition for cornering behaviour. This would be neutral if $\alpha_f \approx \alpha_r$ (Fig. 2.28), understeering if $\alpha_f > \alpha_r$ (Figs 2.29, 2.37 and 1.17), and oversteering if $\alpha_f < \alpha_r$ (Figs 2.29a, 2.38 and 3.57). In both the last two cases, the attitude angle $\pm\beta$ is between the vehicle longitudinal axis and a tangent that is applied in the centre of gravity S on its circular path. In order to represent the driving conditions better, the

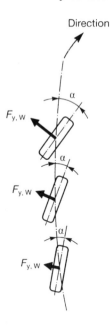

Direction

Fig. 2.27 Increasing lateral forces $F_{y,w}$ during cornering caused by the centrifugal force F_c leads to increasing slip angles α.

wheels of one axle are drawn together in the centre of the vehicle in the three sketches; this is therefore not a single track vehicle (motorcycle).

As shown in Figs 2.29 and 5.0, the steering has to be angled further on an understeering vehicle to force it into the bend, unlike oversteering vehicles (Fig. 2.29a), whose steering input has to be reduced to counteract the tendency of the tail end to swing out. In the case of greater bend radii, where the average steering angle δ_m is less than 5° and it can be assumed that the sine and radius values of the angle are equal, and the angles δ_o and δ_i correspond to this (Fig. 3.70 and Equation 3.17):

$$\sin \delta_m \sim \delta_m \sim \delta_o \sim \delta_i \ [rad]$$

Using Equation 3.12 it is now possible to determine the relationship between steering angle, track circle diameter D_S (Figs 1.51 and 3.68) and slip angles at a constant cornering speed:

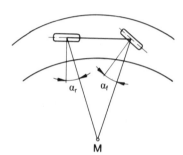

Fig. 2.28 If, during cornering, $\alpha_f \sim \alpha_r$, the handling of a vehicle can be described as neutral.

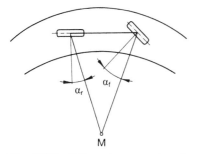

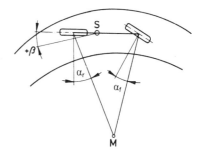

Fig. 2.29 If there is a greater slip angle α_f on the front wheels than α_r on the rear, the vehicle understeers and the attitude angle, which can be seen in Fig. 2.29a becomes $-\beta$ (negative).

Fig. 2.29a If there is a greater slip angle α_r on the rear wheels than on the front (α_f), the vehicle oversteers and there is a positive attitude angle $+\beta$ (see also Fig. 3.57). A tendency to oversteer is particularly prevalent in passenger cars where the centre of gravity S lies towards the back (which is the case when the vehicle is fully laden) or when the tyre pressure on the rear axle is too low.

$$\delta_m = \frac{2l}{D_S} + \alpha_f - \alpha_r \qquad (2.10)$$

The kingpin offset at ground r_S is so unimportant in comparison to D_S that it can be ignored in this case.

2.8.3 Coefficients of friction and slip

To determine the cornering behaviour, the chassis engineer needs the lateral forces (or the coefficient of friction) based on the slip angle and the parameters

- vertical force (or wheel load) in the centre of tyre contact
- tyre pressure
- wheel camber.

The measurements are generally taken on test rigs, up to slip angles of $\alpha = 10°$. The drum surface with its friction values of $\mu_0 = 0.8$ to 0.9 sets limits here, and larger angles hardly give increasing lateral coefficients of friction.

$$\mu_{y,W} = F_{y,W}/F_{z,W} \qquad (2.11)$$

Conditions on the road are very different from those on the test rig, and the type of road surface and its condition play a role here. As can be seen in Fig. 2.30, the coefficient of friction on rough, dry concrete increases to $\alpha = 20°$ and then falls. In precisely the same way as with the longitudinal force the slip S_y (in the lateral direction) is also taken into consideration; this is as a percentage of the sine of the slip angle times 100:

$$S_y = \sin \alpha \times 100 \, [\%] \tag{2.12}$$

In conjunction with the drum value $\alpha = 10°$, this would give a slip of $S_y = 17\%$, and on the street at $\alpha = 20°$ slip values of up to $S_y = 34\%$. If the tyre is further twisted to $\alpha = 90°$, it slides at an angle of 90° to the direction of travel; $\sin \alpha$ would then be equal to one and $S_y = 100\%$. The coefficient of friction then becomes the coefficient of lateral sliding friction $\mu_{y,sl}$, which on average is around 30% lower:

$$\mu_{y,sl} \approx 0.7 \, \mu_{y,w} \tag{2.13}$$

In contrast to dry concrete (as also shown in Fig. 2.30) on asphalt and, in particular on wet and icy road surfaces, no further increase in the lateral cornering forces can be determined above $\alpha = 10°$ (i.e. $S_y \approx 17\%$).

2.8.4 Lateral cornering force properties on dry road

Figure 2.31 shows the usual way in which a measurement is carried out for a series 82 tyre. The lateral force appears as a function of the vertical force in kilonewtons and the slip angle α serves as a parameter. A second possibility can be seen in Fig. 2.32; here, for the corresponding series 70 tyre, $\mu_{y,w} =$

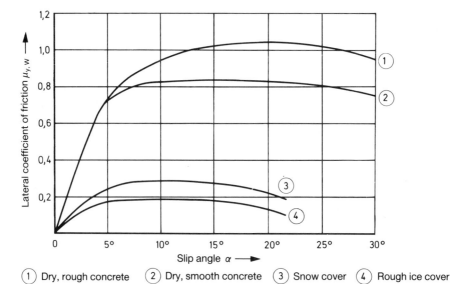

① Dry, rough concrete   ② Dry, smooth concrete   ③ Snow cover   ④ Rough ice cover

Fig. 2.30 Lateral coefficients of friction $\mu_{y,w}$ as a function of slip angle and road condition, shown for an 82 series summer tyre with around 90% deep profile. The ice temperature is around −4°C. The vertical force $F_{z,W}$ was kept constant during the measurements to obtain the dimensionless values of $\mu_{y,w}$. The maximum at $\alpha = 20°$ on a very skid-resistant road can be seen clearly. The further $\mu_{y,w}$ sinks, the further it moves towards smaller angles.

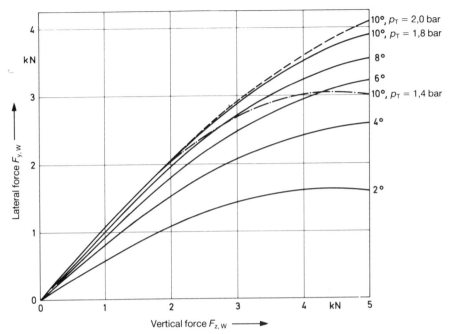

Fig. 2.31 Lateral cornering forces of the 155 R 13 78 S '82 series' steel radial tyre, measured on a dry drum at p_T = 1.8 bar. The load capacity at this pressure is around 360 kg, corresponding to a vertical force $F_{z,w}$ = 3.53 kN. Also shown are the forces at α = 10° and f_T = 1.4 bar and 2.0 bar to indicate the influence of the tyre pressure on the lateral cornering properties.

$F_{y,w}/F_{z,w}$ is plotted against α and $F_{z,w}$ serves as a parameter. The bent shape of the curves in both figures shows that the angle of tangents applied changes as a function of $F_{z,w}$ or $\mu_{y,w}$. The maximum occurs with large angles and small vertical forces. A less stressed tyre in relation to its load capacity therefore permits greater coefficients of friction and higher cornering speeds than one whose capacity is fully used.

This result, which has been used for a long time in racing and sports cars, has also become popular in modern cars, Fig. 2.31, and a mid-range standard car can be taken as an example. The car manufacturer specifies p_T = 2.2 bar/2.5 bar under full load for the front and rear wheels 185/65 R 15 88H. At these pressures the load capacity, in accordance with Tables 2.9 and 2.10, is:

front 505 kg and rear 560 kg

Figure 5.5 contains the authorized axle loads from which the wheel load (divided by two) results. These are

front 375 kg and rear 425 kg

As described in Section 2.2.6, at speeds up to 210 km h^{-1} (H tyres), an

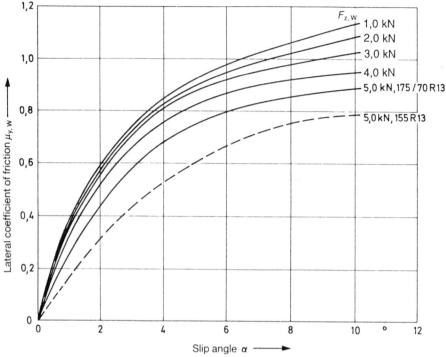

Fig. 2.32   Lateral coefficients of friction $\mu_{y,w}$ as a function of the slip angle α and the vertical force $F_{z,w}$, measured on a dry drum on a 175/70 R 13 82 S tyre at $p_T = 2.0$ bar. The tyre, which has been inflated in such a manner, carries 395 kg or $F_{z,w} = 3.87$ kN. In order to indicate the influence of the cross-section on the transferable lateral forces the 82 series 155 R 13 78 S tyre was also included.

increase in tyre pressure of 0.3 bar is necessary or there is only a correspondingly lower load capacity. This then is, with $p_T = 1.9$ bar at the front or 2.2 bar at the back:

450 kg and 505 kg

Thus, the actual load factor k_m at 210 km h^{-1} becomes:

front $k_{m,f} = (375/450) \times 100 = 83\%$
back $k_{m,r} = (425/505) \times 100 = 84\%$ (2.14)

2.8.5 Influencing variables

2.8.5.1 Cross-section ratio _H/W_   The 185/65 R 15 88H size used as an example in the previous section is a 65 series wide tyre; the 15" diameter also allows a good sized brake disc diameter (Fig. 2.6a). In contrast to the 82 series standard tyre, the sizes of the 70 series and wide tyres ($H/W = 0.65$ and below) generate higher lateral cornering forces at the same slip angles (Figs

2.32, 2.33 and 2.6). As can be seen in Fig. 1.2, these, as $F_{y,w,o} = \mu_{y,w} (F_{z,w} + \Delta F_{z,w})$, are all the greater, the faster the vehicle takes a bend.

2.8.5.2 Road condition   The road's surface and condition have the greatest influence; Fig. 2.30 shows the relationships and Fig. 2.34 shows the influence of tyre tread pattern depth and wetness.

2.8.5.3 Track width change   The track width change that exists, in particular on independent wheel suspensions described in Section 3.3, causes lateral forces at the centres of tyre contact on both wheels when the vehicle is moving unimpeded in a straight line. Figures 3.2 and 3.3 show this, and also what lateral forces can occur if a series 82 radial tyre rolling in a straight line is brought out of its direction by an axle-kinematic dependent change. Wide tyres could cause even higher forces in a lateral direction, whilst diagonal ply tyres could produce lower ones.

2.8.5.4 Variations in vertical force   During cornering, vertical force variations $\pm \Delta F_{z,w}$ in the centre of tyre contact cause a reduction in the transferable lateral forces $F_{y,w}$. The loss of lateral force $\Delta F_{y,W,4}$ depends on the effectiveness of the shock absorbers, the tyre pressure p_T (which can enhance the 'springing' of the wheels, see Equation 5.6) and the type of wheel suspension link mountings. Further influences are wheel load and driving speed. To

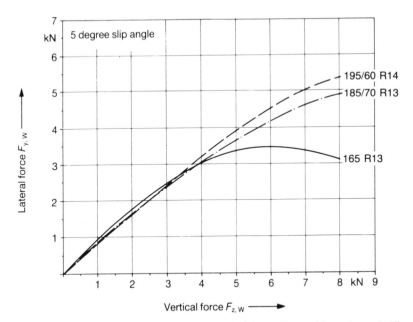

Fig. 2.33   Lateral force $F_{y,w}$ dependent on vertical force $F_{z,w}$ and tyre sizes of different H/W ratios: 165 R 13 82 H, 185/70 R 13 85 H and 195/60 R 14 85 H.

Up to $F_{z,w} = 4000$ N the curves are more or less the same, but at higher loads the more favourable lateral cornering properties of the wide tyre are evident.

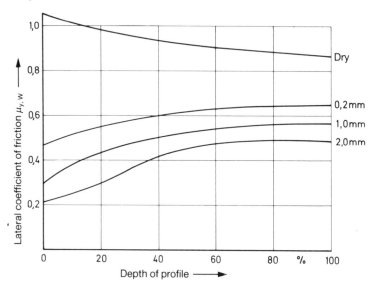

Fig. 2.34 Possible lateral friction coefficients $\mu_{y,w}$ of a steel radial tyre 155 R 13 78 S depending on the depth of the tyre profile as a percentage (starting from 8 mm = 100%) at p_T = 1.8 bar, α = 10°, v = 60 km h^{-1} and varying water film thicknesses in mm.

The improved grip of the profileless tyre on a dry road can be seen clearly as can its significantly poorer grip in the wet; a fact which also applies to the coefficient of friction in the longitudinal direction (see Section 2.7.2).

calculate cornering behaviour, an average loss of lateral force $\Delta F_{y,W,4}$ due to variations in vertical force and dependent only on tyre design and slip angle α, should be considered:

$$\Delta F_{y,W4} \approx 40 \text{ N per degree } \alpha \qquad (2.15)$$

2.8.5.5 Camber change Wheels which incline with the body during cornering have a similar, detrimental influence on the transferability of lateral forces. As can be seen from Fig. 1.2, positive angle (ϵ_w) camber changes occur on the outside of the bend and negative angles on the inside of the bend as a consequence of the body roll. The lateral forces are directed to the centre point of the bend (Fig. 3.9). If a wheel is 'cambered' against this, in other words inclined at the top towards the outside of the bend, the possibility of transferring lateral forces reduces; on a dry road surface, depending on the tyre size, the change is:

$$\Delta F_{y,W,3} = 40 \text{ N to } 70 \text{ N per degree of camber} \qquad (2.16)$$

To counteract this, a greater slip angle occurs and greater steering input becomes necessary for the front wheels. This makes the vehicle understeer more (Fig. 2.29) and appear less easy to handle. Furthermore, the steering aligning moment (see Section 3.10.3) also increases. Negative camber $-\epsilon_w$ on

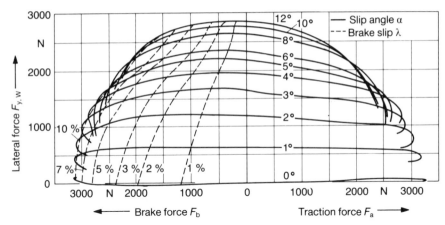

Fig. 2.34a Circumference lateral force performance characteristics with slip angles and brake slip as parameters. The study was carried out on a 165 R 15 86 S radial tyre loaded at 300 kg at p_T = 1.5 bar. The shape of the curves indicates that, with increasing longitudinal forces, those which can be absorbed laterally reduce. At 1.5 bar, the tyre carries a weight of 375 kg, i.e. it is only operating at 80% capacity.

the outside of the bend and positive $+\epsilon_w$ on the inside would have exactly the opposite effect. Wheels set in this manner would increase the lateral forces which can be absorbed by the amount stated previously for $\Delta F_{y,W3}$ and cause a reduction in the tyre slip angle.

2.8.5.6 Lateral force due to camber   The wheels, which tilt due to the body roll inclination on a bend, try to roll towards the outside of the bend, i.e. against the direction in which they are turned (Figs 1.2 and 3.37). An additional force at the centres of tyre contact is necessary to bring them back into the required direction. This force, known as the lateral force due to camber $F_{\epsilon w}$, acts in the same direction as the centrifugal force F_c and therefore increases the lateral forces $F_{y,w,f,o}$ and $F_{y,w,f,i}$ to be absorbed by the tyres. Larger slip angles and greater steering input are required to counteract this.

The average force $F_{\epsilon w}$ with the standard camber values for individual wheel suspensions on a dry road are:

$$F_{\epsilon w} \approx F_{z,W} \, \sin\epsilon_w \tag{2.17}$$

2.9 Resulting friction

Rolling resistance increases when negotiating a bend (see Equation 2.4a), and the vehicle would decelerate if an increased traction force F_A did not create the equilibrium needed to retain the cornering speed selected. In accordance with Equation 6.36, F_A is dependent on a series of factors and the type of drive system (front- or rear-wheel drive); on single-axle drive, the traction force stresses the coefficient of friction (see Sections 1.4 to 1.6 and 2.7.2)

$$\mu_{x,W} = F_{A,f\ or\ r}/F_{V,f\ or\ r} \tag{2.17a}$$

and thus greater slip angles at the driven wheels. With given values for cornering speed and radius (see Equation 2.8) the resulting coefficient of friction μ_{rsl} can be determined:

$$\mu_{rsl} = (\mu^2_{y,W} + \mu^2_{x,W})^{\frac{1}{2}} \tag{2.18}$$

μ_{rsl} cannot be exceeded because the level depends on the road's surface and the condition. When braking in a bend, additional longitudinal forces F_b occur on all wheels (see Section 6.3.1), and act against the direction of travel. In this case Equation 2.18 also applies.

On standard vehicles and front-wheel drives, the front wheels take 70 to 80% of the braking force and the rear wheels only 30 to 20%. This means that the slip angles increase on both axles, but more at the front than the rear and the vehicle tends to understeer (Fig. 2.29 and Equation 6.20). If an axle locks, the friction becomes sliding friction and the vehicle pushes with this axle towards the outside of the bend (Figs 6.8 to 6.10).

Taking into consideration the maximum possible values in the longitudinal and lateral direction of the road – known respectively as $\mu_{x,W,max}$ and $\mu_{y,W,min}$ – the increasing coefficient of friction can be calculated:

$$\mu_{x,W} = \mu_{x,W,max}\left[1 - \left(\frac{\mu_{y,W}}{\mu_{y,W,max}}\right)^2\right]^{\frac{1}{2}} \tag{2.19}$$

Consider as an example a braking process on a dry road at 100 km h^{-1} on a bend with $R = 156$ m. Using Equation 2.9 the calculation gives: $\mu_{y,W} = 0.5$. Figure 2.34a shows a measurement on the tyre in question where the greatest coefficient of friction in the lateral direction at $F_{z,W} = 2490$ N, $s_{x,b} = 10\%$ and $\alpha = 4°$ (see Equation 2.11) amounts to

$$\mu_{y,W,max} = F_{y,W}/F_{z,W} = 2850/2940\ [\text{N/N}] \quad \mu_{y,W,max} = 0.97$$

In the longitudinal direction the possible braking force $F_b = 3130$ N is at $\alpha = 0°$ and therefore (see Equation 2.5)

$$\mu_{x,W,max} = F_b/F_{z,W} = 3130/2940\ [\text{N/N}] \quad \mu_{x,W,max} = 1.06$$

and

$$\mu_{x,W} = 1.06\left[1 - \left(\frac{0.5}{0.97}\right)^2\right]^{\frac{1}{2}} \quad \mu_{x,W} = 0.91$$

The lateral forces that the tyre can absorb during braking can also be calculated:

$$\mu_{y,W} = \mu_{y,W,max} \left[1 - \left(\frac{\mu_{x,W}}{\mu_{x,W,max}} \right)^2 \right]^{\frac{1}{2}} \qquad (2.19a)$$

$\mu_{x,W} = 0.7$ should be given. The lateral coefficient of friction (which can be used) is:

$$\mu_{y,W} = 0.97 \left[1 - \left(\frac{0.7}{1.06} \right)^2 \right]^{\frac{1}{2}} \qquad \mu_{y,W} = 0.73$$

At $S_{x,b} = 10\%$ and $\alpha = 4°$ the transferable lateral force

$$F_{y,W} = \mu_{y,W} F_{z,W} = 0.73 \times 2940 \quad F_{y,W} = 2146 \text{ N}$$

and the available braking force

$$F_b = \mu_{x,W} F_{z,W} = 0.7 \times 2940 \quad F_b = 2058 \text{ N}$$

2.10   Tyre self-aligning moment and caster offset

2.10.1 Tyre self-aligning moment in general

During cornering, the area of tyre contact deforms so that the point at which the lateral force acts alters by the amount $r_{\tau,T}$, known as the caster offset, and comes to lie behind the centre of the wheel (Fig. 3.19a). On the front wheels, the lateral cornering force $F_{y,W,f}$ together with $r_{\tau,T}$ (as the force lever) gives the self-aligning moment $M_{T,f}$ which superimposes the kinematic alignment torque and seeks to bring the input wheels back to a straight position.

The self-aligning moment, lateral force and slip angle are measured in one process on the test rig. $M_{T,f}$ is plotted as a function of the slip angle (Fig. 2.35); the vertical force $F_{z,W}$ serves as a parameter. The higher $F_{z,W}$, the greater the self-alignment and, just like the lateral force, the moment increases to a maximum and then falls again. $M_{T,f,max}$ is, however, already at $\alpha \sim 4°$ (as can be seen in Fig. 2.30) and not (on a dry road) at $\alpha \geqslant 10°$.

2.10.2 Caster offset

Caster offset, $r_{\tau,T}$, is included in practically all calculations of the self-aligning moment during cornering (see Section 3.10.3). The length of this can easily be calculated from the lateral force and moment:

$$r_{\tau,T} = M_{T,f}/F_{y,W} \text{ [m]} \qquad (2.20)$$

This requires two images, one which represents $F_{y,W} = f(F_{z,W}$ and $\alpha)$ or $\mu_{y,W} = f(F_{z,W}$ and $\alpha)$, and another with $M_{T,f} = f(F_{z,W}$ and $\alpha)$. The values of the 175/70R 13 82 S steel radial tyre shown in Figs 2.32 and 2.35 and measured at $p_T = 2.0$ bar serve as an example. At $\alpha = 2°$ and $F_{z,W} = 5.0$ kN the coefficient of friction $\mu_{y,W} = 0.44$ and therefore:

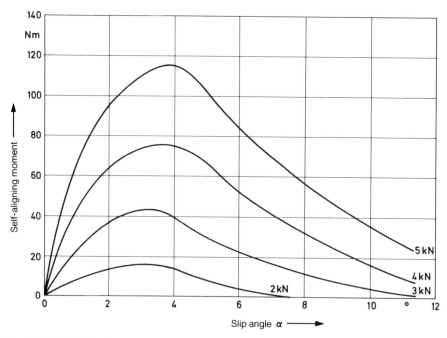

Fig. 2.35 Self-aligning moments of a 175/70 R 13 82 S steel radial tyre measured on a dry drum as a function of the slip angle at p_T = 2.0 bar. The vertical force $F_{z,W}$ in kilonewtons is used as a parameter. The moments increase sharply at low angles, reach a maximum at α = 3° to 4° and then reduce slowly. As the cornering speed increases, the tyre self-aligning moment decreases, whilst the kinematically-determined moment increases (see Section 3.8).

$$F_{y,W} = \mu_{y,W}\, F_{z,W} = 0.44 \times 5.0 = 2.2 \text{ kN}; \; F_{y,W} = 2200 \text{ N}$$

At the same angle and with the same wheel force, the self-aligning moment is $M_{T,f}$ = 95 N m and therefore

$$r_{\tau,T} = M_{T,f}/F_{y,W} = 95/2200 = 0.043 \text{ m}; \; r_{\tau,T} = 43 \text{ mm}$$

Figure 2.36 shows the caster offset (caster trail) calculated in this manner. Higher lateral forces necessitate greater slip angles, and the latter result in smaller self-aligning moments and a reduced caster offset. The explanation for this fact is that, at low slip angles, only the tyre profile is deformed at the area of contact. The point of application of the lateral force can therefore move further back, unlike large angles where, principally, the carcass is deformed. High vertical wheel forces cause the tyre to be severely compressed and therefore an increase both in the area of tyre contact and also in the caster offset occur.

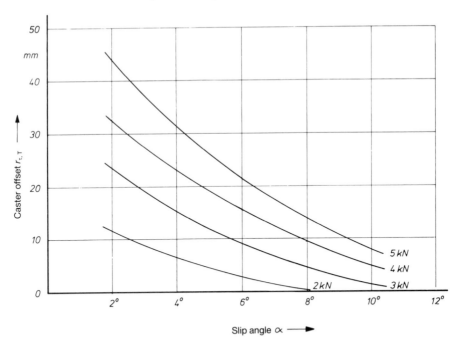

Fig. 2.36   Caster offset of tyre $r_{\tau,T}$ calculated from Figs 2.32 and 2.35 for 175/70 R 13 82 S steel radial tyres at $p_T = 2.0$ bar. The higher the vertical force $F_{z,W}$ (in kN) and the smaller the angle α, the longer is $r_{\tau,T}$.

2.10.3 Influences on the front wheels

The tyre self-aligning moment is one of the causes for the steering forces during cornering; its level depends on various factors.

2.10.3.1 Dry roads   The self-aligning moment is usually measured on a roller test bench with the drum allowing a coefficient of friction of $\mu_0 = 0.8$ to 0.9 between its surface and the tyre. If the resultant self-aligning moment on the open road is required, it is possible to approximate the value $M_{T,f,\mu}$ using a correction factor:

$$k_\mu = \mu_{y,W}/\mu_0 \tag{2.21}$$

A cement block with $\mu_{y,W} \sim 1.05$ (Fig. 2.30) and the 175/70 R 13 82 S radial tyre can be used as an example. In accordance with Fig. 2.35 on this:

$$M_{T,f} = 40 \text{ N m with } F_{z,W} = 3 \text{ kN and } \alpha = 4°$$

As a correction factor this gives

$$k_\mu = \mu \frac{\text{Road}}{\text{Roller}} = \frac{\mu_{y,w}}{\mu_0} = \frac{1.05}{0.80}; k_\mu = 1.31$$

and thus

$$M_{T,f,\mu} = k_\mu \, M_{T,f} = 1.31 \times 40; M_{T,f,\mu} = 52.4 \text{ N m}$$

2.10.3.2 Wet roads   Provided that k_μ is independent of tyre construction and profile, the approximate value for a wet road can also be determined. In accordance with Fig. 2.34, with 1 mm of water on the surface and full profile depth the $\mu_{y,w}$ value reduces from 0.80 to 0.54. Owing to the reduced coefficient of friction, only a smaller value $M_{T,f}$, can be assumed, in other words,

$$k_\mu = \mu_{y,w} \frac{\text{wet}}{\text{Roller}} = \frac{0.55}{0.86} = 0.64, \text{ and}$$

$$M_{T,f,\mu} = 0.64 \times 40 \text{ N m}; M_{T,f,\mu} = 25.6 \text{ N m}$$

A greater water film thickness may cause the coefficient of friction to reduce but the self-aligning moment increases and the water turns the wheel back into the straight position. Furthermore, the self-aligning maximum shifts towards smaller slip angles when the road is wet.

2.10.3.3 Icy roads   Only with greater vertical forces and small slip angles is the smoothness of the ice able to deform the area of tyre contact and generate an extremely small moment, which is nevertheless sufficient to align the tyre. Low front axle loads or greater angles α arising as a result of steering corrections would result in a negative moment $-M_{T,f}$ (in other words in a 'further steering input' of the tyres). The wheel loads at the front, which were only low, were already a problem on rear-engine passenger vehicles.

2.10.3.4 Longitudinal forces   As shown in Fig. 3.91a, traction forces increase the self-aligning moment; the equation for one wheel is:

$$M_{T,f,a} = F_{y,w} \, r_{\tau,T} + F_a \, r_T = F_{z,w} \left(\mu_{y,w} \, r_{\tau,T} + \mu_{x,w} \, r_T \right) \tag{2.22}$$

During braking the moment fades and reduces to such an extent that it even becomes negative and seeks to input the wheels further. The formula for one wheel is:

$$M_{T,f,b} = F_{y,w} \, r_{\tau,T} - F_b \, r_T = F_{z,w} (\mu_{y,w} \, r_{\tau,T} - \mu_{x,w} \, r_T) \tag{2.23}$$

The length of the paths $r_{\tau,T}$ and r_T can be found in the details of Fig. 3.91a.

2.10.3.5 Tyre pressure   When the tyre pressure is increased the self-aligning moment reduces by 6% to 8% per 0.1 bar, and increases accordingly when the pressure reduces, by 9% to 12% per 0.1 bar.

A reduction in pressure of, for example, 0.5 bar could thus result in over a 50% increase in the moment, a value which the driver would actually be able to feel.

2.10.3.6 Further influences   The following have only a slight influence:

- positive camber values increase the moment slightly, whilst negative ones reduce it
- $M_{T,f}$ falls as speeds increase because the centrifugal force tensions the steel belt which becomes more difficult to deform (Table 2.10a)
- widening the rim jaw slightly reduces self-alignment.

2.10.4 Torque steer due to self-aligning moments on the rear wheels

The tyre self-aligning moment $M_{T,f,}$ which is partly responsible for the steering righting moment of the front wheels, is equally present on the rear wheels and can, just as on the front wheels, be increased by longitudinal traction $F_{a,r}$. The moments of the wheel on the outside and inside of the bend, are combined in a central one $+M_r$ (Fig. 2.37), and cause the tail to turn inwards, in other words a tendency towards understeering on standard design or rear-engine vehicles.

Traction forces reinforce the self-aligning moment (Fig. 3.91a). If the driver takes his foot off the accelerator during cornering, the engine brakes. The longitudinal forces $F_{B,r} = 2 F_{b,r}$ which arise (and can be seen in Fig. 6.7) are directed from front to back and cause reducing self-aligning moments at the

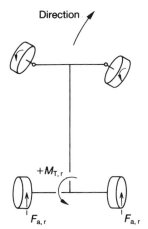

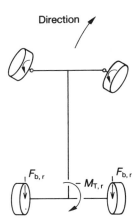

Fig. 2.37 As a result of tyre slip angle, self-aligning moments occur during cornering at the centres of tyre contact of the rear wheels and these can be further reinforced by the traction force F_a. The moments of the outer and inner wheel combined as average $+M_{T,r}$ indicate that the tail is turning inwards, in other words there is a reduction in the tendency to oversteer.

Fig. 2.38 Taking your foot off the accelerator causes the engine to brake which leads to a reduction in the moment $+M_{T,r}$ which can even become negative ($-M_{T,r}$). This makes the tail suddenly swerve out and the vehicle steers towards the inside of the bend; in other words there is momentary oversteering.

centres of contact (see Equations 2.22 and 2.23). The moment $+M_r$, which has so far been moving to the left, becomes zero or (as $-M_r$) even negative. To balance this out, a greater slip angle for transferring the lateral force occurs at the rear wheels, which is equally important when the tail end moves out. If the driver does not spontaneously reduce the steering input, understeering becomes oversteering (Figs 2.29 and 2.29a) and the unpleasant 'torque steer', also known as the turning-in effect occurs (Figs 2.38 and 3.57).

3

Axle kinematics and elastokinematics

'Kinematics' describes the movement caused in the wheels during bottoming out and steering, whilst 'elastokinematics' defines the alterations in the position of the wheels caused by forces and moments between the tyres and the road (Fig. 3.0 and Section 3.6.5), or the longitudinal movement of the wheel required to prevent kinematic changes (Fig. 3.0a). The changes are the result of the elasticity in the suspension parts. The coordinate directions (in which everything is to be considered, Fig. 3.1) and the kinematic formulas are laid down in proposal ISO/TC 22/SC9/WG1, the International Standard ISO 8855, the American Standard SAE J670c and the German Standard DIN 70 000.

3.1 Purpose of the axle settings

To ensure the required road holding and directional stability and to prevent excessive tyre wear, automobile manufacturers specify certain settings, including the permissible tolerances for the front axles of all models and for the rear axles, provided these are not driven rigid axles. Toe-in can be set via the tie rods or eccentric discs (Fig. 3.49a) and camber and caster angles can also be adjusted on some vehicles. The remaining manufacturers' data for kingpin inclination, kingpin offset at ground (scrub radius), caster offset and differential toe angle are design data and not easy to measure and are actually only used for checking the roadworthiness of a vehicle which has been damaged in an accident or has reached a given age. As shown in the figures in the following sections, the axle settings depend on load and load distribution. In order to make the measurements easier for garages to carry out, only the curb weight, in accordance with recommendation ISO/R 1176 or DIN 70 020 (see Section 5.3.1.1) should be used as the basis for measurements.

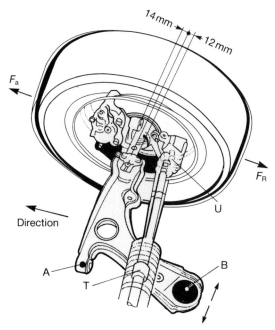

Fig. 3.0 On the McPherson front axle of the VW Passat, the bearing B of the lower sickle-shaped control arm, has a precisely defined lateral compliance to keep the dynamic rolling hardness of the steel radial tyres away from the body; A is the suspension control arm pivot. The wheel can move up to 14 mm forwards under drive-off forces and, in the event of rolling resistance (or braking) forces, up to 12 mm backwards (see also Fig. 3.65). The tie rods (points U and T) must be positioned so that no toe-in alteration occurs (Fig. 4.31).

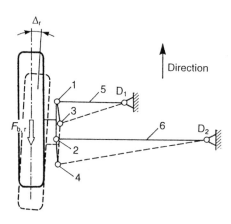

Fig. 3.0a If the front transverse link 5 on the bottom pair of suspension control arms of a rear McPherson axle is shorter than the rear one 6, and if the longitudinal forces are absorbed by a trailing link (not illustrated), its front bearing, which is fixed to the underbody, can comply in a defined manner when braking forces $F_{b,r}$ occur. The outer point 1 of the link 5 then moves in an arc around D_1 to D_3 and point 2 of the link 6 around D_2 to D_4. Due to the different radii of the two arcs, a toe-in angle Δ_r occurs which opposes the returning moment $M_b = F_{b,r}\, r_b$ (Fig. 3.84).

3.2 Wheelbase

The wheelbase l, measured from the centre of the front to the centre of the rear axle (Fig. 6.1), is an important variable in the vehicle's ride and handling properties. A long wheelbase relative to the overall length of the vehicle makes it possible to accommodate the passengers easily between the axles and reduces the influence of the load on the axle load distribution (see Section 5.3.6). The short body overhangs to the front and rear reduce the tendency to pitch oscillations and make it possible to fit soft springing, normally associated with a high level of ride comfort. A short wheelbase, on the other hand, makes cornering easier, i.e. gives a smaller turning circle for the same steering input (see Section 3.7.2). Vehicle designers seek to achieve a long wheelbase on both front-wheel drive passenger cars and on conventional designs. However, this depends on the body shape. A hatchback estate saloon (Figs 1.50 and 1.57) can be of a more compact design, giving a longer wheelbase relative to the vehicle length than notchback saloons and the estate cars developed from them. The ratio

$$k_1 = \frac{\text{wheel base}}{\text{vehicle length}} = 0.60 \pm 0.07 \tag{3.1}$$

can be used as a reference and should be as large as possible.
It is:

$k_1 = 0.57$ to 0.67 on estate saloons and
$k_1 = 0.56$ to 0.61 on notchback saloons.

In coupés the value can be below 0.56 and on small cars it is up to 0.69.
    The wheelbase is quoted in the manufacturers' brochures and the trade press and lies between:

$l = 2160$ and 3040 mm

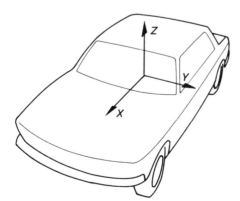

Fig. 3.1 Axis of coordinates in accordance with ISO 4130 and DIN 70000. The positive z direction points upwards and, when viewed into the direction of travel (x direction), the y arrow points left (see also Fig. 3.78).

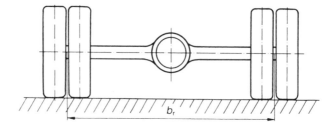

Fig. 3.1a   On twin tyres the tread width specification b_r relates to the mean distance; the lower load capacity of the tyres should be noted here.

3.3   Tread width

The size of the tread width b_f at the front and b_r at the rear (Figs 3.1a and 3.69) has a decisive influence on the vehicle's cornering behaviour and its tendency to body roll (see Section 5.4.3.1). It should be as large as possible but cannot exceed a certain value relative to the vehicle width. On the front axle the compressing, fully turned wheel may not come into contact with the wheel house (Fig. 2.5b) and on the driven axle (regardless of whether front, rear or both) there has to be enough space for snow chains to be fitted. When the wheels compress or rebound, they must not come into contact with any part of the chassis or the bodywork. The tread width on passenger cars is normally:

$b_{f\,\text{or}\,r} = 1210$ to 1602 mm

and k_b can be used as a ratio for the width utilization and should be as large as possible:

$$k_b = \frac{\text{tread width}}{\text{vehicle width}} = 0.81 \text{ to } 0.86 \qquad\qquad (3.1a)$$

When the wheels compress and rebound, the tread width changes on almost all independent wheel suspensions, which may be the result of functional factors or, as the following section shows, unavoidable if a higher body roll centre is necessary. However, the tread width alteration causes the rolling tyre to slip (Figs 3.2 and 3.3) and, on flat cross-sections in particular, causes lateral forces, higher rolling resistance and a deterioration in the directional stability of the vehicle, and may even influence the steering.

Tread width alteration on the front and rear axle must be checked on the drawing when the vehicle is at an early design stage. On a double wishbone suspension, arcs with suspension control arm lengths c and f must be drawn around points C and D (i.e. the suspension control arm axes of rotation), and the centres of the outer ball joints marked as points 1 and 2 (Fig. 3.4). A template can be prepared to show the steering knuckle and wheel (Fig. 3.5) and, in addition to points 1 and 2, must also have holes indicating the centre of

Fig. 3.2 On independent wheel suspensions, the compressing and rebounding of the wheels as they go over a bump can lead to a tread width alteration and this, in turn, to the tyres running at the slip angle α. This causes lateral forces and a deterioration in the directional stability and rolling resistance.

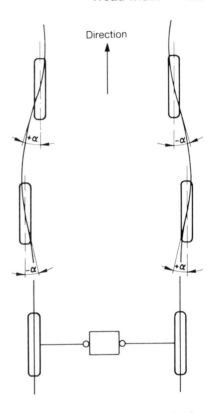

tyre contact N and, if necessary, the central point U of the outer tie rod joint (see Section 4.6.3).

As shown in Fig. 3.4, points 1 and 2 of this template must be drawn upwards along the arcs around C and D until point N of the template has reached the end of the compression travel s_1, previously indicated by a parallel to the ground, and downwards over the rebound travel s_2. The paths from N and U are then filled in step by step with a pencil. The line linking the points, which have been found in this way, gives the alteration of the tread width and the travel of the tie rod joint, but takes no account of any elasticity in the suspension control arm bearings (see Fig. 3.13a). In the case of the longitudinal control arm axle an arc must be drawn around D at the bottom, whilst a vertical line must be drawn on the suspension control arm axis of rotation (Fig. 3.6) and must go through point 1. At the same time a template as per Fig. 3.5 is moved along the arc and the vertical line to determine the tread width alteration.

McPherson struts have a mounting point E (Fig. 1.4) in the wheel house. When the wheel compresses, the distance of the lower ball joint 2 to point C shortens and then lengthens again when the wheel rebounds. The template has to take this length alteration into account (Fig. 3.7) and it is slit in the direction of the strut damper centre line $\overline{EE}$ (only in the direction of the steering axis $\overline{E2}$ if point 2 lies in its extension, see Figs 3.22, 3.23 and 4.31).

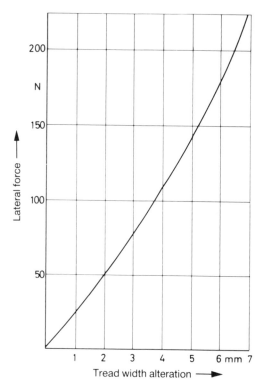

Fig. 3.3 Lateral forces $F_{y,w}$ from the tyre to the road resulting from a tread width alteration – shown on a radial 175/65 R 14 82 H tyre inflated to 1.9 bar under a load of 380 kg and at a speed of 80 km h^{-1}.

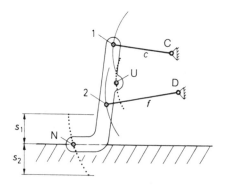

Fig. 3.4 Calculation by drawing of the tread width alteration of a wheel (in the centre of tyre contact N) and the track of the outer tie rod joint U on the double wishbone suspension, using the template shown in Fig. 3.5.

Using point 2, which also has to appear on the template, a movement is made along the arc around D, whilst the slit is shifted over point C (Fig. 3.8). A needle should mark this point on the drawing board.

If an arc is drawn around poles P, the tread width alteration of the dual

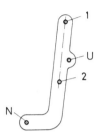

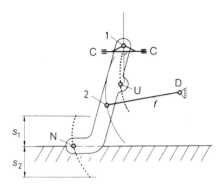

Fig. 3.5 Template for easy calculation of tread width alteration, can be used on double wishbone suspensions (Fig. 3.4) and longitudinal link axles (Fig. 3.6).

Fig. 3.6 Determination of the tread width alteration and the track of the outer tie rod joint U using the template shown in Fig. 3.5 on the longitudinal link axle.

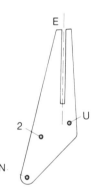

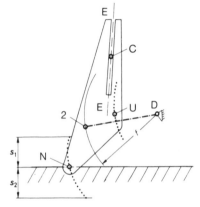

Fig. 3.7 The template needed to calculate, by drawing, the tread width alteration on the McPherson strut and strut damper must have a slot in the direction of the damper centre line E.

Fig. 3.8 Calculation by drawing of the tread width alteration of one wheel and the track of the outer tie rod joint U on the McPherson strut and strut damper using the template shown in Fig. 3.7. C is the centre of the upper strut mount; this point is marked as E in Figs 1.4 and 3.107.

joint swing axle can easily be drawn. Figure 3.8a shows both this and the advantages of lowering the tail end of the vehicle.

On all independent wheel suspensions the position of the pole P determines the momentary alteration $\pm \Delta b$ (present in a small springing range). Tread width alteration is avoided completely if P is at ground level and the lengths of the suspension control arms on a double wishbone suspension have

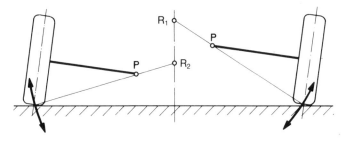

Fig. 3.8a Lowering the suspension control arm pivots P reduces the tread width alteration on the dual swing axle, causes the body roll centre to drop from R_1 to R_2 and a wider tread width. With two people in the vehicle, there is already negative camber on the wheels – giving the advantage of the tyres accepting more of the lateral forces, but the disadvantage of reduced compression travel.

been determined so that the pole moves horizontally from side to side on it when the wheels compress and rebound (Fig. 3.9). This can be demonstrated up to spring travel $s = \pm70$ mm using a drawing, calculation or a simulator, whereby any elasticity has been ignored (Fig. 3.13a). The tread width alteration can be measured as a function of the compression and rebound travel (s_1 and s_2) on the finished vehicle by determining the lateral shift of two parallel plates on which the two wheels of an axle are standing. It is necessary to run them parallel because a kinematic toe-in alteration when the wheels bottom out (see Section 3.6.2) could turn the plates slightly and distort the measurement results.

When this is represented on a graph, the wheel travel should be plotted on the y-axis (Fig. 3.10) and – in accordance with the direction in which the axle is moving – compression can be shown as positive and upward (s_1), and rebound as downward (s_2). The zero position should correspond to the design weight (see Section 5.3.4), in other words the weight when three (or even two) people, each weighing 68 kg, are in the vehicle. An empty vehicle would be unrealistic.

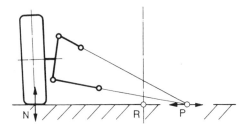

Fig. 3.9 An almost zero tread width alteration requires a body roll centre at ground level (or at infinity, Fig. 3.20). Better kinematic properties are also obtained if the pole is on the ground.

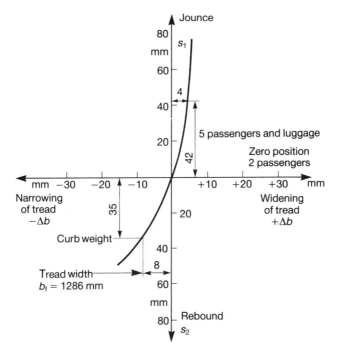

Fig. 3.10   The tread width (b_f or b_r) between the two wheels of an independent wheel suspension depends on the loading.

The tread width alteration Δb of the two wheels (or $\Delta b/2$ of one wheel) appears on the x-axis, with the increase (as a positive value) entered to the right and the reduction (as a negative value) to the left. The existing tread width $b_{f\,or\,r}$ in the zero position is an important dimension that should be stated. The tread width difference Δb to fully laden (or empty) can be determined using the springing curve. The spring travel Δs_1 from the zero position to the permissible axle load (or the compression travel Δs_2 to the 'empty status') can be read off this to obtain the tread width alteration curve Δb as a function of Δs.

Figure 5.4a shows the front wheel springing of a front-wheel drive vehicle, where the dimension 80 mm must be deducted from 115 mm to get the rebound travel, $\Delta s_2 = 35$ mm starting from the zero position (here, two people each weighing 68 kg). The vehicle compresses (from the zero position) by $\Delta s_1 = 92 - 50 = 42$ mm at the permissible wheel load (half the axle load). The paths are marked in Fig. 3.10; Δs_1 gives $\Delta b_1 = +4$ mm and Δs_2 gives $\Delta b_2 = -8$ mm. The tread width should be specified for the kerb weight: $b_f = 1286$ mm.

Figures 3.4, 3.11 and 3.13a show the tread width alteration of double wishbone suspensions and McPherson struts and the lower alteration values at compression can be clearly seen. As described in more detail in Section 3.4.1, the shape of the curve determines the level of the body roll centre. On all three passenger vehicles W_f is above the ground and falls perceptively (with

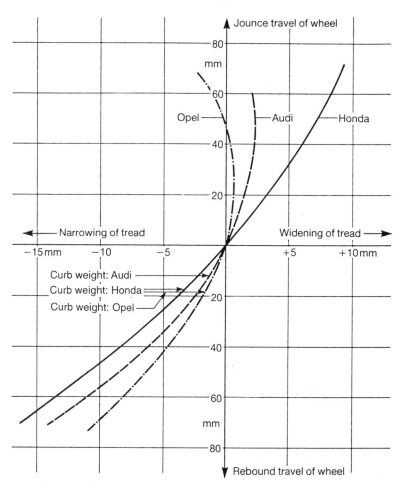

Fig. 3.11 Tread width alteration of one wheel measured on the front axle of a front-wheel drive Audi 100, Vauxhall Astra and Honda Accord (Figs 1.40, 5.37 and 1.38). The Honda is the only passenger car to have double wishbone suspension; the kinematic advantages can be seen clearly.

The 'body roll centre height' $h_{Ro,f}$ in mm is:

	Design position	Permissible axle load
Opel/Vauxhall	40	15
Audi	77	30
Honda	138	111

the exception of the Honda, Fig. 3.11) when the vehicle is laden. If the vehicle manufacturer has designed it at ground level as standard and the vehicle is subsequently lowered (Fig. 3.12) the body roll centre then moves into an adverse position; W_f drops below ground level and directional stability is likely to be impaired, particularly with wide tyres.

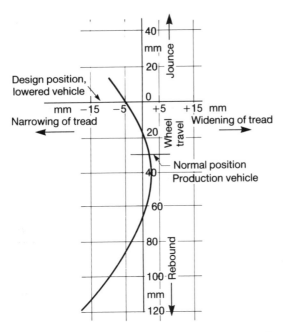

Fig. 3.12 Tread width alteration of both wheels measured on the front axle of a lowered VW Golf GTi. In the normal position, specified by the manufacturer, the body roll centre is around road level. Lowering the vehicle by 30 mm means the body roll centre moves 115 mm below ground, resulting in a longer body roll lever and a theoretically increased roll pitch. However, due to the early acting jounce bumper and virtually non-existent compression travel, the cornering inclination is greatly reduced (see Fig. 5.11 and Section 5.5.3).

In double wishbone suspensions, the springs sit on the upper or lower suspension control arms and, in both cases, a force pair arises (Figs 3.13 and 1.3) which, as a result of the elasticity in the suspension control arm bearings, causes the tread width alteration curve to take on a slightly different shape, thereby slightly altering the position of the roll centre (Fig. 3.13a). The alteration curve, determined by measurements on the vehicle (with springs), gives the correct height in any case.

The tread width alteration curves of typical rear wheel suspensions are shown in Figs 3.14, 3.15, 3.58a and 3.8a. Non-driven rigid and compound crank axles experience an increase or decrease in the tread width as a result of the elastic camber alteration (Fig. 3.46).

3.4 Roll centre and roll axis

In all independent wheel suspensions, there is a direct correlation between the tread width alteration and the height of the roll centre, so the two should be examined together.

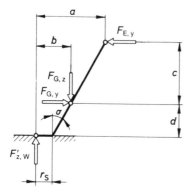

Fig. 3.13 The force $F_{z,W}$ at the centre of the tyre contact and $F_{G,z}$ on the lower supporting ball joint form a moment, which is absorbed laterally on the suspension control arms causing the force pair $+F_{E,y}$ and $-F_{G,y}$ here. For reasons of simplification, upper and lower suspension control arms are assumed to be horizontal.

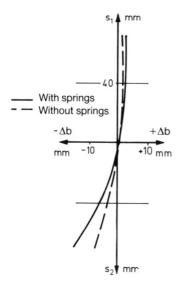

Fig. 3.13a Tread width alteration of both wheels, measured with and without springs as a function of the spring travel on a double wishbone suspension. The curvature differs, being equivalent to a higher body roll centre on the drivable vehicle than the theoretical value (without opposing spring force) calculated or drawn on the drawing board (see Fig. 3.4).

3.4.1 Definitions

In accordance with the German standard DIN 70 000, the body roll centre is the point in the vertical plane which passes through the wheel centre points (Fig. 3.16), and in which transverse forces (y-direction) can be exerted on the sprung mass, in other words the body, without kinematic roll angles occurring. The body roll centre is therefore the point in the centre of the vehicle (from the front) and in the centre of the axle (when viewed from the side) around which the body begins to roll when a lateral force acts, and at which reaction forces are absorbed between axle and body. Based on the existing tread width alteration curve of a wheel, the body roll centre is the point R in the centre of the vehicle (Fig. 3.17), with which a vertical, drawn on the tangent $\overline{AB}$ laid on the alteration curve in the centre of tyre contact, intersects.

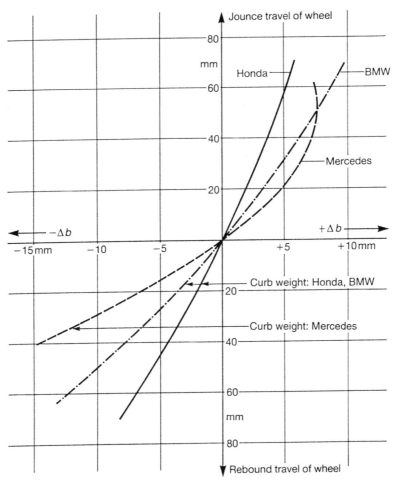

Fig. 3.14 Tread width alteration of one wheel, measured on the driven rear axle of a Mercedes, of a BMW 3-series (Fig. 1.0) and the non-driven axle of a Honda Accord (Fig. 1.38). The shape of the curve indicates that, with the multi-link axle of the Mercedes, the body roll centre falls under load (Fig. 3.17). The levels $h_{Ro,r}$ are as follows:

	Design position	Permissible axle load
BMW	122	92
Honda	74	58
Mercedes	65	–

The height $h_{Ro,f}$ of point R at the front (or $h_{Ro,r}$ at the rear) can be determined in this way using the paths Δs and Δb drawn at the tangents, considering all elasticities in the suspension control arm bearings (Fig. 3.13a). It behaves as follows:

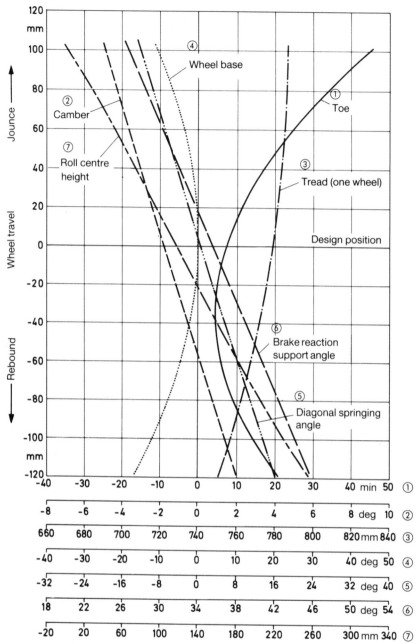

Fig. 3.15 Kinematics of the semi-trailing rear axle of the BMW 7-series shown in Fig. 1.28. This measurement shows the tread width of one wheel only. The toe-in alteration curve indicates a roll-steer effect on the rear axle tending towards understeering, which was achieved by matching camber (in design position) and the rear view angle of the control arm (Fig. 3.29). The lowering of the rear body roll centre under load favourably reduces the dynamic wheel load transfer on the bend at permissible axle load (relative to that on the front): it allows the vehicle to understeer more.

Brake reaction support angle ϵ and diagonal springing angle κ are shown in Fig. 3.124.

Fig. 3.16 The body roll centre R is in the centre of the vehicle (viewed from the front) and in the centre of the axle (viewed from the side).

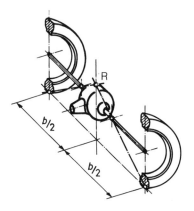

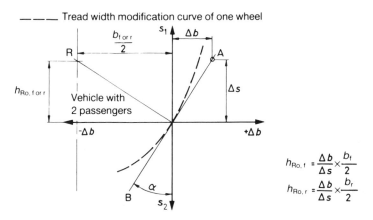

Fig. 3.17 The height $h_{Ro,f\ or\ r}$ of the body roll centre can be determined using a tangent from the measured tread width alteration curve in the respective load condition.

$$\frac{\Delta b}{\Delta s} = \frac{h_{Ro,f\ or\ r}}{0.5 b_{f\ or\ r}} = \tan \alpha \qquad (3.2)$$

and therefore the height of the body roll centre related to one wheel is:

$$\text{front } h_{Ro,f} = \frac{\Delta b}{\Delta s}\frac{b_f}{2} \quad \text{and rear} \quad h_{Ro,r} = \frac{\Delta b}{\Delta s}\frac{b_r}{2} \qquad (3.2a)$$

Where $b_f = 1400$ mm, $\Delta b = 6$ mm per wheel and $\Delta s = 40$ mm:

$$h_{Ro,f} = \frac{6}{40}\frac{1400}{2} = 105 \text{ mm}$$

The greater the tread width alteration in the point corresponding to the respective load (Fig. 3.10), the steeper the vertical on the tangent becomes and the higher the body roll centre lies above ground. However, in the case of small tread width alterations, R is only slightly above, or on, the ground if the tangent $\overline{AB}$ is parallel to the y-axis (Fig. 3.17). If (as partly shown in some figures in Section 3.3) the tread width alteration of both wheels is entered, the height of the body roll centre can be determined in the same way but only half the alteration travel, i.e. $\Delta b/2$ has to be considered. The equation is therefore related to both wheels:

$$h_{\mathrm{Ro,f\,or\,r}} = \frac{\Delta b}{\Delta s}\frac{b_{\mathrm{f\,or\,r}}}{4} \tag{3.3}$$

In Fig. 3.11, in the Audi and Opel, tangents drawn on the upper curve are always parallel to the y-axis when the wheels compress, this being the equivalent of an (undesirable) drop in the body roll centre under load, a disadvantage of McPherson struts. However, on the double wishbone suspension the tangent angle, and therefore the height of point R, alters less under load (Honda and Fig. 3.13a). The same applies to this type of rear axle (Figs 3.14 and 3.15).

3.4.2 Body roll axis

The position of the theoretical body roll axis C in the vehicle, i.e. the line connecting the body roll centres front and rear (Fig. 3.18), is just as important to the handling properties as the alteration in height under load. On independent wheel suspensions this axis should be largely parallel to the ground but as high as possible – parallel to achieve approximately equal wheel force alteration on the front and rear axles (with the aim of neutral behaviour) during cornering, and as high as possible to limit the tendency of the body to roll (Fig. 1.2). However, the level at the front is restricted by the permissible

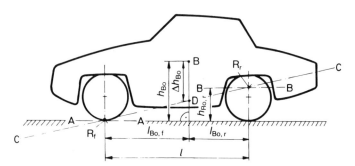

Fig. 3.18 Connecting line C between the front and rear body roll centre represents the theoretical roll axis (here at an angle). The path Δh_{Bo} is the body roll lever pointing vertical to the ground between this line and the body centre of gravity B. If the passenger car has a rigid rear axle, this angled disposition is beneficial. The body roll axis of a vehicle with independent wheel suspensions front and rear should only be at a slight angle (h_{Ro} see Equations 6.7 and 6.24).

tread width alteration and it is rarely acceptable to exceed $h_{Ro,f} = 150$ mm. Furthermore, on front-wheel drive vehicles, due to the high front axle load and the presence of the drive, the wheel force alteration at the front should be as low as possible. The body roll centre levels in the design position on independent wheel suspensions are (except with trailing link axles):

front $h_{Ro,f} = 30$ to 100 mm
rear $h_{Ro,r} = 60$ to 130 mm

Depending on the curvature of the tread width alteration curve, the body roll centres sink under load to a greater or lesser degree (Figs 3.11, 3.14, 3.15 and 3.17).

The design of a chassis firstly requires the determination of the height $H_{ro,f}$ of the front body roll centre (dependent on the tread width alteration) so that, in a second step, an appropriate rear axle can be provided with a slightly higher $H_{ro,r}$ in the case of independent wheel suspensions.

If the vehicle has been fitted with a rigid axle, the body enjoys less anti-roll support on bends ($i_\varphi = b_{Sp}/b_r$, Fig. 1.13) as a result of the shorter effective distance b_{Sp} of the springs relative to the tread width b_r. To balance this out, it is recommended that the body roll centre be designed slightly higher at the back (as shown in Fig. 3.18).

The lines A and B, also drawn in Fig. 3.18, are the actual body roll axes, which are mostly parallel to the ground. The precise location depends on the angular position of the steering control arms. The body inclines around A and B under the influence of a lateral force.

3.4.3 Body roll centre on independent wheel suspensions

The height of the pole P determines the position of the body roll centre R (Fig. 3.19). If P is above ground level, R will also be above ground. As can be seen in Fig. 3.17, the tangent drawn at the zero point on the tread width alteration curve varies by the angle α from the vertical. However, the shape of the curve at this point depends on the distance between pole P and the centre of tyre contact N. The further the two are apart (i.e. the longer the path q, Fig. 3.23), the less pronounced the curvature. This can result in a less favourable kinematic camber alteration (see Section 3.5.2). The following figures show the determination of height h_{Ro} of the body roll centre and path p by drawing. The pole distance q can be measured or calculated simply:

$$q = \frac{p \, b_{f\,or\,r}}{h_{Ro}\,2} \tag{3.4}$$

As can be seen in Figs 3.19 and 3.4, on the double wishbone suspension only the position of the steering control arms is important (i.e. the sizes of the angles α and β). The lines connecting the inner and outer steering control arm pivots need to be extended to get pole P and, at the same time, its height p. P linked with the centre of tyre contact N gives the body roll centre R in the intersection with the vehicle centre plane. In the case of parallel control

The following dimensions have to be known: c, d, b_f, r_s, α, β, σ

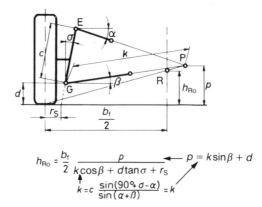

$$h_{Ro} = \frac{b_f}{2} \; \frac{p}{k\cos\beta + d\tan\sigma + r_s} \quad \longleftarrow \quad p = k\sin\beta + d$$

$$k = c \; \frac{\sin(90°\; \sigma - \alpha)}{\sin(\alpha+\beta)} = k$$

Fig. 3.19 Determination by drawing and calculation of the paths h_{Ro} and p on double wishbone suspensions and a multi-link as well as a longitudinal transverse axes (Figs 3.0 and 3.25).

arms, P is at ∞, and a parallel to them needs to be drawn through N (Fig. 3.20).

Where the pole is a long way from the wheel centre of contact, it is recommended that the distances p and h_{Ro} be calculated using the formulae in Fig. 3.19. Steering control arm axes of rotation, which are sloped when viewed from the side (designed this way to obtain a vehicle pitch pole – Fig. 3.120), need E_1 and G_1 to be moved perpendicularly up or down (Fig. 3.21). The points E_2 and G_2 obtained in this way – linked with E_1 and G_1 when viewed from the rear – give the pole P, and the line from this pole to the centre of tyre contact (as shown in Fig. 3.19) gives the body roll centre. If the axle is controlled by transverse leaf springs, where these are held in the middle (Fig. 3.21a), the kinematic lever L_3 is important for calculating the body roll centre and, if the springs are attached at two points, the distance L_2 to the spring attachment point is important (Fig. 3.21b). Further details are given in Section 4.6.3.1.

On McPherson struts, or strut dampers, a vertical must be created in the body side fixing point E to the centre line of the shock absorber piston rod, and the lower steering control arm must be extended. The intersection of the two lines will then give P (Fig. 3.22). The illustration also shows how increasing the tread width out from b_{f1} to b_{f2} results in the body roll centre being

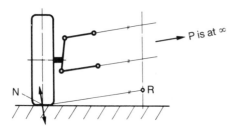

Fig. 3.20 Determination of the body roll centre on parallel double wishbones; the pole is at infinity.

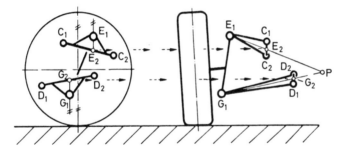

Fig. 3.21 If the suspension control arm axes of rotation are at an angle to one another when viewed from the side, a vertical should first be drawn to the ground through the points E_1 and G_1; the intersections with the axes of rotation C_1C_2 and D_1D_2 yield the points E_2 and G_2 needed for determining the pole when viewed from the rear.

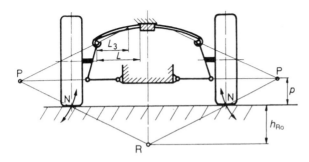

Fig. 3.21a Determination of R and P on a high, centrally-anchored transverse leaf spring.

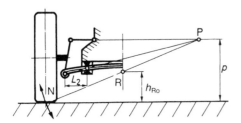

Fig. 3.21b Determination of R and P on a low transverse leaf spring supported in two places.

raised from R_1 to R_2. A negative kingpin offset at ground makes it necessary to shift the lower guide joint in to the wheel (Fig. 3.79) which separates the kingpin axis from the shock absorber centre line. Figure 3.23 shows the path $\overline{EP}$, which is then vertical to the shock absorber centre line and also that h_{Ro} is not dependent on the steering control arm length, which is the decisive factor for the kinematic properties. Where the suspension control arm lies flat, it is recommended that the heights h_{Ro} and p be calculated because, if drawn,

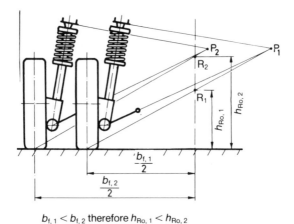

$$b_{f,1} < b_{f,2} \text{ therefore } h_{Ro,1} < h_{Ro,2}$$

Fig. 3.22 The greater the tread width b_f, the higher the body roll centre R, shown using the example of a McPherson strut (Fig. 1.39).

the pole would be too far outside the drawing board (Fig. 3.24). Section 4.6.3.2 contains further details.

On the longitudinal link axle (Fig. 3.25), the direction of movement of the upper point E (vertical to the suspension control arm axis of rotation) plays a role. A parallel to $\overline{CF}$ must be drawn through E to obtain P and W. The calculation can be seen in Fig. 3.19. Whilst on the McPherson strut, the height of the body roll centre can only be influenced by placing the lower suspension control arm at an angle and only marginally by changing the angle between steering axis $\overline{EG}$ and the McPherson strut centre line (Fig. 3.23), which is a disadvantage of this type of suspension. On the longitudinal control arm axle it is possible to increase the angle of the axis of rotation $\overline{CF}$ further and therefore to raise R. At the same time, the pole moves closer to the wheel, giving the additional advantage that the compressing wheels move more strongly into negative camber.

The heights $h_{Ro,f}$ of the front body roll centres determined in accordance with Figs 3.19 to 3.25 only agree in the case of bearings which, although they can be rotated, are otherwise not flexible, and only at body roll angles up to $\varphi = 2°$. The elasticity of the rubber elements used slightly alters the height available on the vehicle (Fig. 3.13a). Furthermore, calculations and studies have both shown that, in the case of larger body roll angles, the left and right poles take on a different position, but that the body roll centre in the centre of the vehicle experiences an alteration of only $\Delta h_{Ro} = \pm 10$ mm. Parallel measurements carried out on passenger cars showed a deviation of up to $\Delta h_{Ro} = 20$ mm.

In contrast to the front independent wheel suspensions, rear ones sometimes have only one control arm on each side; here too the position of the pole determines the height of the body roll centre, with the direction of movement of the wheel providing additional information. If the axis of rota-

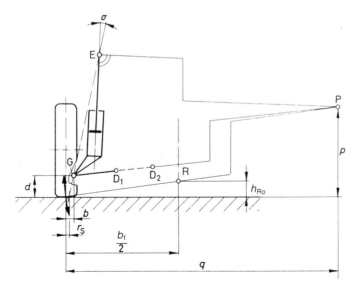

Fig. 3.23 The more vertical the McPherson struts and dampers and the more horizontal the lower control arm GD_1, the closer the body roll centre R is to the ground. This results in an adverse camber alteration when the wheels compress. Lengthening the lower suspension control arm (point D_1 to D_2) improves the kinematic properties.

To achieve a small or negative kingpin offset at ground r_S, point G must be drawn outwards into the wheel, giving the benefit of a shorter lever b for the vertical force $F_{z,W}$. The shorter can be path b, the less friction occurs between the piston rod and rod guide, as well as at the piston, and the smaller the forces in bearing points D, E and G (see also Fig. 1.6). A long path q means tread width alteration can be restricted. Fig. 1.4 shows the precise position of points E and G.

The lever b is easy to calculate:

$$b = r_S + d \tan \sigma \qquad \text{(Equation 3.4a)}$$

Depending on the design, either $+r_S$ or $-r_S$ has to be included in the equation.

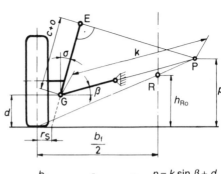

Fig. 3.24 Calculation of the paths h_{Ro} and p in the standard configuration of a McPherson strut and strut damper.

$$h_{Ro} = \frac{b_f}{2} \frac{p}{k\cos\beta + d\tan\sigma + r_S}$$

$$k = \frac{c+o}{\sin(\sigma+\beta)} = k$$

$$p = k \sin\beta + d$$

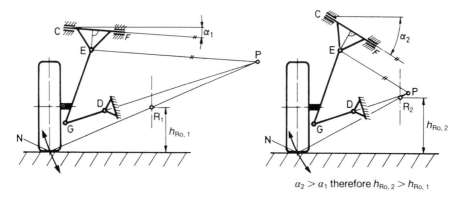

$$\alpha_2 > \alpha_1 \text{ therefore } h_{Ro,2} > h_{Ro,1}$$

Fig. 3.25 With the longitudinal transverse axle, a parallel to $\overline{CF}$ should be drawn through E and this made to intersect with the extension of the path $\overline{GD}$ to determine the pole P. It is then connected to N to give R in the vehicle centre plane. The greater the angle of the upper suspension control arms, when viewed from the rear (α_2 right), the closer P moves to the vehicle centre; tread width and camber alteration increase and W_1 becomes W_2 at a higher level (see also Fig. 4.34).

tion lies horizontal (Fig. 3.26) on the link axle, the wheel moves vertically and the roll centre R is at ground level. If the axis of rotation is inclined (Fig. 3.27), R moves above ground or, if the angle is in the other direction, below ground.

The single joint swing axle (Fig. 3.28) has its point of rotation in the centre of the vehicle. The pole is, at the same time, the body roll centre, unlike the dual joint swing axle on which point P is to the side next to the differential and R is therefore disproportionally high. Figure 3.8a shows how R is calculated, with the fall in the body roll centre in the case of negative camber ϵ_W (left) clearly indicated.

In the case of the semi-trailing link axle, the movement of the wheel vertical to the three-dimensional axis of rotation $\overline{EG}$ plays a role (Fig. 3.29). The point at which the extension of the axis of rotation intersects a vertical plane in the centre of the axle gives the pole $P_{1,2}$, from which the height h_{Ro} of the body roll centre in the middle of the vehicle can be concluded. To determine

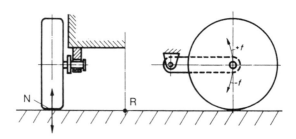

Fig. 3.26 If, with longitudinal links, the axis of rotation is horizontal, the body roll centre is at ground level and P is at ∞; the size of the diagonal springing $\pm f$ depends on the suspension control arm length (diagonal springing angle κ see Fig. 3.122).

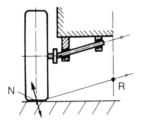

Fig. 3.27 If, with longitudinal links, the axis of rotation is at an angle, the body roll centre will lie above ground (or below it, if the angle is reversed); P is at ∞ in both cases.

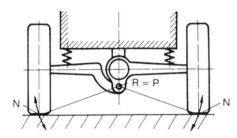

Fig 3.28 On the single-joint swing angle, the suspension control arm pivot, which is approximately at the centre of the vehicle, is both the pole and roll centre.

The following dimensions have to be known: $e, f, k, b, \alpha, \beta$

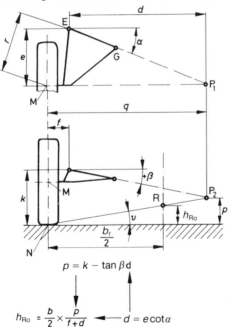

$$p = k - \tan \beta \, d$$

$$h_{Ro} = \frac{b}{2} \times \frac{p}{f+d} \longleftarrow d = e \cot \alpha$$

Fig. 3.29 On the semi-trailing axle, the positions of the pole P and roll centre R are determined by the length r of the suspension control arms and the top view angle α and rear view angle β. The equations are used for calculating the height h_{Ro} in the vehicle centre. When the vehicle is laden, points E and G (and therefore also P and R) move down. The momentary tread width alteration results from an arc around P_2 (see also Figs 3.15 and 3.124).

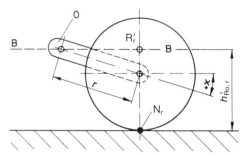

Fig. 3.30 On the compound crank axle, lateral forces are supported on two trailing links, which point forwards and are stiff to torsion and bending; the distance $h'_{Ro,r}$ of the pivot points O to the ground determines the position of the kinematic body roll centre R'_r. Only the length r of the suspension control arms and their angled position, i.e. the angle $\pm\kappa$, influence the level $h'_{Ro, r}$.

this, first draw the top view, taking into account the angle α, and in it the extension of the suspension control arm axis of rotation made to intersect with the axle centre. The pole P_1 obtained in this way is moved perpendicularly down in to the rear view and made to intercept with the extension of the axis of rotation – this time using the angle β. Finally, the pole P_2 found in the rear view must be linked with N. With small angles α and β, it may be sensible to calculate h_{Ro} and p as a function of the dimensions specified by the designer. Figure 3.29 also contains the formulas for these relationships.

3.4.4 Body roll centre on compound crank axles

The kinematic or static body roll centres of this axle are the bearing points O (Figs 3.30 and 1.42) at which – as specified in DIN 70 000 and described in Section 3.4.1 – the lateral forces are absorbed. The elastokinematic body roll centre, on the other hand, determines the alteration to toe-in and camber on reciprocal springing. Owing to the low torsion resistance of the transverse members the wheels swing during cornering, as on the semi-trailing link suspension, around the line connecting the points O_l/O_{rs} with the thrust centre point SM (Fig. 3.31). Toe-in and camber alteration are shown in Figs 3.45 to 3.46b.

3.4.5 Body roll centre on rigid axles

As shown in Figs 1.14 and 1.15, on rigid axles the lateral forces are absorbed in only one or two places. The body roll centre can therefore only occasionally be determined using the theory of transmission kinematics. It is the laws of statics which mainly apply, and the centre axle point – at which the forces are transferred between body and axle – which should be observed.

  If longitudinal leaf springs are used as the suspension, the lateral force is concentrated on the main leaves, and R is at their centre within the clamp (Fig. 3.32). To keep it flat for a low underbody on a passenger car, the spring

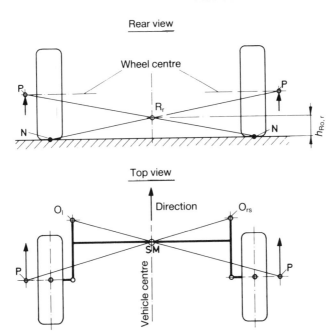

Fig. 3.31   Determination of the height $h_{Ro,r}$ of the elastokinematic roll centre R_r around which the body inclines under the influence of the centrifugal force acting on the body centre of gravity for the compound crank axle. The thrust centre point SM of the cross-member, which must be linked in the top view with the bearing points O and intersected with a straight line through the wheel centres, must be known. The resulting poles must be moved vertically upwards to the wheel centre axis in the rear view and linked with the centres of tyre contact N to obtain point R_r in the vehicle centre.

The position of the 'thrust centre point' also determines camber and caster alteration on reciprocal springing (Figs 3.45 and 3.46).

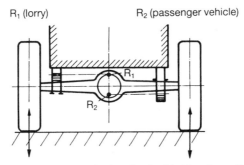

Fig. 3.32   If the rigid axle is carried by longitudinal leaf springs, the lateral forces are concentrated in its main bearings. The body roll centre is on the axle mounting in the middle of the main leaf, regardless of whether the spring is fixed above (left side and Figs 1.13a and 1.22) or below the axle (right and Fig. 1.15).

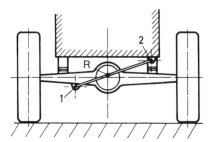

Fig. 3.33 If a panhard rod provides lateral force reaction support, the body roll centre is at the intersection of the rod with the vehicle centre line.

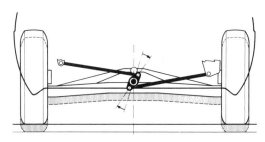

Fig. 3.33a Watt linkage on a passenger car rear axle. This allows the axle to be carried without any lateral deviation. When the springs compress and rebound, the linkage turns around the mounting point on the axle housing, which is also the roll centre.

is fitted below the axle (right-hand side of the picture), whilst commercial vehicles need a high body roll centre to reduce the body inclination. The spring is then above the axle (left-hand side, see also Fig. 1.22) with the advantage that the fixing bolts are not subject to further tensile forces.

If the lateral force is supported by a panhard rod (Fig. 3.33), the body roll centre will be at the intersection of the panhard rod with the vehicle centre line (and not, as sometimes thought, in the centre of the bar). During cornering, the position of the bar changes and therefore so does the height of R. However, if a watt linkage supports the forces in a lateral direction, the point at which it is fixed to the axle housing is the decisive point of reference (Fig. 3.33a).

The upper pair of longitudinal control arms and the panhard rod can be replaced by an A-arm (Fig. 3.34), which transfers lateral and longitudinal forces to the body. The body roll centre R is then the fixing point on the axle. In contrast to the panhard rod, point R maintains its height h_{Ro} when subjected to load.

Instead of the upper A-arm, two suspension control arms at an angle to one another can be used (Fig. 3.35). In this case, the intersection of the extension of the suspension control arm from the top view gives the pole P which must be brought down perpendicularly in the side view. In the case of parallel lower suspension control arms, a line drawn in the same direction as the arms intersects with the axle centre in the body roll centre R.

Fig. 3.34 If a longitudinal A-arm supports the rigid axle, its fixing position on the axle housing is also point R.

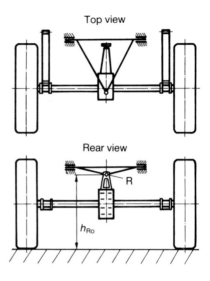

Top view

Rear view

R

h_{Ro}

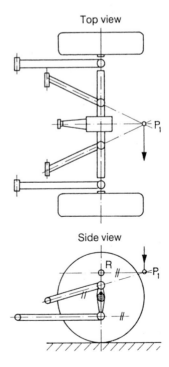

Top view

P_1

Fig. 3.35 If the two upper suspension links, which lie at an angle to one another in the top view, absorb the lateral forces, their extensions give pole P_1. To determine R in the side view, a parallel must be drawn to the lower suspension control arms through P_1. Since – as can be seen in the top view – these two suspension links point in the same direction, their pole is at ∞.

Side view

R

P_1

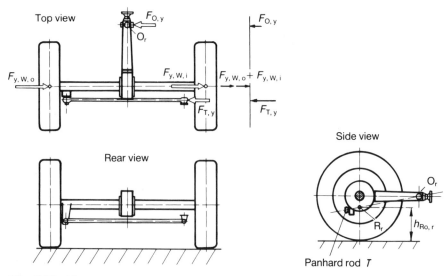

Fig. 3.36   The lateral forces $F_{y,w,o}$ and $F_{y,w,i}$ are transferred from the axle to the body at the front drawbar mounting and the rear panhard rod. The reaction forces $F_{O,y}$ and $F_{T,y}$ occur. The body roll centre R_r must therefore lie on the line connecting points T and O_r from the side view.

Unlike the rigid axle suspensions discussed so far, on the drawbar axle (also known as the A-bracket axle) lateral forces can be absorbed jointly on the front bearing point O_r and two lateral struts (Fig. 1.43a). The body roll centre is then at the height at which these three parts are attached to the body. If, instead of the two struts, there is a panhard rod, the forces are supported on this and point O_r. The side view shown in Fig. 3.36 next to the top view clearly shows both reaction forces $F_{O,y}$ and $F_{T,y}$. The body roll centre is therefore on the line linking the two points, which can be seen in the side view. If (as shown in Fig. 3.33) the panhard rod is at an angle, the mean height of the rod in the rear view must be determined and then transferred to the side view.

3.5   Camber

3.5.1 Camber values and data

In accordance with the standards ISO 8855 and DIN 70 000, camber is the angle between the wheel centre plane and a vertical to the plane of the road. It is positive if the wheel is inclined outwards (Fig. 3.37) and negative, as $-\epsilon_W$, when inclined inwards. When a vehicle is loaded with two three persons (design weight, see Section 5.3.4), a slightly positive camber would be useful on passenger cars to make the tyres roll as upright as possible on the slightly

Fig. 3.37 Positive camber $+\epsilon_w$ is the deviation of the wheel plane outwards from the vertical.

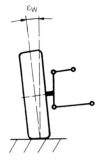

curved road surface and give more even wear and lower rolling resistance. As Fig. 3.38 shows, the optimum value for this purpose would be:

$\epsilon_w = 5'$ to $10'$, i.e. around $0.1°$

To give better lateral tyre grip on bends and improve handling, nowadays this rule is generally no longer adhered to and, on passenger cars, the setting is negative even when the vehicle is empty. Front axle values are as follows on newer production vehicles:

$\epsilon_{w,f,ul} = 0°$ to $-1° \, 20'$

In addition to the absolute camber, the tolerance values are important, i.e. both the deviation from the permitted value and also the difference between the left and right wheel. A $\pm30'$ deviation is usual to enable the components of the front axle to be manufactured economically. This is why it is not always possible to adjust the camber on front wheel suspensions.

To avoid the steering pulling to one side when the vehicle is moving in a straight line, the difference in the kingpin inclination angle between left and

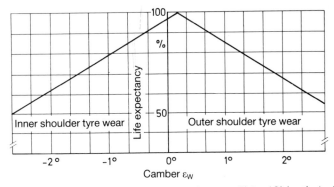

Fig. 3.38 Studies have shown that a camber of $\epsilon_w = +5'$ to $10'$ leads to the most even tyre wear; more positive camber would lead to more pronounced wear on the outer shoulder and negative camber to more pronounced wear on the inside of the tyre tread.

right wheels should not exceed $\Delta\sigma = 30'$. As can be seen in Fig. 3.80, camber and kingpin inclination are directly related, i.e. if the camber deviation is too great, so is the kingpin inclination angle. This is why no camber difference greater than $\Delta30'$ should be allowed as a factory setting. The information in the sub-assembly drawing of the front axle would then be as follows, for example:

Camber $-40'\pm30'$ (3.4b)
Max difference between left and right 30'

The measurement condition, which must relate to the kerb weight (i.e. the unoccupied vehicle, see DIN 70 020), must also be added. In the case of rear independent wheel suspensions and compound crank axles, designers prefer to use negative camber to increase lateral tyre grip; the mean value for the kerb weight can be:

Camber $-1°\,30'\pm20'$ (3.4c)
Max difference between left and right 20'

The existing setting options allow tighter tolerances here. On semi-trailing link axles there is a danger of too negative a value in the fully laden condition (Fig. 3.40); this could lead to the risk of the tyres becoming excessively warm and the protective cover coming free. This is the reason why passenger car manufacturers have reduced the kinematic camber alteration on this type of suspension by means of the angles α and β of the control arm axis of rotation (see Fig. 3.29 and Section 2.2.6.3).

3.5.2 Kinematic camber alteration

As described in Section 1.2.1, one disadvantage of independent wheel suspension is that the wheels incline with the body on a bend, i.e. the wheel on the outside of the bend goes into positive camber relative to the ground, and the lateral grip of the tyre under the greatest load (unlike the one on the inside of the bend) reduces (Figs 3.45 and 3.46). To balance this out, manufacturers tend to design the suspension on passenger cars such that the wheels go into negative camber as they compress and into positive camber as they rebound (Figs 3.38a and 3.39).

On the x-axis, negative camber is given in degrees on the left and positive camber on the right, whilst on the y-axis wheel compression travel s_1 is plotted in mm upwards and rebound travel s_2 downwards. The curve for the double wishbone suspension, which bends sharply into the negative during the compression, shows the advantage of this axle. For the McPherson strut or strut damper the curve bends (unfavourably) in the other direction. However, the wheel on the strut dampers takes on more positive camber during rebound, this being the equivalent of better lateral force absorbtion on the (less loaded) wheel on the inside of the bend.

The camber alteration curves for rear independent wheel suspensions are

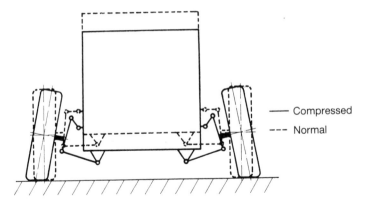

— Compressed

--- Normal

Fig. 3.38a In independent wheel suspensions, the wheels incline with the body when the vehicle is cornering (Fig. 1.2). To even this out, the compressing wheels should go into negative camber and the rebounding ones into positive camber.

shown in Figs 3.40, 3.15 and 3.58a, where better properties can be seen than on front ones. As there is no steering input to be considered, the semi-trailing links or transverse links can adopt a better position. From the zero position shown, as can be seen in Fig. 5.9, the Mercedes compresses by 53 mm under full load. The camber is then $\epsilon_{w,lo} = -2°50'$ and remains above the critical value $\epsilon_w = -4°$, which should not be exceeded.

3.5.3 Camber alteration calculation by drawing

From a design point of view, the camber alteration on the front wheels can easily be determined as a function of the wheel travel over the angle of alteration $\Delta\sigma$ of the kingpin inclination if elasticities are ignored. On double wishbone suspensions, arcs with the suspension control arm lengths e and f must be drawn around the points C and D (in other words the suspension control arm axes of rotation) and, in the normal position, the centres of the outer ball joints marked as points 1 and 2 (Fig. 3.41). A point 3 is determined on the upper arc and an arc with the path $\overline{1,2}$ drawn around it to give point 4. The line connecting them, $\overline{3,4}$ then has the alteration angle $\Delta\sigma$ to the path $\overline{1,2}$, if the wheel compresses by the path s_1. If it goes into negative camber (as in the example), $\Delta\sigma$ must be subtracted from the camber angle $\epsilon_{w,0}$ in the normal position i.e.

$$\epsilon_w = \epsilon_{w,0} - \Delta\sigma \text{ (e.g. } -40' -2° = -2° 40') \tag{3.4d}$$

In the case of positive camber, $\Delta\sigma$ would have to be added:

$$\epsilon_w = \epsilon_{w,0} + \Delta\sigma$$

On McPherson struts and strut dampers, the distance 1,2 is shortened when

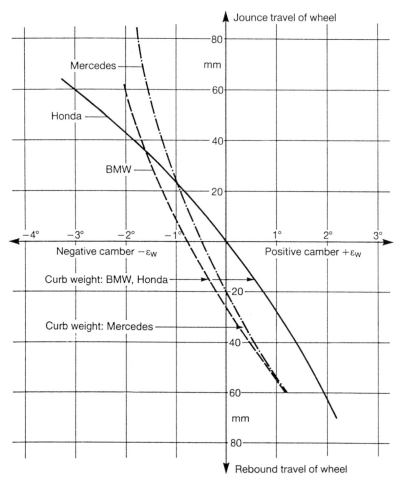

Fig. 3.39 Camber alteration on the front double wishbone suspension of a Honda Accord (Fig. 1.38) as a function of the wheel jounce travel s_1 and rebound travel s_2 in comparison with the McPherson suspension of a 3-series BMW (Fig. 1.24) and the strut damper axle of a Mercedes.

the wheel compresses, the upper mounting point is in the wheel house and only the lower point 2 moves to 3. $\Delta\sigma$ is again the angle between the two connecting lines (Fig. 3.42). The upper suspension control arm of the longitudinal link axle (Fig. 3.43) requires a vertical to be created on the axes of rotation $\overline{CC}$ through the point 1 so that point 4 can be obtained using an arc around 3 and the length $\overline{1,2}$. If the axes $\overline{CC}$ were to deviate more from the horizontal, $\Delta\sigma$ (and therefore the camber alteration, Fig. 3.25) would improve.

An arc around pole P must be drawn on the swing axle (Fig. 3.8a). The tangents drawn to this one after the other give the camber alteration which must

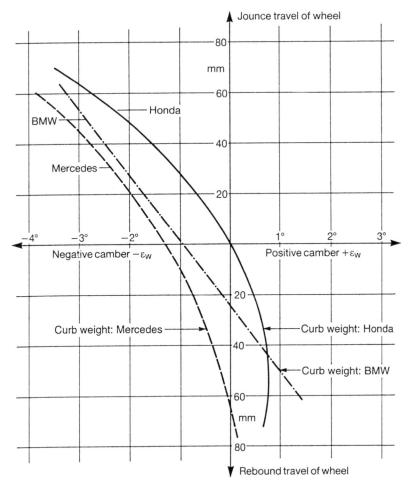

Fig. 3.40 Camber alteration on the rear wheels of a Mercedes, a 3-series BMW and a Honda Accord. The multi-link independent suspension of the Mercedes has a fairly precise camber setting. In the empty condition this was $\epsilon_{W,O,l} = -55'$ and $\epsilon_{W,O,rs} = -35'$ and increased to around $-1°30'$ when there were three people in the vehicle. When the springs compress, the curve shape is slightly progressive. The manufacturer's specification for the empty condition is: $\epsilon_W = -50' \pm 30'$.

The multi-link axle of the BMW (Fig. 1.0) exhibits a straight-line curve; when the springs compress, the negative camber is less than on the Mercedes. The double wishbone suspension of the Honda (Fig. 1.38) has zero camber in the design position, but the wheels take on higher alterations (negative values) when the springs compress than on the two other suspensions.

be subtracted from or added to $\epsilon_{W,0}$. The same applies to the semi-trailing link axle where the compass needs only to be inserted into P_2 (rear view, Fig. 3.29).

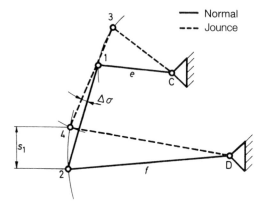

Fig. 3.41 Design determination of the kingpin inclination alteration Δσ on double wishbones which is equal to the camber alteration.

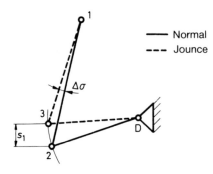

Fig. 3.42 Design determination of the camber and kingpin inclination alteration on the McPherson strut and strut damper.

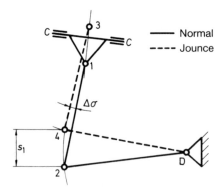

Fig. 3.43 Design determination of the camber and kingpin inclination alteration on the longitudinal transverse axle.

3.5.4 Roll camber during cornering

When the body roll-pitches, the camber of individually suspended wheels also changes, on the outside of a bend by the angle $\Delta\epsilon_{W,o}$ and on the inside by $\Delta\epsilon_{W,i}$ (Fig. 1.2). The mean value of the two $\Delta\epsilon_{W,\varphi} = 0.5\,(\Delta\epsilon_{W,o} + \Delta\epsilon_{W,i})$ together with the kinematic body roll angle $\varphi_{V,k}$ gives the

roll camber coefficient $k\epsilon_{W,\varphi} = d\epsilon_W/d\varphi_V$ $\qquad\qquad$ (3.5)

A wheel that is cambered positively to the ground on the outside of a bend by the angle $+\epsilon_{W,o} = \epsilon_{W,0} + \Delta\epsilon_{W,o}$ and one that is inclined on the inside of the bend by the angle $\epsilon_{W,i} = \epsilon_{W,0} - \Delta\epsilon_{W,i}$ can experience an additional camber due to the vertical force elements:

$$F_{\epsilon W,o} = F_{z,W,o} \sin \epsilon_{W,o} \text{ and } F_{\epsilon W,i} = F_{z,W,i} \sin \epsilon_{Wi}$$

(Fig. 3.44). The softer the suspension control arm bearings have to be, and the shorter the path c on double wishbones (Fig. 1.1b) or the distance l–o between piston and rod guide on McPherson struts and strut dampers (Fig. 1.6), the worse the roll camber becomes. The diameter of the piston rod (see Section 5.8.1) and the basic kinematics of the suspension also have an influence.

The body roll camber factor can be determined by heeling the body over to both sides and measuring the body roll angle and the camber angle. The

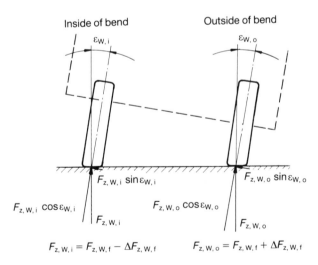

Fig. 3.44 When the body (and therefore also the wheels) incline, the vertical force element $F_{z,W,o} \sin \epsilon_{W,o}$ on a left-hand bend pushes the wheel on the outside of the bend (here the right-hand one) further into positive camber and the force $F_{z,W,i} \sin \epsilon_{W,i}$ pushes the one on the inside into a (equally unfavourable) negative camber (see also Fig. 1.2).

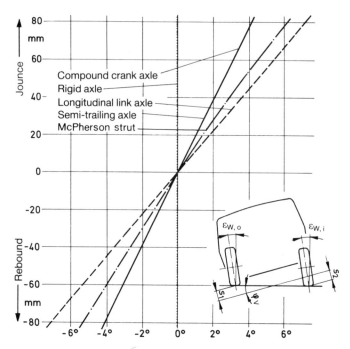

Fig. 3.45 Camber alteration relative to the ground of various rear-wheel suspensions in the case of reciprocal springing; with the exception of the rigid axle, the wheel on the outside of the bend goes into positive camber and the one on the inside into negative camber on all configurations. Wheel travel when the wheels compress and rebound is entered on the axis of the ordinate. The body roll angle φ_V is easy to calculate using the path differences Δ_{s_1} and Δ_{s_2} (see Equation 3.6).

wheel travel in compression and rebound can be plotted on the y-axis instead of the body roll angle (Figs 3.45 and 3.46), and the body roll angle can be easily calculated from this using the tread width $b_{f \, or \, r}$:

$$d\varphi_V = \frac{s_1 + s_2}{b_{f \, or \, r}} \text{ (rad) and } d\varphi_V = 57.3 \, d\varphi_V \text{ (degree)} \tag{3.6}$$

The compound crank axle of the VW Golf has a tread width $b_r = 1444$ mm and where the path is $s_1 + s_2 = 80$ mm, the body roll angle is

$$d\varphi_V = 80/1444 = 0.00554 \text{ rad} = 3.17° = 3°10'$$

The progressive springing curve of this passenger car means the wheel on the outside of the bend only compresses a little relative to the amount by which the opposite wheel rebounds (see Section 5.4.2). Given the permissible axle load, the following paths are assumed:

$$s_1 = 27 \text{ mm and } s_2 = 53 \text{ mm}$$

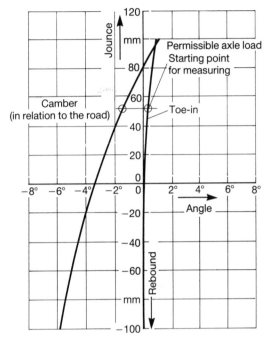

Fig. 3.46 Values for toe-in and camber angle measured on the compound crank axle of a Golf with reciprocal springing, entered as a function of the wheel travel relative to the body. The compressing wheel on the outside of the bend goes into positive camber and the rebounding one on the inside of the bend goes into negative camber relative to the ground. The vehicle was measured with permissible rear axle load. Toe-in does not alter favourably. Fig. 1.2 shows the body roll angle φ_V and Fig. 3.31 the relevant thrust centre point SM.

The following values arise:

Camber $\epsilon_{W,o} = -0.1° \; \epsilon_{W,i} = -3.55°$,

Camber alteration $d\epsilon_W = [\epsilon_{W,o} - \epsilon_{W,i}]/2$ (3.7)

$d\epsilon_W = [-0.1 - (-3.5)]/2 = 1.7°$

and (referring to Equation 3.5) as a

body roll camber factor $k\epsilon_W = d\epsilon_W/d\varphi_V = 1.7/3.44 = 0.49$

The average roll camber factors for the following axles are:

longitudinal link axles	1.05
McPherson struts	0.85
double wishbone suspensions	0.80
compound crank axles	0.55
rigid axles	0.0

3.5.5 Elasticity camber

In addition to the body roll camber, the camber alteration caused by the lateral forces must also be taken into consideration. In accordance with DIN 70 000, $\Delta\epsilon_W$ is the proportion of the camber of a wheel that can be ascribed to the elasticity in the suspension and the steering, and is caused by forces acting between the tyre and road or by their moments.

Figure 3.46a shows the values calculated on the McPherson strut front axles of two passenger cars and Fig. 3.46b those measured on various rear axles. If there are no test results available, the following can be taken as the elasticity camber coefficient (per kiloNewton):

$$d\epsilon_W/dF \approx 22'/1 \text{ kN} \tag{3.7a}$$

3.6 Toe-in and self-steering

3.6.1 Toe-in and crab angle, data and tolerances

In accordance with the standards ISO 8855 and DIN 70 000, the static toe-in angle Δ is the angle that results in a standing vehicle (reference status), between the vehicle centre plane in the longitudinal direction and the line intersecting the centre plane of one wheel with the road plane. It is positive, when the front part of the wheel is turned towards the vehicle longitudinal centre plane and negative ('toe-out') when it is turned away.

The total toe-in angle is obtained by adding the toe-in angle of the right and left wheels. The total value is sometimes given in millimetres. The toe-in is then the dimensional difference $r_{\Delta,t} = b - c$ (Fig. 3.47), by which the rim flanges are further apart at the back than at the front. The toe-in should be measured at the height of the wheel centre, when the vehicle is empty, with the wheels pointing straight forward. $r_{\Delta,t}$ therefore relates to both wheels of one axle. Expressed in degrees, the toe-in angle Δ of a wheel corresponds to the tyre slip angle $\alpha_{f\,or\,r}$ (see Section 2.8.1); i.e. where there is toe-in, the front wheels of a vehicle are set to slip, with the disadvantage of an increase in rolling resistance (Equation 2.4) of:

$$\Delta F_R \approx 0.01 \; F_R \; \text{ per } \; \Delta = 10' \tag{3.7b}$$

The toe-in dimension r_Δ of just one wheel is included in determining the toe-in angle Δ (i.e. $r_{\Delta,t}/2$):

In radians $\Delta = r_\Delta/D$ 　　　　　　　　　　　　　　　　　　　　(3.8)

In minutes $\Delta_i = r_\Delta/D \times 57.3 \times 60$ 　　　　　　　　　　　　　(3.8a)

r_Δ should be taken at the rim flanges, which is why its distance D must be considered. With a given toe-in dimension, e.g. $r_\Delta = 2$ mm there is a larger angle on small 12" rims than on ones with a 15" diameter. Figure 3.47a shows the

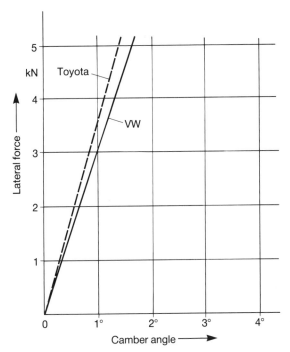

Fig. 3.46a Camber alteration measured on the driven McPherson front suspension of a lower mid-size passenger car with lateral forces directed inwards and applied statically at the centre of tyre contact. Disc wheel elasticity was eliminated on the measurements, and caster, which exercises no influence here, was ignored.

influence of the rim diameter and Fig. 2.7 the individual dimensions: $D = d + 2\,b$.

A tyre moving in a straight line has the lowest tyre wear and rolling resistance. When it rolls, a rolling resistance force F_R, directed from front to back, arises at the centre of tyre contact, which generates a moment with the lever arm r_a, which is absorbed via the tie rod to the steering (Figs 3.48 and 3.86 and Equation 2.4).

As a result of existing elasticity, particularly in the suspension control arm bearings, this moment pushes the wheel backwards slightly and, in order to make it run straight when the vehicle is moving, 'slip' is set as toe-in when it is stationary. In front-wheel drive vehicles, the traction forces directed from back to front attempt to push the wheels together at the front edge (Fig. 3.49), so toe-out (i.e. negative toe-in) alignment can be beneficial. As a result of the built-in elastokinematics (Figs 3.65 and 3.66a) and in order not to cause a deterioration in the driving stability in the overrun condition (i.e. when the driver removes his foot from the accelerator), front-wheel drive vehicles may also be set with toe-in.

In addition to the absolute value of the toe-in, tolerances must be specified which, because they can be adjusted by changing the tie rod length on the

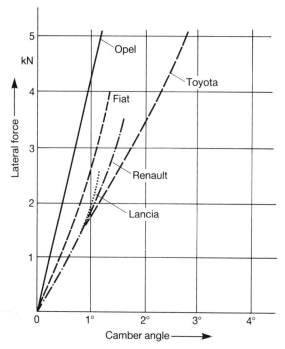

Fig. 3.46b Elastic camber change measured on various non-driven rear axles of mid-size passenger cars with lateral forces introduced statically in the middle of the centres of tyre contact. The type of axles were:

Vauxhall: compound crank axle
Fiat: compound crank axle
Lancia: McPherson strut
Toyota: McPherson strut
Renault: trailing link suspension

The low elasticity of the compound crank axles is clearly visible. Considering the caster would also give the same results.

front wheels (Fig. 4.11), only need to be $\Delta\Delta_f = 5'$ per wheel. Average values in factory information for toe-in are:

on rear-wheel drive vehicles $\Delta_f = +15' \pm 10'$ (3.8b)

on front-wheel drive vehicles $\Delta_f = 0° \pm 10'$ (3.8c)

With semi-trailing links it is possible to alter the toe-in on the rear axle by swivelling the axis of rotation of the suspension control arms (Figs 3.49a, 1.9 and 1.9a) and, on 'double wishbone suspensions', by a lateral length alteration on one suspension control arm (Figs 1.0 and 1.45). Tolerances of $\Delta\Delta_r =$

Fig. 3.47 The toe-in $r_{\Delta,t}$ of both wheels in accordance with the German standard DIN 70 020 is the difference in dimension b–c in mm, measured on the rim flanges at the level of the wheel centre.

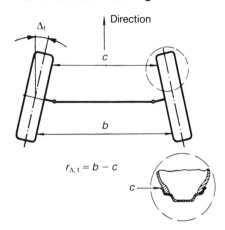

$$r_{\Delta,t} = b - c$$

±5' can be maintained where there is a setting. If this has not been provided in the design, values of $\Delta\Delta_r = \pm 25'$ are almost inevitable, if tight component tolerances are not to render manufacturing uneconomical. Regardless of whether the rear axle is steered or not, toe-in angles of the same size, both left and right, are required to ensure that the direction of movement x'–x' of the vehicle corresponds to its longitudinal axis X–X (Figs 3.49b and 3.1). The German standard DIN 70 027 therefore specifies that the so-called crab angle β', must be quoted, i.e. half the total toe-in angle of the rear axle. It is:

$$\beta' = (\Delta_{r,rs} - \Delta_{r,l})/2 \qquad\qquad (3.8d)$$

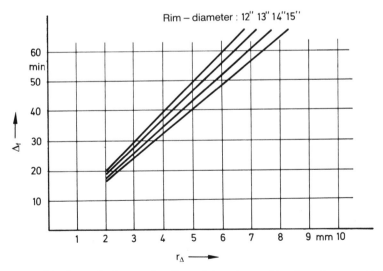

Fig. 3.47a Toe-in angle Δ_f as a function of rim size and toe-in r_Δ in mm, measured on one front wheel.

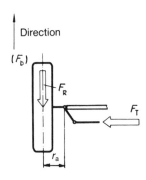

Fig. 3.48 The rolling resistance causes a longitudinal force F_R in the wheel centre, and tries to push the wheel backwards into toe-out via the lever r_a; for reasons of simplification, the steering axis $\overline{EG}$ (Fig. 3.80) is assumed to be vertical in this and the next illustration. The moment $M_R = F_R\, r_a$ causes the force F_T to arise in the tie rod. Braking force F_b operates in the same direction as F_R but has a different lever (Figs 3.83 and 3.84).

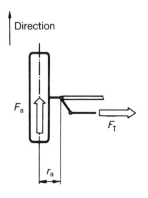

Fig. 3.49 On front-wheel drive vehicles, tractive forces F_a attempt to push the wheels into toe-in. The tie-force F_T arises on both sides; the same applies to driven rear axles (Fig. 3.50).

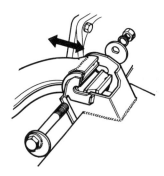

Fig. 3.49a Hexagonal bolts with eccentric discs, which come into contact with lateral collars, can be provided for setting camber and toe-in on both semi-trailing links (illustration: Ford).

Where it is possible to set the toe-in, $\beta' = \pm 10'$ can be maintained, and if there is no facility for setting toe-in on independent wheel suspensions or the vehicle is fitted with a compound crank axle or a rigid axle (Fig. 3.58b), up to $\beta' = \pm 25'$ must be allowed to enable economical production.

Taking as an example a passenger car with $\Delta_{r,l} = -10'$ and $\Delta_{r,rs} = +5'$ in accordance with Fig. 3.49b:

$$\beta' = [+5' - (-10')]/2 = +7.5'$$

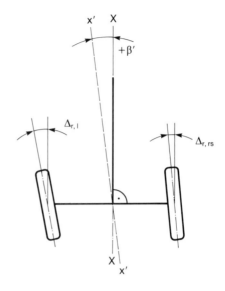

Fig. 3.49b The difference between the toe-in angle $\Delta_{r,l}$ on the left and $\Delta_{r,rs}$ on the right rear wheel determines the size of the crab angle $\pm\beta'$. It is positive if the median points forward and left (see also Fig. 3.58b).

This means the angle is positive. The toe-in on the rear axle of passenger cars is $\Delta_r = 10'$ to $20'$; the drawing information for a vehicle with independent wheel suspension would then, for example, be:

toe-in $15' \pm 10'$, crab angle max $\pm\ 15'$

BMW already specifies this condition for all models:

geometrical crab angle $0° \pm 15'$

and on vehicles with compound crank axles VW specifies:

maximum permissible deviation from the direction of travel $25'$.

3.6.2 Kinematic toe-in alteration

Even more important than a toe-in which has been correctly set on the stationary vehicle, is whether this is maintained when the vehicle is moving or whether it changes as a consequence of the wheels compressing and rebounding. This can be the fault of inadequate steering kinematics (see Section 4.6) or deliberately introduced to achieve certain handling properties. To avoid increased tyre wear and rolling resistance or impeding directional stability (as shown in Fig. 3.50 and curve 1, Fig. 3.51) no toe-in change should occur when the wheels compress or rebound. The wheel travel upwards (s_1) and downwards (s_2) is plotted on the y-axis of the figures, whilst on the x-axis positive toe-in is plotted to the right for one wheel each time, and negative toe-in (i.e. toe-out) plotted to the left. The ideal curve 1 would be difficult to achieve at

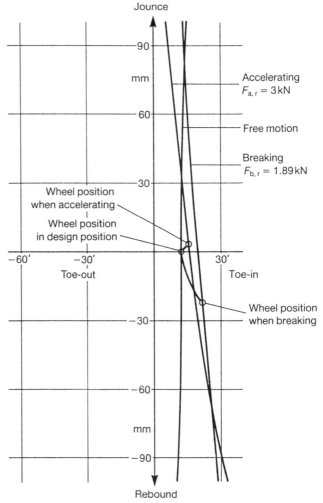

Fig. 3.50 Kinematic toe-in alteration of one wheel on the multi-link independent rear suspension of the Mercedes Benz S class with barely any deviation from the static value $\Delta_r = 12'$. The illustration also shows the behaviour of the wheel when subjected to a constant drive-off force $F_a = 3$ kN (Fig. 3.87) introduced in the wheel centre and an opposed braking force $F_b = 1.89$ kN acting at the centre of tyre contact (Fig. 3.83), all beginning in the design position (see Section 5.3.4). As the wheel compresses when the vehicle moves off, it goes $+\Delta r = 3'$ further into toe-in and, for elastokinematic reasons, further into toe-in by $+\Delta r = 10'$ when the brakes are applied. The rear axle stabilizes the braking process (see Section 3.6.5.1).

the design stage and certain deviations from the ideal shape have to be accepted.

A toe-in alteration can be the result of incorrect tie rod length or position. Provided that the steering arms are behind the front axle (Figs 3.48, 4.2 and

Fig. 3.51 Possible alteration of toe-in of one wheel (in minutes) as it compresses and rebounds due to an incorrect tie rod length or position.

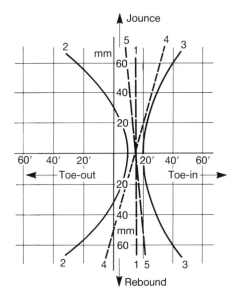

4.3), the example of a double wishbone suspension can be used to explain how different length tie rods act (Fig. 3.52). If they are too short (point 2), they pull the wheels together at the back both during compression and rebound, and go into toe-out as shown in curve 2 of Fig. 3.51 and Fig. 3.52a. Tie rods, which are too long, push the wheels apart in the direction of toe-in, curve 3; in both cases the graph displays a high curvature.

If, when the tie rods are the correct length, the inner joint 4 is too high (or the outer one too low, Fig. 3.53), when the wheel rebounds, the back of the wheel is drawn inwards and toe-out occurs; whilst, when it compresses, the wheel goes into toe-in. This results in approximate straight line running but at an angle (curve 4 in Fig. 3.51). A tie rod joint, 5. which sits too low on the inside or too high on the outside has, as the corresponding curve shows, the opposite effect – as do steering arms, which point forwards in all observed cases (Figs 4.22 and 4.24).

Fig. 3.52 Too short a tie rod (point 2) causes both the compressing and the rebounding wheel to go into toe-out. However, too long a tie rod (point 3) causes toe-in in both directions (see Fig. 3.51).

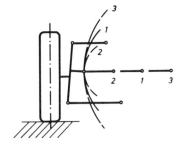

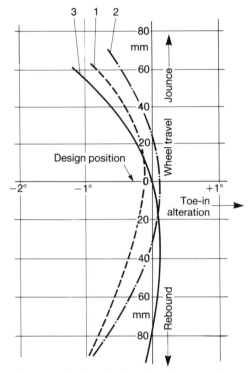

Fig. 3.52a As can be seen in Figs 3.52 (point 2) and 4.31, too short tie rods on McPherson struts cause the toe-in curve to be bent. If the steering arms are behind the axle, both compressing and rebounding wheels go into toe-out. The Fig. shows values measured on the left front wheel on three front-wheel drive vehicles of the lower mid-size range. Curve 3 shows a 'roll–steer effect' on the front axle. This measure, which tends towards understeering, is achieved due to the difference in height between the inner and outer tie rod joint (which can be seen in Fig. 3.53 under point 5).

3.6.3 Toe-in alteration due to roll steering

As a further example, Figs 3.54 and 3.55 show a deliberate toe-in alteration. When cornering, the outer compressing wheel goes into toe-out and the inner rebounding one into toe-in. The steering angle is reduced slightly under the

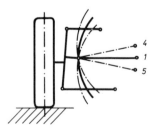

Fig. 3.53 An inner tie rod joint that is too high (point 4) produces curve 4 (Fig. 3.51), and one that is too low generates curve 5.

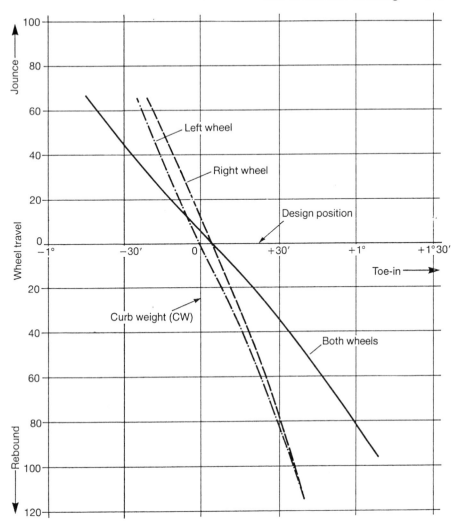

Fig. 3.54 Toe-in alteration recorded on an Opel/Vauxhall Omega indicating body roll understeering on the front axle. The individual wheels were measured to obtain the total toe-in. The design position relates to the vehicle with three passengers each weighing 68 kg; the height of the unladen vehicle is also marked.

influence of the body inclination in order to improve the tendency to oversteer by means of body roll understeering on the front axle or to improve handling when changing lanes (Fig. 3.56, see also curve 3 in Fig. 3.52a).

As described in Section 3.6.4, rear axles can tend to lateral force oversteer – which can lead to an overswing of the vehicle's rear end (Fig. 3.57). To compensate for this and make the overall handling of the vehicle neutral, designers like to make the rear axle body roll understeer (Fig. 3.58). On indi-

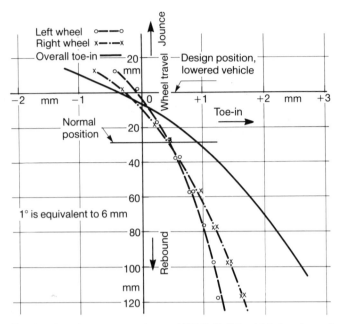

Fig. 3.55 Toe-in alteration measured on a VW Golf GTi which has been lowered by $\Delta_s = 30$ mm. In the normal position (also marked as specified by the manufacturer), as the wheels compress and rebound, the alteration values (which have a negative influence on directional stability and tyre wear) are less than in the lowered condition. The (now minimal) residual small compression spring travel can be seen clearly.

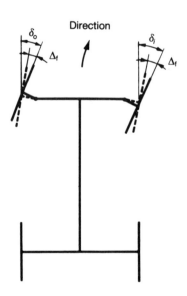

Fig. 3.56 If the compressing wheel on the outside of the bend goes into toe-out and the rebounding one on the inside of the bend into toe-in under the influence of the body roll inclination or due to lateral forces, the steering input is slightly reduced by the angle Δ_f. The axle understeers.

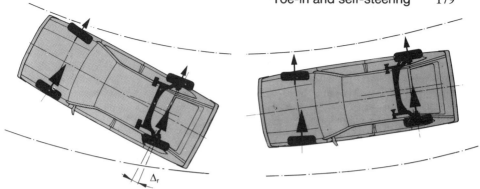

Fig. 3.57 Under the influence of a lateral force, the rear axle can take on the angle Δ_r and oversteers the vehicle to the inside edge of the bend (left and Fig. 2.29a). For this reason, VW install track-correcting bearings to the Golf, Vento and Passat models, which largely prevent oversteering. Another possibility is to allow body roll understeering of the axle, (see Figs 1.41, 1.42, 3.60 and 3.61).

vidual wheel suspensions the compressing wheel on the outside of the bend in this case must go into toe-in and the rebounding inner one into toe-out; Figs 3.15 and 3.58a show this type of alteration curve (see also Section 2.10.4).

As they are directly linked to one another, the wheels of rigid and compound crank axles have no toe-in alteration where the springing is parallel. However, due to design tolerances or incorrect installation, the axle can sit at an angle in the vehicle, i.e. one wheel has toe-in and the other toe-out in respect of the longitudinal axis of the vehicle. In this case, the direction of

Fig. 3.58 To reduce the tendency to oversteer, the rear wheel suspension can be designed so that body roll or lateral force understeering of the axle is possible, i.e. under the influence of the body pitch (or lateral forces) the compressing wheel on the outside of the bend goes into toe-in and the rebounding one on the inside of the bend into toe-out.

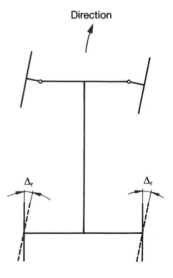

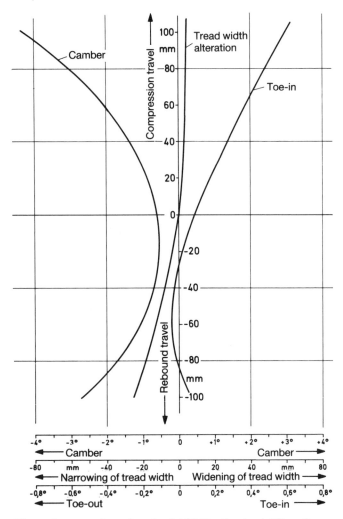

Fig. 3.58a Kinematic properties of an Audi 100 Quattro (and A6) as the rear wheels compress and rebound. The relatively small tread width alteration of the two wheels, the favourable negative camber as the springs compress and the toe-in alteration (of one wheel), which points to roll understeering of the rear axle, are clearly visible.

movement x'–x' of the vehicle and its longitudinal axis X–X deviate from one another by the crab angle (Figs 3.49b and 3.58b).

Even with rigid axles, body roll understeering can be achieved by – as shown in Figs 3.59, 1.16 and 1.17 – the axle being drawn forwards on the outside of the bend when the body inclines and backwards on the inside of the bend. The alteration dδ of the steer angle in the axle as a whole, divided by the alteration d$\varphi_{V,k}$ of the kinematic roll inclinations, is termed the 'roll steering factor' (Fig. 3.60).

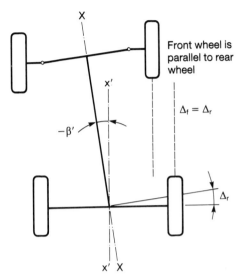

Fig. 3.58b If the rigid rear axle is not fitted at a right angle to the vehicle's longitudinal axis X–X – i.e. if the vertical on it deviates from the direction of movement x'–x' by the crab angle β' – a slight steering input is necessary to make the vehicle move in a straight line. The figure also shows how the self-steering of the rear axle makes it necessary to turn the front wheels if the vehicle is to move in a straight line on an uneven road surface under reciprocal springing (Fig. 1.11). The axle can displace by the angle $\Delta_r = \beta'$ (Figs 1.16 and 3.49b).

A rigid rear axle which self-steers when the body inclines also self-steers when going in a straight line on an uneven road. The steering effect this causes occurs not only on reciprocal bend springing (Fig. 1.11) but, to a certain extent, also on unilateral springing. This is the reason why 'self-steering', which can only be compensated for by spontaneously turning the front wheels (see Fig. 3.58b), is generally limited.

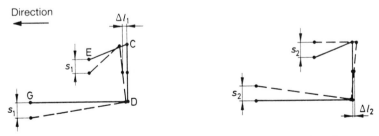

Fig. 3.59 If the body jounces by the path s_1 on the outside of the bend – caused by trailing links at an angle to one another and of different lengths, as can be seen in Fig. 3.125 – the axle centre is drawn forwards slightly by the path Δl_1 (left) and pushed backwards by Δl_2 on the rebounding inside. As a result of this, the rigid axle moves by an angle and roll understeers. This reduces the tendency to oversteer of standard vehicles.

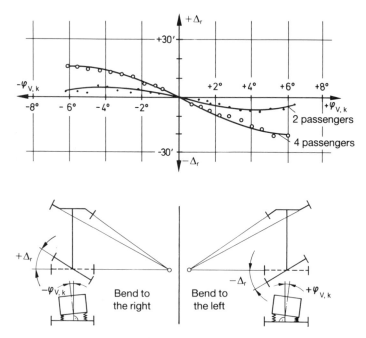

Fig. 3.60 Angled position by the angle Δ_r as a function of the roll angle $\varphi_{V,k}$ measured on the driven rigid rear axle of a conventional passenger car occupied with two and four people. When there are two people in the vehicle and $\varphi_{V,k} = 4°$, $\Delta_r = 6'$. The body roll-steering factor would then be $d\Delta/d\varphi_{V,k} = 0.1°/4° = 0.025$.

When there are four people in the vehicle, this increases to 0.075; the tendency of this vehicle to oversteer is therefore reduced, regardless of load.

On compound crank axles, the bearing points O shown in Fig. 3.30 move under the centre of the wheel when the vehicle is loaded, resulting in negative angles κ. This results in increasing body roll understeering, respectively, decreasing roll oversteering with load and therefore an improved roll-steer factor (Fig. 3.61).

3.6.4 Toe-in alteration due to lateral forces

Increasing lateral forces try to push the turned-in front wheels with the lever of the kinematic caster $r_{\tau,k}$ and the caster offset $r_{\tau,T}$ (Fig. 3.92) into the straight running position. As a result of the elastic compliance in the system this reduces the steering angle and lateral force understeering takes place.

To achieve this on the rear wheels (as shown in Fig. 3.58), the wheel on the outside of the bend has to go into toe-in and the one on the inside of the bend in the direction of toe-out.

To some extent, exactly the opposite of this can be seen in Fig. 3.62. The rear wheels of the compound crank axle (Opel and Fiat) on the outside of the bend are pushed into toe-out by the lateral force $F_{y,W,o}$ and the ones on the

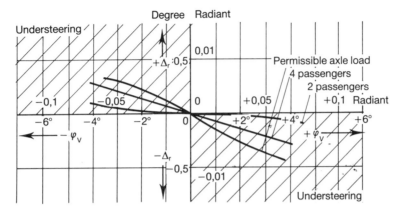

Fig. 3.61 Roll-steering measured on a VW Polo; increasing the load increases the understeering of the compound crank axle. At $\varphi_{V,k} = 4°$, the roll-steer factor is 0.025, 0.07 and 0.1 depending on the load.

inside of the bend into toe-in by $F_{y,W,i}$. The result is lateral force oversteering (Fig. 3.57), which is also noticeable on the longitudinal link axle of the Renault (Fig. 1.45a) and is also slightly evident on the McPherson struts of the Lancia (Fig. 1.7). Toyota moves the two transverse links 1 and 2 (Figs 3.0a and 3.63) backwards in parallel, therefore achieving the elastokinematic toe-in angle Δ_r on the outside of the bend and (as shown in Fig. 3.62) toe-out on the inside.

For the measurement, the lateral force was applied statically in the centre of the tyre contact; shifting it backwards by the caster offset $r_{\tau,T} = 10$ to 40 mm would cause all toe-in curves to turn counter-clockwise. The Toyota Corolla would then have a slight tendency to lateral force understeer, whilst there is an increased tendency on all other passenger cars to oversteer.

Another way of reducing this would be to give the rear wheels negative caster $-r_{\tau,k}$ (Figs 3.90 and 3.111a); however this must be greater than that of the tyre, which itself reduces as the slip angles α increase (Figs 2.36 and 3.91a). This can be achieved on double wishbone suspensions (Fig. 3.111b); the negative caster increases on the compressing wheel on the outside of the bend and under load. Even with rigid axles lateral force understeering is possible. If the panhard rod is behind the axle casing (Figs 3.63a and 1.44), the effective distance a between the lateral forces $F_{y,W,r,o \text{ or } i}$ on the two rear wheels and the rod force $F_{T,y}$ results in a pair of forces which generates the forces $\pm F_x$ in the trailing links and – due to the elasticity in the rubber bearings – causes the desired self-steering.

3.6.5 Toe-in alterations due to longitudinal forces

3.6.5.1 Toe-in during braking Toe-in leads to stabilization of the vehicle breaking. This means better straight-running behaviour and it can be

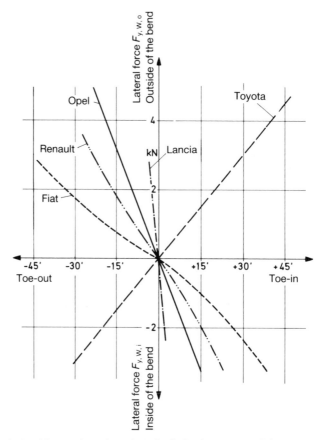

Fig. 3.62 Lateral forces introduced statically in the centre of the tyre contact of different rear axles give the Toyota toe-in on the outside of the bend, but toe-out on the other vehicles tested; these exhibit a lateral force steering tending towards oversteering. If the lateral force operates in the other direction (i.e. from the inside out), there is toe-in instead of toe-out on the vehicles. The toe-in alteration in minutes appears on the x-axis and the force in kilonewtons on the y-axis.

The vehicles are fitted with: compound crank axle (Opel/Vauxhall and Fiat), McPherson strut (Lancia and Toyota) and trailing link suspension (Renault, Fig. 1.45a).

achieved both by negative kingpin offset (Fig. 6.12) and by an elastokinematic toe-in alteration.

The front end of the vehicle jounces when the brake is activated. If (as shown in Fig. 3.54) the body roll has been kinematically designed to understeer, both front wheels go into toe-out, i.e. with a positive kingpin offset, and they continue to travel in the same direction in which they were already being pushed by braking forces F_b (Figs 3.48 and 6.11). To limit this effect, the necessary counter-steering in the direction of toe-in can be achieved, $r_s =$

Fig. 3.63 Under the influence of the lateral force $F_{y,W,o}$ acting on the outside of the bend behind the wheel centre by the tyre caster $r_{\tau,T}$, the mountings of the transverse link 1 flex more than the brace 2, which is offset backwards; point 6 moves to 7 and the toe-in angle Δ_r occurs (see also Fig. 3.0a).

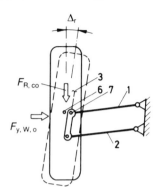

0 or there is a small positive kingpin offset at ground. The only prerequisite for this is a top view angle ξ between transverse link 1 and tie rod 7 (Fig. 3.64).

Using a Mercedes model as an example: the front of the longitudinal rod 4 is anchored at point G on the suspension control arm, and the back carries the supporting bearing 6. Under the influence of the braking force F_b the defined longitudinal elasticity of part 5 yields, the lower guiding joint G moves out to 4 and the outer tie rod joint U moves to 9. As points G and U move in different arcs and the tie rod joints are also less laterally elastic than the bearing D of the transverse link 1, both front wheels are pushed into toe-in in spite of the opposing moment $M_b = F_b \, r_b$ seen in Fig. 3.84.

In the same way, individually suspended rear wheels can experience an elastokinematic toe-in alteration during braking (Figs 3.0a and 3.50).

3.6.5.2 Absorbtion of the radial tyre dynamic rolling hardness without toe-in alteration Nowadays, manufacturers fit only steel radials to series production vehicles. However, unlike the diagonal tyres used in the past, these have the disadvantage of dynamic rolling hardness (see Section 2.2.2). The very

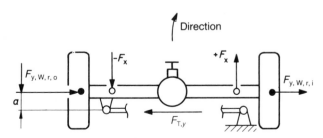

Fig. 3.63a The effective distance a between the lateral forces $F_{y,W,r,o\ or\ i}$ on the wheels of the rigid axle and the force $F_{T,y}$ on the panhard rod at the back, leads to a force pair which generates the forces $\pm F_x$ in the longitudinal links and can cause lateral force understeering due to the elasticity in the rubber mountings. If the rod is in front of the axle, oversteering is possible.

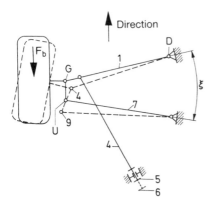

Fig. 3.64 A positive top view angle ζ between the tie rods 7 and the transverse links 1 close to them (mostly the lower ones) can cause an elastokinematic toe-in alteration during braking.

stiff belt causes longitudinal oscillations which, on independent wheel suspension, are transferred to the body via the steering knuckle and the tie rod and – particularly on cobbles, rough concrete and at speeds below 80 km h^{-1} – can cause an unpleasant droning noise inside the vehicle. The vibrations can be stopped if the steering knuckle is given a precisely defined longitudinal mobility. This is a task that is not easy to fulfil at the design stage because neither a toe-in alteration may occur, nor a lateral force at the centre of tyre contact (Fig. 3.3) under the influence of the paths of $s \leqslant \pm 10$ mm (Fig. 3.0), as straight rolling ability and rolling resistance would deteriorate. On the front axle it can be solved using a transverse link, which has an outrigger pointing backwards (or forwards) (Figs 3.65 and 3.65a), and which is supported at the side in a rubber bearing with a highly progressive, precisely defined spring rate. The important thing is that stiff bearing elements, which only yield a little under cornering lateral forces and braking forces, sit in the pivots D and G.

If a transverse link anchored at point D controls the wheel, it can have a hole in which a longitudinally elastic rubber bearing sits (Figs 3.66, 3.66a and 1.40). The inner tube of this part is supported on the anti-roll bar 5 or a tension or compression rod strut 4, pointing either backwards or forwards.

On driven independent rear wheel suspensions it is especially important that the trailing or semi-trailing arms be controlled as well as possible to avoid elastic camber and toe-in alterations. The three or four rubber bearings, which link the suspension subframe and the differential with the body, then have to be designed so that the dynamic rolling hardness of the radial tyres is absorbed (points 2, 3 and 4 in Fig. 1.9). This task is carried out by the bearings in the longitudinal struts on rigid axles and by the rubber elements sitting in the pivot points O on compound crank axles (as shown in Figs 3.67, 1.41 and 1.44).

3.6.5.3 Due to front-wheel tractive forces   As can be seen in Fig. 1.35, on a transverse engine the differential is relocated from the middle of the vehicle to the manual transmission that is sited at the side, resulting in drive shafts of different lengths. When the vehicle moves off in the lower gears, the front end rebounds and the shorter (left-hand) shaft takes on a steeper working

Fig. 3.65 BMW fits the sickle-shaped control arm 1 (shown separately in the next figure) on the front axle of their 3-series.

Under the effect of longitudinal forces, it rotates around the (only slightly compliant) ball joint D and is supported with the outrigger 4 by means of a large rubber mounting on the body. In the lateral direction this bearing has an initially soft, but then highly progressive, springing curve.

Tie-rod 7 lies at the height of the control arm and is almost parallel to the line linking the bearing points $\overline{GD}$; points U and G therefore move on an arc of around the same radius and longitudinal wheel movements do not cause any toe-in alteration. As shown in Fig. 3.86, the rolling resistance F_R, which varies in size, must be observed in the wheel centre as F'_R.

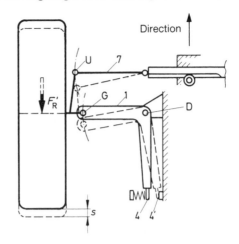

angle α to the wheel axis than the longer one (right-hand, Fig. 3.67a). The left or right-hand moments $M_{S,a,rs\ or\ l}$ moving around the steering axis $\overline{EG}$ result from α and – due to the direction of rotation of the drive shafts – try to push both wheels into toe-in:

$$M_{S,a,rs\ or\ l} = \tfrac{1}{2} F_A\, r_{stat}\, \tan \tfrac{1}{2}\alpha_{rs\ or\ l} \tag{3.8e}$$

(F_A see Equation 6.36 and r_{stat} Section 2.2.5).

As the angle α is larger on the left, a slightly higher moment can arise there than on the other side, with the risk of the vehicle pulling to the right. If the driver takes his foot off the accelerator quickly, a braking moment is generated by the engine, the front end dips and a steering effect in the other direction is inevitable. This is the main reason why (as shown in Figs 1.36 and 1.40) front-wheel drive vehicles with powerful engines necessarily have an intermediate shaft, and drive shafts of equal lengths.

3.7 Steering angle and steering ratio

Section 4.6 covers steering kinematics.

3.7.1 Steering angle

When the vehicle is moving slowly, it will only corner precisely when the verticals drawn in the middle of all four wheels meet at one point – the centre of the bend M. If the rear wheels are not steered, the verticals on the two front

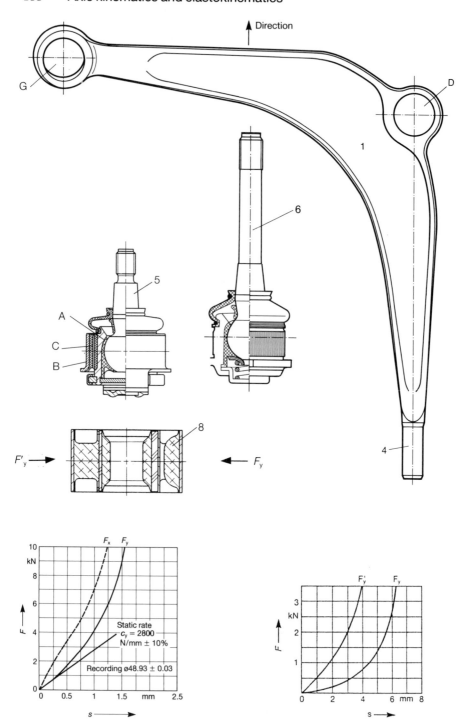

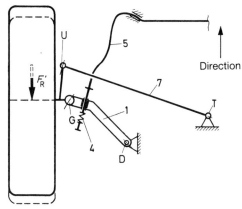

Fig. 3.66a An A arm can be replaced by two individual suspension arms: one is transverse (position 1) and carries lateral forces; the other is longitudinal (position 5) and transfers forces in this direction. A longitudinally elastic bearing (position 4) in a hole in part 1 absorbs the dynamic rolling hardness of the radial tyres. As in the Audi 100 (Fig. 1.40), part 5 can also be the arm of the anti-roll bar.

wheels must intersect with the extension of the rear axle centre line at M (Figs 3.68 and 1.51) whereby different steer angles δ_i and $\delta_{A,o}$ occur on the front wheels on the inside and outside of the bend. The nominal value $\delta_{A,o}$ of the outer angle – also known as the Ackerman angle – can be calculated from the larger inner angle δ_i.

$$\cot\delta_{A,o} = \cot\delta_i + j/l \qquad (3.9)$$

whereby l is the wheelbase and j the distance between the two steering axis extensions (Figs 3.69 and 3.80), measured at the ground, i.e.

$$j = b_f - 2\,r_S \qquad (3.10)$$

Where the kingpin offset r_S is negative, the integer is positive (Fig. 3.87).

The differential steer angle $\Delta\delta_A$ included in Fig. 3.68 (also known as the toe difference angle) must always be positive for the nominal values calculated (nominal curve in Fig. 3.71).

$$\delta_A = \delta_i - \delta_{A,o} \qquad (3.11)$$

Fig. 3.65a Front sickle-shaped suspension control arm of the BMW 3-series. The guiding joint 5 links the suspension control arm 1 with the suspension strut and is press-fitted from below into hole G; the inner joint 6 sits in hole D. The suspension control arm rotates around this part under the influence of longitudinal forces and is supported by outrigger 4 on the transversely elastic bearing 8. Its progressive compliance in the y direction is shown by the illustration on the right. In the vertical direction (z) the bearing is stiffer.

In part 5 the rubber ring C is vulcanized in between the joint housing A and the outer ring B, and is – as can be seen from the illustration on the left – laterally more compliant (F_y) than in the longitudinal (x) direction (illustration: Lemförder Metallwaren).

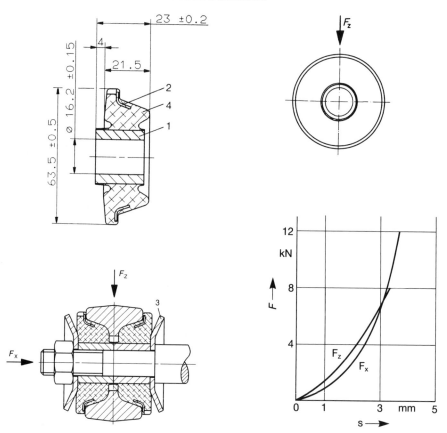

Fig. 3.66 Mounting of the anti-roll bar fitted at the front in the transverse links on the Audi 100 (Fig. 1.40). The two rubber parts in the suspension control arms are vulcanized to the inner tube 1 and ring 2. Under the influence of longitudinal forces F_x one part comes into contact at the dome-shaped washer 3 and the other part relaxes. As can be seen on the left, the rubber part 4 projects beyond the sleeve 1; when fitted this achieves the necessary pre-tensioning. Ring 2 ensures that it sits firmly in the suspension control arm, so that the mounting can transmit vertical forces F_z without complying too much. The diagram shows the longitudinally progressive characteristic curve of the two bearings and the almost vertical linear characteristic curve when fitted (illustration: Lemförder Metallwaren).

The theoretical track circle diameter D_S can be calculated using the angle $\delta_{A,o}$ (Fig. 3.68), i.e. the diameter of the circle which the outer front wheel traces with the largest steering angle (see also Equation 2.10). The track circle of a vehicle should be as small as possible to make it easy to turn and park. The formula

$$D_S = 2 \left(\frac{l}{\sin \delta_{A,\sigma,max}} + r_S \right) \tag{3.12}$$

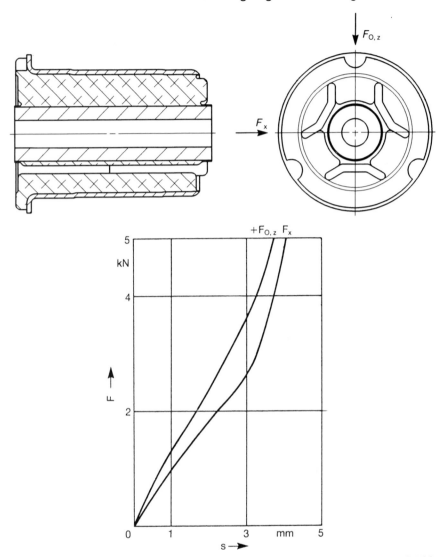

Fig. 3.67 Elastic bearing in the front eyes of the torsion crank axle of the Audi 100 (Fig. 1.44). The indents in the rubber part give the necessary elasticity. The bearing must be soft enough in the longitudinal direction to absorb the dynamic rolling hardness of the tyres (when the axle shown in Fig. 1.44 is controlled precisely) and not very compliant in the vertical direction to be able to absorb safely the forces $-F_{O,z}$ which occur during braking (Fig. 3.123). (illustration: Lemförder Metallwaren).

derived using the illustration shows that this requirement necessitates a short wheelbase and a large steer angle $\delta_{A,o}$ on the outer wheel of the bend. This in turn requires an even larger inner wheel angle, which is limited by the fact that the fully turned compressing wheel may not come into contact with the

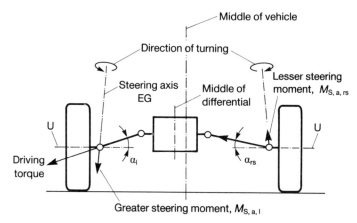

Fig. 3.67a   When the engine is transverse, the differential is no longer in the centre of the vehicle and an intermediate shaft is necessary (Fig. 1.36), otherwise the drive shafts are not the same length. If they are at different angles α, different moments can occur around the steering axes, causing the steering to pull to one side. $\alpha_l = \alpha_{rs}$ can be achieved by tipping the differential by up to 2°.

wheel house or the front axle components. The wheel house cannot be brought too far into the sides of the front foot well as the pedals (on both left and right-hand drive vehicles) would then be at an angle to the direction in which the driver faces and foot-space would be restricted. In front-wheel drive vehicles, room also needs to be allowed for snow chains (Figs 2.5b and 3.79) and the largest working angle of the drive joints (Figs 1.1a and 1.36a).

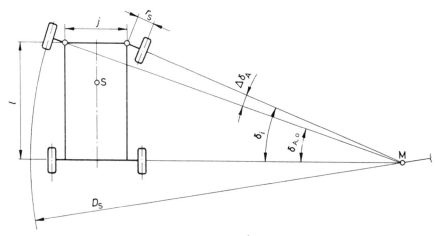

Fig. 3.68   Kinematic relationships in accordance with Ackermann between the steering angle $\delta_{A,o}$ on the wheel on the outside of the bend and δ_i on the inside of the bend. The illustration also shows the $\Delta\delta_A$ and the track circle diameter D_S (see Fig. 1.51).

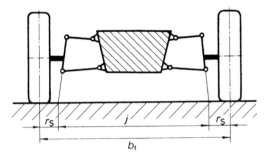

Fig. 3.69 Path designations on the front axle; b_f is the tread width on the front and r_S the (in this case) positive kingpin offset on the ground (scrub radius).

3.7.2 Track and turning circles

The inner angle δ_i is therefore limited, whilst the wheel angle on the outside (for functional reasons a smaller angle) is not. This may be the same size as the inner one. The disadvantage is that it impairs the cornering behaviour of the vehicle (Fig. 3.70), but with the advantage that the track circle becomes smaller and the lateral tyre force capacity on the outside of the bend increases. For this reason, the outer steering angle is larger on most passenger cars, i.e. the actual value δ_o (without index A) is greater than the nominal angle $\delta_{A,o}$ calculated according to Ackerman (Fig. 3.71) by the steering flaw $\Delta\delta_F$. In other words, the required steering deviation is as follows:

$$\Delta\delta_F = \delta_o - \delta_{A,o} = \Delta\delta_A - \Delta\delta \qquad (3.13)$$

whereas $\Delta\delta = \delta_i - \delta_o$ expresses the so-called differential steer angle.

The track circle diameter D_S shown in Fig. 3.68 can be reduced by deliberately accepting a steering deviation. In addition to $\Delta\delta_F$, the angle $\delta_{A,o,max}$, in other words the largest outer nominal angle according to Ackerman calculated using Equation 3.9, must also be known. A series of test measurements has shown that a reduction by $\Delta D_S \sim 0.1$ m per $1°$ steering deviation can be

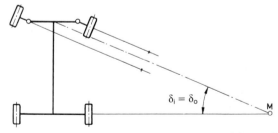

Fig. 3.70 To use the space available in the wheel house, it is an obvious idea to turn the wheel on the outside of the bend inwards by as much as the wheel on the inside of the bend; the wheels are then turned parallel and $\Delta\delta$ is zero. It is possible to increase the cornering force by turning the outer wheel more (compared with the wheel on the inside of the bend, Fig. 3.71).

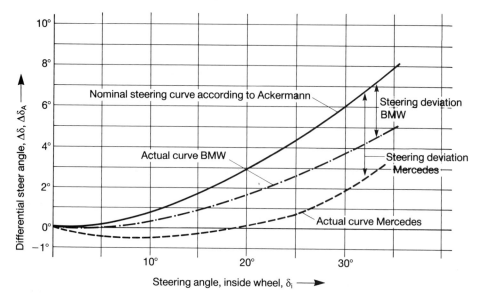

Fig. 3.71   Required, nominal steering curve for two standard passenger cars with the same wheelbase and approximately the same tread width calculated in accordance with Equation 3.9. The mean value of the actual curve measured when the wheels are turned to the left and right is included, and the steering deviation $\Delta\delta_F$ (also known as the steering error) is also marked. The steering angle δ_i of the wheel on the inside of the bend is entered on the x-axis, and the differential steer angle $\Delta\delta = \delta_i - \delta_o$ (which relates to the actual curve) and $\Delta\delta_A = \delta_i - \delta_{A,o}$ (which is valid for the nominal curve according to Ackermann) are marked on the y-axis.

In the workshop manuals $\Delta\delta$ must be indicated with a tolerance at $\delta_i = 20°$; on the BMW 3-series it would be $\Delta\delta = 1° \, 35'$ and on the Mercedes $\Delta\delta = 10'$. The differential steer angle of the Mercedes, which is negative up to $\delta_i \approx 20°$ indicates that the wheel on the outside of the bend is turned more than the one on the inside and so the lateral force absorbed by the front axle when it corners – and with it the steering response – is increased.

achieved; the formula which should include all dimensions in metres would then be:

$$D_S = 2 \left(\frac{l}{\sin \delta_{A,\sigma,max}} + r_S \right) - 0.1\Delta\delta_F \, [m] \tag{3.14}$$

A front-wheel drive vehicle with normal steering flaw can be used as an example.

The data when the wheels are turned to the right is:

$$l = 2.677 \text{ m}; \; b_f = 1.47 \text{ m}; \; r_S = -0.015 \text{ m}; \; \delta_{i,max} = 42°; \; \delta_{o,max} = 35°40'$$
$$j = 1.47 - [2\,(-0.015)] = 1.5 \text{ m}$$
$$\cot\delta_{A,o} = \cot 42° + 1.5/2.677 = 1.671; \; \delta_{A,o} = 30°55'$$
$$\Delta\delta_F = 35°40' - 30°55' = 4°45'$$
$$D_S = 2\,[2.677/\sin 30°55' + (-0.015)] - 0.1 \times 4.75°; \; D_S = 9.91 \text{ m}$$

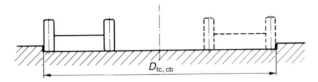

Fig. 3.72 Turning circle kerb to kerb $D_{tc,cb}$; an important dimension for the driver when turning the vehicle.

The track circle diameter measured on the passenger car was $D_{s,t} = 9.92$ m

The track circle radius is basically only a theoretical value which can be calculated at the design stage; for the driver it is the turning circle kerb to kerb that is important, in other words the distance between two normal height kerbs (Fig. 3.72) standing parallel to one another, between which the driver can just turn the vehicle. This circle diameter $D_{tc,cb}$ can be measured but can also be calculated easily using the track circle diameter D_s and the actual width of the tyre (Figs 2.7 and 2.10):

$$D_{tc,cb} = D_s + B \; [m] \tag{3.15}$$

However, the turning circle, the diameter of which D_{tc} is greater than that of the track circle by the front overhang length $L_{Ov,f}$ (see the caption to Fig. 1.18) is probably a more important dimension.

According to DIN 70 020, D_{tc} is the diameter of the smallest cylindrical envelope in which the vehicle can turn a circle with the largest steering input angle (Fig. 3.73). The smallest turning circle can be calculated at the design stage, but is easier to measure and appears as manufacturer's information in the specifications or as a measurement value in test reports.

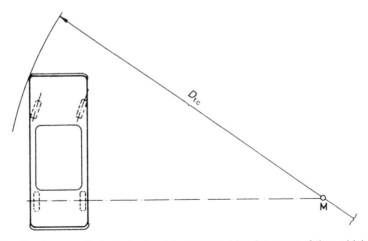

Fig. 3.73 The turning circle D_{tc} is the arc described by the parts of the vehicle protruding furthest outwards when the wheels are turned in at the largest steering angle.

The radius $R_{r,o}$ of the track circle, which the rear wheel on the outside of the bend traces, or $R_{r,i}$ – that traced by the wheel on the inside of the bend – can be calculated from the known track circle diameter D_{tc}. These are:

$$R_{r,o} = \left((D_S/2 - r_S)^2 - l^2\right)^{\frac{1}{2}} + \frac{b_r - j}{2} \tag{3.16}$$

$$R_{r,i} = R_{r,o} - b_r \tag{3.16a}$$

The formulae indicate that the longer the wheelbase l, the shorter $R_{r,o}$ and $R_{r,i}$ become (relative to D_S), i.e. the vehicle requires more width when cornering slowly (Fig. 1.51).

3.7.3 Kinematic steering ratio

. The kinematic steering ratio i_S is the ratio of the alteration $\Delta\delta_H$ of the steering wheel angle to $\Delta\delta_m$ of the mean steering angle, of a pair of steered wheels, where steering is operated free of moments and begins from the on-centre (straight ahead) position. Initially, the steering elasticity and the alteration of the ratio during steering are ignored. It is:

mean steering angle $\delta_m = (\delta_o + \delta_i)/2$ $\hspace{2cm}$ (3.17)

kinematic ratio $i_S = \Delta\delta_H/\Delta\delta_m$ $\hspace{2cm}$ (3.18)

The equations are only valid when there is a greater input range (e.g. $\delta_m = 20°$) or a ratio which remains approximately constant over the whole steering range (Fig. 3.74). However, if this changes (Fig. 3.75), the steering wheel

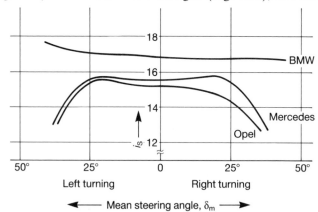

Fig. 3.74 Overall steering ratio i_S (see Section 4.3.3), measured on three conventional passenger cars with power-assisted recirculating ball steering. Whilst the BMW has a ratio which remains almost constant throughout the turning range it reduces on both sides from around $\delta_m = 20°$ on the Vauxhall/Opel and the Mercedes, so the driver needs less turns of the steering wheel to park. These two model groups have an opposed steering square positioned behind the axle (Figs 1.25, 4.10 and 4.23), whilst the BMW uses a synchronous one which also sits behind the axle (Fig. 4.2).

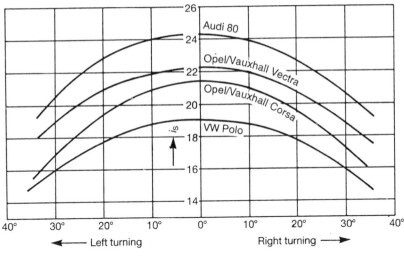

Mean steering angle, δ_m

Fig. 3.75 Total steering ratio i_S (Equation 3.19) measured on four front-wheel drive passenger cars with manual (non power-assisted) rack and pinion steering, entered over the mean steering angle δ_m of the wheels (Equation 3.17). It is important to note the relatively severe drop in ratio as the wheels are turned more (due to the steering kinematics, see Section 4.6). To limit the forces on the steering wheel when the vehicle is being parked, heavy vehicles, such as the Audi 80 and the Opel/Vauxhall Vectra, have the larger ratio $i_{S,0} = 24.2$ or 22.2. All vehicles have a constant steering gear ratio i'_S, i.e. not the varying split seen in Fig. 3.76.

angle proportions $\Delta\delta_{H,h}$ as a result of steering must be assumed (index h = hand) as well as the mean steering wheel proportion $\Delta\delta_{H,S}$ (index S = steering) related to both wheels:

$$i_S = \Delta\delta_{H,h}/\Delta\delta_{H,S} \tag{3.19}$$

If the overall steering ratio relates to the on-centre position a zero should be given as the index: i_{S0}.

As shown in Figs 4.2 and 4.21 to 4.23, steering gears with a rotational movement need a steering square arrangement of the linkage, in which the length and position of the tie rods and steering arms allow almost every type of steering ratio as a function of the input angle. However, the entire steering system has more component parts and is more expensive (see Section 4.3).

The more economical design is rack and pinion steering, although this has the disadvantage that – as can be seen in Fig. 3.75 – for kinematic reasons the ratio reduces as steering angles increase. On power-assisted steering systems, the reduction in ratio has a favourable effect on the handling properties. In the straight running position, a more generous ratio is desirable on passenger cars at high speeds in order not to make the steering too sensitive, whilst a reducing ratio could be better for cornering and making parking and manoeuvring possible with less turns of the steering wheel.

The hydraulics (or electrics) support the increasing activation forces at greater steering angles, however this is not the case on vehicles without power-assisted steering. Here, forces can become disproportionately high because the fall in ratio cannot be reduced, especially on front-wheel drive vehicles. The reasons for this are:

- the steering gear is located in the narrow space available between the dashpanel and engine
- the fixing points have to be laterally stiff
- toe-in alteration (Fig. 3.52a) must be avoided
- the need to produce the actual steering curve (Fig. 3.71) .

The design position of the tie rods in the top view is also a consideration. It makes a difference whether these – as shown in Figs 4.3 and 4.24 to 4.26 – are situated in front of or behind the centre of the axle (or intersect with it) and whether the inner joints are screwed into the sides of the steering rack (outer take-off) or must be fixed in the centre (centre take-off). The influence of the kingpin inclination angle and caster offset angle and the size of the steering arm angle (Fig. 4.2) also have to be taken into account.

Series measurements have shown that, on front-wheel drive vehicles, the reduction in ratio from the 'on-centre position' to full lock is 17 to 30%.

Standard passenger cars have space under the engine–gearbox–block; this is the reason for the significantly lower reduction (Fig. 3.77b) of only 5 to 15%.

Rear-engine vehicles offer even more space under the front-end boot. Of these, passenger cars were found with rack and pinion steering systems in which the ratio does not change throughout the entire input range.

The curve shown in Fig. 3.75 of the steering ratio of the Vauxhall Cavalier exhibits $i_{SO} = 22.2$, with the wheels in the straight ahead position and at a mean steering angle of $\delta_m = 35°$ the value $i_{S,min} = 17.7$ and $i_{S,min}/i_{SO} = 0.80$, i.e. the reduction is 20%.

The steering gear manufacturer ZF has developed a system to counteract the disadvantage of the reduction in ratio on non-assisted steering systems. For this purpose the steering rack varies its pitch from t_1 to t_2 (Fig. 3.76). This causes the rolling circle diameter of the pinion gear to reduce on both sides from d_1 to d_2 when the wheels are turned to the off-centre range position. The path s_2 shortens as the wheels are turned more and therefore the ratio i_S in the steering gear itself increases. The consequence is more turns of the steering wheel from stop to stop, but a reducing steering wheel moment (Fig. 3.77).

3.7.4 Dynamic steering ratio

The true steering ratio, as experienced by the driver, would be the dynamic ratio i_{dyn}; this comprises the proportions resulting from steering $\Delta\delta_{H,h}$ and the

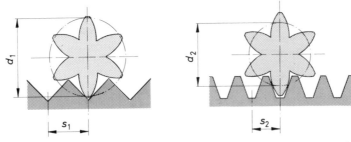

Fig. 3.76 If the steering rack is designed in such a way that the pinion gear is given a larger pitch circle (d_1, left) in the middle than on the outside (d_2, right), the rack travel reduces from s_1 to s_2 as the wheels are turned more, which leads to a more generous ratio (illustration: ZF).

elastic $\Delta\delta_{H,e}$ (Fig. 3.77a). To calculate the group of curves a given steering angle range $\Delta\delta_{H,S}$ must be assumed on both wheels (e.g. 0° to 5°, 0° to 10°, 0° to 15°, etc.) and the respective mean value determined in each case (here $\Delta\delta_m$ = 2.5°, 5°, 7.5°, etc.) to be able to take the kinematic steering ratio i_S at these points on the respective curves. The dynamic ratio depends on the height of the steering wheel moment M_H, so that only one point of a given curve can be considered in each instance. The equation is:

$$i_{dyn} = i_S + \Delta\delta_{H,e}/\Delta\delta_{H,S} \tag{3.20}$$

Figure 3.77b shows the dynamic steering ratio measured on a standard passenger vehicle. As an example i_{dyn} at $M_H = 5$ N m in the range $\Delta\delta_{H,S} = 0° - 5°$ can be calculated. Taken from the lower curve (for i_S) the overall steering ratio is $i_S = 21$. In accordance with Fig. 3.77a, the mean value of the steering wheel proportions as a result of elasticities is $\Delta\delta_{H,e} = 17°$. This means:

$$i_{dyn} = 21 + 17/5 = 24.4$$

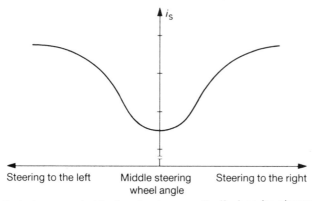

Fig. 3.77 Ratio i_S, generated in the steering gear itself when (as shown in Fig. 3.76) the steering rack has a varying split (illustration: ZF).

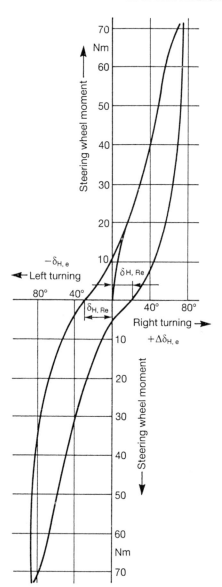

Fig. 3.77a Characteristic result of a steering elasticity measurement on a passenger car with rack and pinion steering that records the steering wheel angle as a result of elasticities in accordance with DIN 70 000. It shows the compliance $\delta_{H,e}$ when the wheels are turned to the left and right and the steering wheel moments M_H increase; the wheels were locked during the measuring process.

If the curve is steep, there is a high $C_H = M_H/\Delta\delta_{H,e}$ value, i.e. low steering elasticity. The greatest moment $M_H = \pm 70$ N m corresponds to a force of $F_H = 184$ N per hand with a steering wheel diameter of 380 mm. This should be enough to permit conclusions about the elasticity behaviour during driving. The hysteresis also shows the residual angles $\delta_{H,Re}$ which remain when the wheels are turned and the vehicle is stationary.

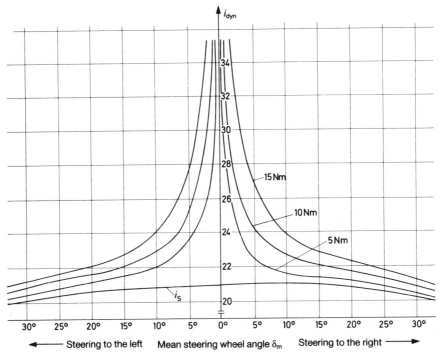

Fig. 3.77b Typical curve of the dynamic steering ratio i_{dyn} of a vehicle with rack and pinion steering entered as a function of the mean steering angle $\Delta\delta_m$ and the steering wheel moments M_H = 5, 10 and 15 Nm. The kinematic total ratio i_S measured on the same vehicle was entered for comparison; this falls from $i_{S,0}$ = 21 (in the centre position) to $i_{S,min}$ = 19.7 (where δ_m = ±35°) in other words only by 6°.

This value should then be entered at $\Delta\delta_m - 2.5°$. The smaller the steering angle range, and the greater M_H becomes, the more the dynamic ratio increases; if, for example, M_H is 15 N m, i_{dyn} already has a value of 31.

3.8 Steering righting – general

If there were no self-righting torque on the front wheels of a vehicle, straight ahead driving would be impaired and only a small force would be needed to turn it into the bend. When the bend had been negotiated, the steering wheel would have to be turned back and would not go back of its own accord to the straight ahead position. The driver would have no feel for cornering speed and handling with the risk that he/she would not be able to return the steering to the normal position fast enough when coming out of a bend. Sections

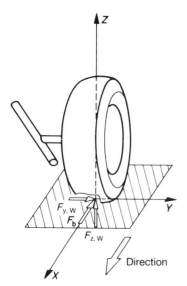

Fig. 3.78 The forces occurring between tyres and road are transferred from the suspension to the body. This is shown on the left front wheel for the vertical force $+F_{z,W}$, the rolling resistance or braking force $-F_b$ and the lateral force $+F_{y,W}$ (see also Fig. 3.1) acting from the inside out.

1.4.1, 1.5 and 1.6.2 refer to the correlations with the various types of drive and Fig. 1.20 shows the differences.

There are several ways of righting the steering at the end of a bend with, in each instance, one of the three forces acting on the centre of tyre contact (vertical force $F_{z,W}$, lateral force $F_{y,W}$ or longitudinal force F_l) having a lever to generate moments. They have been given indices to differentiate them and these indicate the direction of the righting force (Fig. 3.78) or other associated aspects:

$M_{T,f}$ moment from lateral force $F_{y,W}$ and caster $r_{\tau,T}$ (Figs 3.91a and 2.35)

$M_{S,z}$ moment from vertical force $F_{z,W}$, kingpin offset r_S and kingpin inclination σ (Figs 3.81 and 3.82)

$M_{S,y}$ moment from lateral force $F_{y,W}$ and lateral force lever $n_{\tau,k}$ (Figs 3.93 and 3.98)

$M_{S,x}$ moment from rolling resistance force F_R, and lateral force lever $n_{\tau,k}$ (Fig. 3.94)

In addition, there can be self-righting torques on front-wheel drive vehicles caused by the tractive forces (Fig. 3.100), by the body roll when drive shafts lie diagonally (Figs 1.2 and 3.67a) and drive joints, whose centres lie outside the steering axis (Fig. 3.79). Braking forces would also right the wheel on the outside of the bend, whilst turning the (lesser loaded) wheel on the inside of the bend further (see Equation 3.26a).

In accordance with the German standard DIN 70 000, the steering moment M_S is the sum of all moments around the steering axis of the steered wheels. This moment is introduced by the driver, whereas righting after the vehicle has negotiated a bend is a question of the driving condition and the coeffi-

Fig. 3.79 Left front axle of an Audi with negative kingpin offset on the ground $r_s = -18$ mm and an almost vertical damper unit; the spring was angled to reduce the friction between the piston rod and rod guide. For reasons of space, the CV-joint centre Q had to be shifted inwards; the space for snow chains has to be considered (see Fig. 2.5b and position 10 in Fig. 1.39).

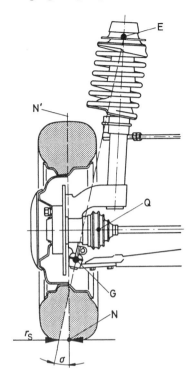

cients of friction. Reference is merely made to this difference. The vertical force $F_{z,W}$, which influences all righting moments and is sometimes called wheel force, is half the weighed front axle force $F_{V,f}$ determined in the design position (see Section 5.3.4), i.e. when there are three people each weighing 68 kg in the vehicle.

$$F_{z,W} = F_{V,f}/2 \text{ and } F_{V,f} = m_{V,f} g \text{ [kN]} \tag{3.21}$$

As can be seen, the level of the front axle load $m_{V,f}$ is also a consideration here and so we sometimes speak of 'weight righting'.

Using $F_{z,W}$ we can obtain:

the lateral force	$F_{y,W} = \mu_{y,W} F_{z,W}$
the rolling resistance force	$F_R = k_R F_{z,W}$, sometimes also
the tractive force	$F_a = \mu_{x,W} F_{z,W}$ (see Equations 6.36 and 6.37a)
the braking force	$F_b = \mu_{x,W} F_{z,W}$

The values for $\mu_{y,W}$, k_R and $\mu_{x,W}$ are given in Sections 2.8.3, 2.6.1 and 2.7, and Section 3.10.3 contains a summary of all righting moments.

The opinion still sometimes expressed that the steering is righted by the vehicle front end lifting when the wheels are turned would only apply at zero caster. As Fig. 3.118 shows, at $\tau = 0°$ the body lifts on both wheels (ΔH), but

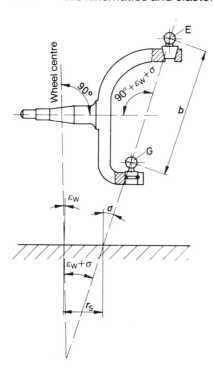

Fig. 3.80 The precise position of the steering axis – also known as kingpin inclination axis – can only be determined if the centre points E and G of the two ball joints are known. The total angle of kingpin inclination and camber ($\sigma + \epsilon_W$) must also be included when dimensioning the steering knuckle as an individual part.

if there is caster, the wheel on the outside of the bend moves upwards, the most highly loaded side of the body sinks and instead of righting it, the weight would turn the steering further. However, the less loaded side on the inside of the bend lifts.

3.9   Kingpin inclination and kingpin offset at ground

3.9.1 Relationship between kingpin inclination and kingpin offset at ground (scrub radius)

In accordance with the standards ISO 8855 and DIN 70 000, the kingpin inclination is the angle σ which arises between the steering axis $\overline{EG}$ and a vertical to the road (Figs 3.79 and 3.80). The kingpin offset is the horizontal distance r_S from the steering axis to the intersecting point of line $\overline{N'N}$ in the wheel centre plane with the road. Values on present passenger cars are:

$\sigma = 11°$ to $15°30'$ and
$r_S = -18$ mm to $+ 20$ mm

As shown in Fig. 2.5b, r_S can also depend on the tyre width.

Larger kingpin inclination angles are necessary to give the vehicle a small or negative kingpin offset. In commercial vehicles, tractors and building-site lorries, the inclination of the kingpin is often equivalent to the angle σ (Fig. 1.1a), whilst the wheels are controlled by ball joints on the front axles of passenger cars. On double wishbones suspensions, the steering axis therefore goes through the centres of the ball sockets E and G indicated (Figs 1.23, 3.80 and 3.92); the engineering detail drawing must show the total angle of camber and kingpin inclination.

The McPherson strut and strut damper have a greater effective distance between the lower ball joint G and the upper mounting point E in the wheel house (Figs 1.4 and 3.79); however the upper axle parts are next to the wheel, so attention should be paid to creating enough clearance for the rotating tyre (possibly for snow chains). Resulting from this measure, a higher inclination of the steering axis and a higher angle σ have to be accepted. In addition, as can be seen in the illustrations, point G has been shifted to the wheel to obtain a negative kingpin offset. The steering axis then no longer matches the centre line of the suspension strut (Figs 1.4, 3.23 and 3.80a).

Due to the relationship between camber and kingpin inclination shown in Fig. 3.80, the angle σ does not need to be toleranced on double wishbone suspensions. The permissible deviations on the overall angle $\epsilon_w + \sigma$ are given in the detailed drawing of the steering knuckle. If the camber has been set correctly on this type of suspension, the kingpin inclination angle will also be correct. However, the important thing is (as specified in the camber tolerance) that the deviation between left and right does not exceed 30', otherwise the steering could pull to one side if the caster angle τ on the left and right-hand sides differ (see Section 3.10.7).

On McPherson struts and strut dampers, the steering knuckle is usually bolted to the damping unit (Figs 1.39 and 5.39). In this case there may be play between the bolts and holes or the position may even be used for setting the camber (Fig. 3.80a). In this case it is sensible to tolerance the kingpin inclination angle because, provided the camber is correct, the kingpin inclination does not have to be.

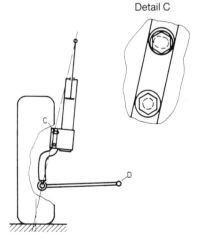

Detail C

Fig. 3.80a Camber can be set at the bracket between steering knuckle and suspension strut using an excenter on the upper bolt C; the lower screw is then used as a pivot. The kingpin inclination, which is important for driving behaviour, cannot be corrected in such cases. The steering axis entered here does not lie in the damper centre line.

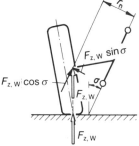

Fig. 3.81 For static observation, the vertical force $F_{z,W}$ must be shifted to the wheel axis and resolved into its components. The distance to the steering axis is equivalent to the vertical force lever r_n, the size of which depends on the kingpin offset r_S and the angle σ.

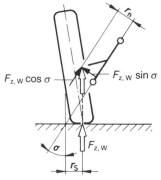

Fig. 3.81a The negative kingpin offset reduces the vertical force lever r_n. However, its length is one of the determining factors in the self-aligning torque $M_{S,z}$. To maintain its level, the kingpin inclination angle σ would have to be increased.

There is also a direct correlation between the alteration to camber and kingpin inclination when the wheels compress and rebound. As described in Section 3.5.2, the aim is to make the compressing wheel go into negative camber, as this leads to small changes in camber at body roll, but an increase in kingpin inclination by the same angle. Strictly speaking, the calculation by drawing of the camber alteration, shown and described in Figs 3.41 to 3.43, relates to the kingpin inclination, and for this reason the angle alteration $\Delta\sigma$ is also entered.

To obtain the self-aligning torque $M_{S,z}$, which is important for righting, the vertical force $F_{z,W}$, which is always present on the centre of tyre contact, must, for static consideration, be shifted up to the wheel axis and resolved there in the direction of the steering axis;

$F_{z,W} \cos \sigma$ and, vertical to it, $F_{z,W} \sin \sigma$ (Figs 3.81 and 3.81a).

The vertical force lever r_n at the resolution point is:

$$r_n = (r_S + r_{dyn} \tan\sigma) \cos \sigma \qquad (3.21a)$$

The equation will apply provided that $\cos \epsilon_W = 1$, a condition that applies to normal camber angles. If the vehicle has caster, the force components $F_{z,W} \sin \sigma$ must be further resolved by the angle τ (see Equation 3.33). r_{dyn} can be calculated using Equation 2.2.

When the wheels are turned, the force $F_{z,W} \sin \sigma$ is at the angle δ to the wheel axis (Fig. 3.82) and the component $F_{z,W} \sin \sigma \sin \delta$ will, with smaller steering angles, give the approximate righting moment based on the whole axle:

$$M_{S,z} = F_{V,f} \sin \sigma \sin \delta_m \, r_n \qquad (3.22)$$

($F_{V,f}$ see Equation 3.21 and δ_m Equation 3.17)

Fig. 3.82 When the wheels are turned by the angle δ, the vertical force component $F_{z,W} \sin \sigma$ gives the self-aligning torque $M_{S,z}$; the extent of this weight related self-alignment depends on the kingpin inclination angle σ, the lever r_n, the front axle load $m_{V,f}$ and the caster (Fig. 3.113).

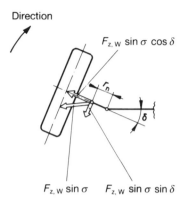

Direction

$F_{z,W} \sin \sigma \cos \delta$

$F_{z,W} \sin \sigma \qquad F_{z,W} \sin \sigma \sin \delta$

The exact solution has to take the changing kingpin inclination angle (due to lateral forces when the wheels are turned and due to the body roll) into account as well as the positive and negative caster that occurs (Figs 3.39, 3.44, 3.46a and 3.102). The influence of the paths $r_{\tau,T}$ and r_T in the tyre contact area (Figs 3.91a and 3.92) also has to be considered. Both can have a significant influence on the size of M_{Sz} during cornering. On the outside of the bend, r_T reduces the kingpin offset (or causes it to become more negative) while on the inside of the bend, it increases or becomes less negative (Fig. 3.98).

There is also a load alteration during cornering, whereby $F_{z,W,f,o} > F_{z,W,f,i}$ and also δ_i and δ_o are not always of the same size, so that different moments always occur on individual wheels. The kingpin offset r_S, which appears in the equations, influences the level of the self-aligning torque $M_{S,z}$; if this offset is large, the righting increases, if r_S decreases or even becomes negative (owing to the shorter lever r_n), the moment reduces (Fig. 3.81a).

The more $M_{S,z}$ increases, the more the front axle becomes longitudinally sensitive. There is, therefore, a clear tendency towards a small positive or negative kingpin offset.

If $M_{S,z}$ is to remain at the same level, the kingpin inclination angle has to be enlarged with the disadvantage that, when the wheels are turned, the wheel on the outside of the bend goes in the more positive camber direction, which makes more space necessary because the brake disc has to be shifted into the disc wheel (Figs 3.79 and 2.14). With a given path r_{S1} the necessary angle σ_1 can be calculated from the existing values r_S (in mm) and σ_O:

$$\tan \sigma_1 = \frac{r_{S1}}{2B} + \sqrt{\left[\frac{r_{S1}}{2B} \right]} + A/B \qquad (3.23)$$

It is:

$$A = (r_S + r_{dyn} \tan \sigma_O) \sin \sigma_O \cos \sigma_O$$
$$B = r_{dyn} - A$$

The dynamic tyre radius r_{dyn} can be determined from the rolling circumference C_R (or $C_{R,dyn}$, see Section 2.2.8 and equation 2.2)

$$r_{\text{dyn}} = C_R/(2\pi) \qquad (3.24)$$

Taking as an example the old Ford Granada with the tyre size 185 R 14 90 S, which has a rolling circumference of 1965 mm; the axle settings were: $\sigma_O = 5°54'$ and $r_S = 73$ mm.

The aim is to find the kingpin inclination angle σ_1 with a negative kingpin offset $r_{S1} = -18$ mm:

$r_{\text{dyn}} = 1965/2\pi = 313$ mm
$A = (+73 + 313 \tan 5°54') \sin 5°54' \cos 5°54'$
$A = 11$ mm: $B = 311 - 11 = 302$ mm

$$\tan \sigma_1 = \left(\frac{-18}{2 \times 302}\right) + \sqrt{\left[\frac{-18}{2 \times 302}\right]} + \frac{11}{302} = 0.0298 + 0.1912$$

$\tan \sigma_1 = 0.211; \sigma_1 = 12.46° = 12°28'$

The following would then appear on the drawing and in the workshop manual:

kingpin inclination 12°30'

a normal value for a negative kingpin offset. r_{S1} can be more easily calculated as a function of the amended kingpin inclination angle σ_1:

$$r_{S1} = \frac{A_T}{\sin \sigma_1 \cos \sigma_1} - r_{\text{dyn}} \tan \sigma_1 \qquad (3.25)$$

3.9.2 Braking force lever

During a braking process carried out with the brake mounted on the steering knuckle or wheel carrier, the braking force F_b tries to turn the wheel with the brake force lever (Fig. 3.83)

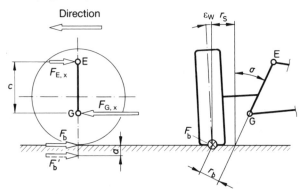

Fig. 3.83   The braking force F_b has the lever $r_b = r_S \cos \sigma$ to the steering axis $\overline{EG}$; shifted vertically on this axis, F_b acts by the amount a below ground and causes the greatest force in point G: $F_{G,x} = F_b + F_{E,x}$ (see also Fig. 3.120). When there is caster, F_b must be resolved at the centre of tyre contact around the angle τ (Fig. 3.88).

Fig. 3.84 If the brake is in the wheel, the braking force F_b causes the moment $M_{S,b} = F_b\, r_b$, which tries to push the wheel into toe-out and causes the tie rod force F_T. The steering axis is assumed to be vertical to simplify the calculation.

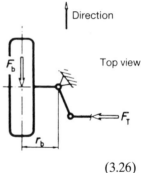

$$r_b = r_S \cos \sigma \tag{3.26}$$

around the steering axis, i.e. the moment

$$M_{S,b} = F_b \cos \tau\, r_b \tag{3.26a}$$

is generated, which, as Fig. 3.84 shows, results in the tie rod force F_T and, where r_S is positive, pushes the wheel into toe-out (for caster angle τ see Fig. 3.88).

The longer the path r_S, the more the moment $M_{S,b}$ increases and the larger the influence of uneven front brake forces on the steering – which is the reason for keeping r_S as low as possible or even making it negative (Figs 3.79 and 3.81a). Thus, as shown in Fig. 6.12, brakes that do not respond equally cause a counter-steering effect, which can reduce or eliminate the yaw response of the body, which is also true for an elastokinematic toe-in alteration (Figs 3.0a and 3.64). The longitudinal force F_b arising on the ground produces the reaction forces $F_{E,x}$ and $F_{G,x}$ in the pivot points of the steering knuckle. In order to be able to determine their size, F_b must be shifted towards the braking force lever on the extension of the steering axis $\overline{EG}$. Therefore, with positive kingpin offset F_b lies below the ground by the amount a and is shown in the side view of Fig. 3.83 as F'_b:

$$a = r_b \sin \sigma = +r_S \cos \sigma \sin \sigma \tag{3.27}$$

If r_S is negative, F'_b moves above ground (Fig. 3.121) and $F_{G,x}$ becomes smaller.

If the brake is on the inside on the differential, the braking moment is transmitted via the universal joints to the engine and causes the bearing reaction forces ΔF_z in the engine mounting (Fig. 3.85):

$$\pm \Delta F_z = (F_b\, r_{dyn})/c$$

The smaller the wheel (r_{dyn}) and the larger the effective distance c, the lower the forces and therefore also the compliance in the rubber mountings. The braking force F_b which occurs at the centre of tyre contact must, in such cases, be shifted to the centre of the wheel (like the rolling resistance force F'_R in Fig. 3.86), because a shaft bearing can only transfer forces, and not

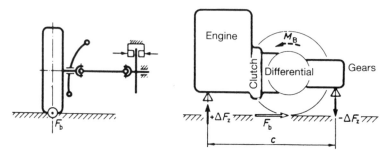

Fig. 3.85 If a front-wheel drive vehicle has an inside brake, the engine mounting must absorb not only the drive-off moment, but also the braking moment; the reaction forces $\pm\Delta F_z$, the size of which depends on the effective distance C, occur in the rubber buffers.

moments, in its effective direction. Just like F'_R, F'_b acts on the longitudinal force lever r_a, also known as the disturbance and traction force lever:

$$r_a = r_S \cos \sigma + r_{dyn} \sin (\sigma + \epsilon_W) \tag{3.28}$$

With it, F_b causes the moment:

$$M_{S,b} = F_b \cos \tau\, r_a \tag{3.28a}$$

which also occurs when $r_S = 0$. In the equations, note must be taken of the plus or minus signs; in the case of a negative kingpin offset, the first element $(-r_S \cos \sigma)$ must be subtracted from the second. Figure 3.88 contains $\cos \sigma$. Because $r_a > r_b$, there is a higher moment when the brake is on the inside at the differential, which has a more pronounced influence on the steering. The

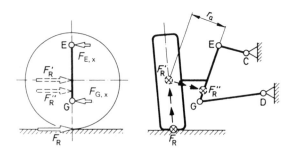

Fig. 3.86 When the wheel is rolling in a straight line, the rolling resistance force F_R must be observed as F'_R in the wheel centre; its distance to the steering axis is r_a. This so-called longitudinal force lever depends on the kingpin offset r_S and the smaller this can be, the further up F_R acts as F''_R on the steering axis and the more evenly points E and G are stressed in the longitudinal direction. The same static conditions apply to the braking force if the brake is located on the inside on the differential (see also Fig. 3.119a).

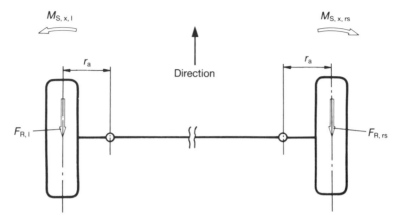

Fig. 3.86a The rolling resistance force F_R pushes the wheels backwards via the longitudinal force lever r_a, i.e. into toe-out $-\Delta_f$. A moment arises on both sides, which is absorbed and cancelled out at the tie rods. In the case of caster the angle τ must also be observed (see Fig. 3.88).

reaction force $F_{G,x}$ in the lower ball joint, however, becomes much smaller. To determine the forces $F_{E,x}$ and $F_{G,x}$, F'_b has to be shifted vertical to the steering axis and, in the side view, comes to lie below the wheel centre as F''_b (Fig. 3.119a) by:

$$a = r_a \sin \sigma \tag{3.28b}$$

3.9.3 Longitudinal force lever

Figure 3.86 shows the rolling resistance force F_R, which always occurs when the vehicle is moving. It generates the same moment left and right:

$$M_{S,x} = F_R \cos \tau \, r_a \tag{3.28c}$$

which is absorbed at the tie rods (Fig. 3.86a); any caster angle τ must be considered. If the moments are of the same size, the vehicle moves in a straight line, but if they are different it can pull to one side. Tyres that have a different rolling circumference (Fig. 2.10) or front axles where the angles $\sigma + \epsilon_W$ differ to the left and the right can be the reason for this. The factor $r_{dyn} \sin (\sigma + \epsilon_W)$ primarily determines the length of the lever r_a (see Equation 3.28). On a bend, the outer wheel experiences a force increase ($F_{z,W} + \Delta F_{z,W}$, Fig. 1.2) and the inner one a reduction equivalent to $F_{R,o} > F_{R,i}$. Where there is no caster, the wheel on the outside of the bend rights itself more than the one on the inside is trying to turn into the bend.

The tractive forces F_a, which occur at the contact points of the front wheels on front-wheel drive vehicles, cause moments acting in the opposite direction, but also have to be taken into consideration in the centre of the wheel

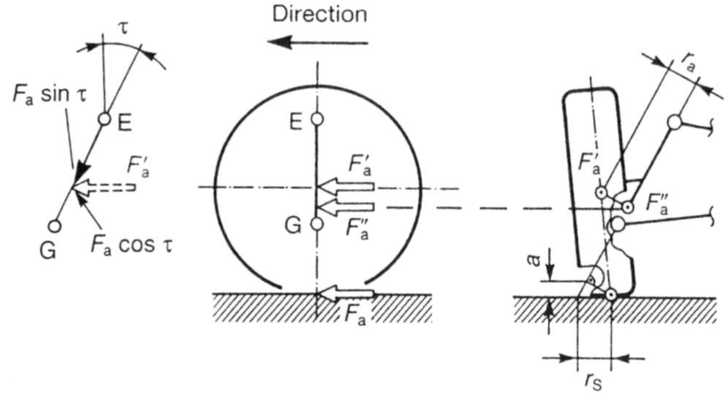

Fig. 3.87 The negative kingpin offset on the ground favourably shortens the longitudinal force lever r_a. The tractive force F_a relating to one wheel must be resolved around the angle τ in the wheel centre in the case of caster.

(Fig. 3.87), i.e. in vehicles of this design, a smaller longitudinal force lever r_a will be particularly important. Citroën has achieved this by shifting the ball joints E and G in to the wheel centre plane (Fig. 3.87a). This means that:

$$\epsilon_W + \sigma = 0, \ r_S = 0 \ \text{and therefore also} \ r_a = 0$$

The longitudinal force lever should be as short as possible. Comparison with the formula for the vertical force lever r_n (Equation 3.21a) shows the difficulties:

$$r_n = r_S \cos \sigma + r_{dyn} \sin \sigma$$
$$r_a = r_S \cos \sigma + r_{dyn} \sin(\sigma + \epsilon_W)$$

If, for example, the camber is $\epsilon_W = 0°$, $r_n = r_a$ if there is only a small or no caster angle τ and the vehicle moves unimpaired in a straight line.

However, during cornering, r_n changes significantly whilst r_a remains virtually unchanged (Fig. 1.2).

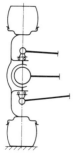

Fig. 3.87a Section through the centre axle steering of the model GSA, which Citroën no longer builds; guide and supporting joints are in the wheel centre plane.

3.9.4 Alteration to the kingpin offset

To improve cornering behaviour, disc wheels with lower wheel offsets e are sometimes used (Fig. 2.14) or (in the past) spacer rings were laid between the wheel and brake disc to give the advantage of a slightly wider tread width (around 2 to 4%), but with the disadvantage of up to 100% greater kingpin offset at ground. The result is a more noticeable disturbance effect on the steering when the road is uneven and particularly when the front brakes do not pull evenly.

If, as on almost all passenger cars with negative kingpin offset, the two brake circuits are designed to be diagonal, these measures cannot be implemented. The negative kingpin offset would either change from negative to positive or become too positive when there is an elastokinematic toe-in alteration on the front axle (Fig. 3.64), and toe-out would occur during braking instead of toe-in.

Figure 2.5b shows the alteration of r_S depending on the tyre width, using the example of the VW Golf III.

3.10 Caster

3.10.1 Caster trail and angle

We differentiate between the kinematic caster trail $r_{\tau,k}$ of the wheel, the caster angle τ, the caster offset $r_{\tau,W}$, the tyre caster $r_{\tau,T}$ the lateral force lever $n_{\tau,k}$ and the elastokinematic caster $r_{\tau,e}$.

In accordance with the standards ISO 8855 and DIN 70 000, τ is the angle between the steering axis $\overline{EG}$ projected onto an xz-plane and a vertical, drawn through the wheel centre (Case 1, Fig. 3.88), and $r_{\tau,k}$ the distance between the points K and N on the ground. The castering of the wheel centre of contact N behind the intersection K can also be achieved by shifting the

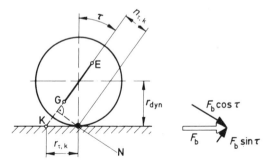

Fig. 3.88 If the extension of the steering axis goes through the ground at point K in front of the wheel centre, the distance arising is the kinematic caster trail $r_{\tau,k}$ (Case 1). A vertical to $\overline{EG}$, drawn through the centre of tyre contact N, when projected onto the xz-plane, gives the lateral force lever $n_{\tau,k}$ (Equation 3.30).

Longitudinal forces which arise, such as the braking force F_b (or the rolling resistance F_R), must be resolved at the centre of tyre contact (or as $F3'_R$ in the wheel centre, Fig. 3.86) by the angle τ.

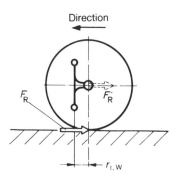

Fig. 3.89 Caster can also be achieved by shifting the wheel centre behind the steering axis (Case 2); if this is vertical, as shown, the (here) positive caster offset is equal to the lever: $r_{\tau,w} = r_{\tau,k} = +n_{\tau,k}$. Rolling resistance forces F_R acting at the centre of tyre contact must be observed as F'_R in the wheel centre.

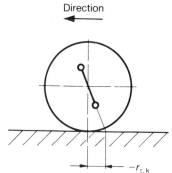

Fig. 3.90 Caster (Case 3): a steering axis, which is inclined opposite by the angle $-\tau$, results in negative caster $-r_{\tau,k}$, and the associated disadvantage of a more positive camber on the outside of the bend when the wheels are turned. However, where the angles $-\tau$ are small, the tyre caster $r_{\tau,T}$ balances out the negative caster trail (Fig. 3.93). On the independent rear wheel suspensions the steering knuckle (here not the steering axis) can be given negative caster to achieve lateral force understeering (see Figs 3.111a and 3.111b).

axes of rotation in front of the wheel centre: $+ r_{\tau,k}$, (Case 2 Fig. 3.89). On some front-wheel drive vehicles, owing to the increased self-righting moment caused by tractive forces, negative caster was designed, which can be achieved with a reversed angled steering axis (Case 3 Fig. 3.90) or by positioning an axis $\overline{EG}$ behind the wheel centre and inclining it by the angle τ, which leads to negative caster offset $-\Delta r_{\tau,w}$ (as can be seen in Figs 3.91 and 3.107). For the following reasons, linking a positive caster angle and $-r_{\tau,w}$ is popular with designers:

- the kinematic caster trail $r_{\tau,k}$ is smaller, i.e. the influence on the steering resulting from uneven road surfaces reduces

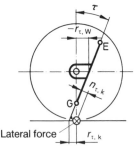

Fig. 3.91 Front axle properties can be improved by a negative caster offset $r_{\tau,w}$; the caster trail $r_{\tau,k}$ on the ground shortens by this amount and the camber alteration when the wheels are turned becomes more favourable.

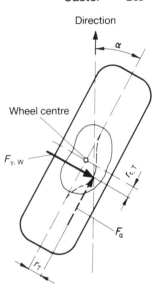

Direction

Fig. 3.91a The tyre contact area (also known as the 'tyre print', Fig. 3.26) of a tyre rolling at an angle under the influence of lateral forces deforms in the shape of a kidney; this means the point of application of the vertical force $F_{z,W}$ and the lateral force $F_{y,W}$ moves by the trail $r_{\tau,T}$ – the tyre caster – behind the wheel centre and the tyre self-aligning torque $M_{T,f}$ = $F_{y,W} \times r_{\tau,T}$ occurs. If the vehicle has front-wheel drive, F_a acts at a point in the tyre contact area, offset by r_T from the wheel centre plane, as does the rolling resistance force $F_{R,co}$ on a bend. Tyre caster is between $r_{\tau,T}$ = 10 mm and 40 mm; lateral offset is $r_T \sim$ 3 mm per $\mu_{y,W}$ = 0.1. (See Section 2.10.2 and Figs 3.98 and 3.99.)

If the slip angle α is specified instead of the coefficient of friction $\mu_{y,W}$, Equation 2.4c will apply.

- the camber alteration is increased when the wheels are turned (Fig. 3.102).

The trail $r_{\tau,k}$ and the lever $n_{\tau,k}$ of the lateral force (i.e. the path projected onto the vertical plane xz) both with and without negative offset $-r_{\tau,W}$ can be easily determined using the dynamic rolling radius r_{dyn} (see also Section 3.10.7):

$$r_{\tau,k} = r_{dyn} \tan \tau \tag{3.29}$$
$$r_{\tau,k} = r_{dyn} \tan \tau - r_{\tau,W}$$
$$n_{\tau,k} = r_{dyn} \sin \tau \tag{3.30}$$
$$n_{\tau,k} = r_{dyn} \sin \tau - r_{\tau,W} \cos \tau$$

During a bend, the area of tyre contact deforms due to the slip angle α (Fig. 3.91a). The lateral force $F_{y,W}$ therefore acts offset by the amount $r_{\tau,T}$ – known as tyre caster – behind the wheel centre (Figs 3.92 and 2.36). The tyre caster of practically $r_{\tau,T}$ = 10 mm to 40 mm must therefore be included in all static and elastokinematic observations. Without and with caster offset the overall path $r_{\tau,t}$ is then as follows (Fig. 3.93):

$$r_{\tau,t} = r_{\tau,k} + r_{\tau,T} \tag{3.31}$$

$$n_{\tau,t} = n_{\tau,k} + r_{\tau,T} \cos \tau \tag{3.32}$$

$$n_{\tau,t} = n_{\tau,k} + (r_{\tau,T} - r_{\tau,W}) \cos \tau \tag{3.32a}$$

If precise calculations are required, the elastokinematic caster $r_{\tau,e}$ must be

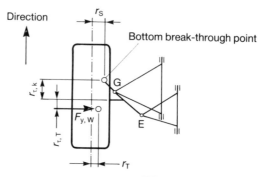

Fig. 3.92 The extension of the steering axis $\overline{EG}$, which is 3-dimensionally at an angle due to kingpin inclination and caster, penetrates the ground in front of the wheel centre and gives (in the example) the positive kingpin offset r_S and the kinematic caster trail $r_{\tau,k}$. On a bend, the lateral force acts offset by the tyre caster $r_{\tau,T}$ in the tyre contact area. The total trail (index t) is therefore $r_{\tau,t} = r_{\tau,k} + r_{\tau,T}$ and – in accordance with Fig. 3.91a – the kingpin offset (overall on the outside of the bend) $r_{S,t} = r_S - r_T$.

used instead of $r_{\tau,k}$, although this can only be determined by experiment on the vehicle.

In order to make them clearer, the path $r_{\tau,T}$ is not shown in some of the following figures.

3.10.2 Caster and straight running

Caster can be compared with the tea trolley effect, where the pulled wheel takes on the direction of pull and the wheel centre adopts a position behind the axis of rotation 1 (Fig. 3.93a). The tensile force and the opposed force F_R generated by the rolling resistance are on an effective line, in other words in a stable ratio to one another because the guiding and wheel axis lie behind one another. The same effect also exists (despite kingpin offset and kingpin inclination) on the wheels of a vehicle if these can be rotated around axes. The wheels are set to caster on both sides and are linked by tie rods.

If unevenness in the road surface or a steering input pushes the wheels out

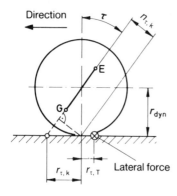

Fig. 3.93 Due to the tyre caster $r_{\tau,T}$, which is always present during cornering, the lateral force lever is extended and becomes $n_{\tau,k,t} = n_{\tau,k} + r_{\tau,T} \cos \tau$.

Fig. 3.93a If the rolling resistance force F_R acts behind the steering axis 1, the wheel follows in a stable manner in the direction in which it is pulled.

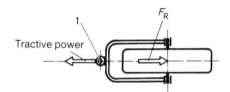

from the straight ahead direction, by the angle δ (as shown in Fig. 3.94), the rolling resistance components $F_R \sin δ$ move both wheels back via the force lever $n_{\tau,k}$ (or the overall lateral force lever $n_{\tau,t}$) until they roll in a straight line again. The components $F_R \cos δ$ (left and right) compensate, and only subject the tie rods to pressure. Negative caster on the wheels could have the opposite effect and the vehicle would become unstable.

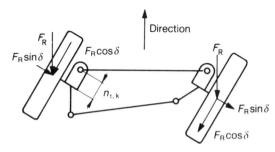

Fig. 3.94 When the vehicle is travelling in a straight line, caster has a stabilizing effect. Fig. 3.113 shows the necessary further resolution of the force components by the angle τ.

On a vehicle moving in a straight line, caster would not only have advantages but also disadvantages. Uneven road surfaces cause alternating lateral forces on the centres of tyre contact and these, together with the lever in $n_{\tau,k}$ (or $n_{\tau,t}$, Equations 3.30 to 3.32a) cause moments around the steering axis (Fig. 3.95), which are supported on the tie rods and can cause steering disturbances and vibrations. Furthermore, there is increased wind sensitivity due to the fact that a wind force acting on the body (Fig. 3.96) causes lateral forces $F_{y,w}$ in the opposite direction, on the centres of tyre contact. In addition the

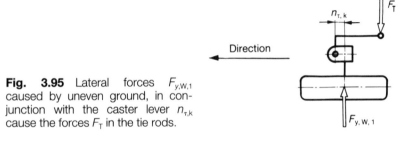

Fig. 3.95 Lateral forces $F_{y,W,1}$ caused by uneven ground, in conjunction with the caster lever $n_{\tau,k}$ cause the forces F_T in the tie rods.

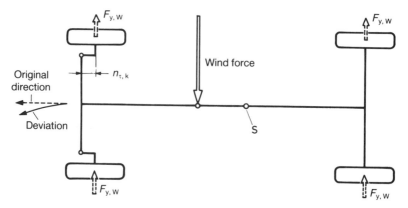

Fig. 3.96 Caster can increase the wind-sensitivity of a vehicle. The point at which the wind acts is usually in front of the centre of gravity S; a moment arises which seeks to turn the vehicle and causes the wheels to steer in the same direction.

front forces $F_{y,W,f}$ together with the caster lever $n_{\tau,k}$ (or $n_{\tau,t}$) result in moments that turn the vehicle in the direction of the wind, i.e. further in the direction in which the body is already being pushed by the wind.

3.10.3 Righting moments during cornering

The alteration to the caster and kingpin inclination (or camber) angle, which is influenced by the body roll inclination (Figs 3.44 and 3.111) and is caused by the steering angle of the wheels (Figs 3.102 and 3.104), results in an alteration to all levers on which vertical, lateral and longitudinal forces are acting. Observation of each individual wheel would mean looking at these very complicated kinematic relationships and errors would almost be inevitable because of the additional elastokinematic movements. Determination of the righting moments based on the whole axle and on the position of the vehicle parallel to the ground is – particularly in the case of small steering angles and low cornering speeds – sufficiently precise. The caster $r_{\tau,T} = 10$ mm to 40 mm (Fig. 3.92) (not present when the vehicle is moving in a straight line) must, nevertheless, be included in the equation:

$$M_{S,z} = F_{V,f} \sin \sigma \cos \tau \sin \delta_m \, r_n \tag{3.33}$$

The caster angle in the lever $n_{\tau,k}$ is considered in the case of the righting moment due to lateral force. Here, the kingpin inclination must also be included in the calculation (Fig. 3.97).

$$F_{y,W,f,o} + F_{y,W,f,i} = \mu_{y,W} \, F_{V,f} = F_{y,W,f,t} \tag{3.33a}$$

$F_{y,W,f,t}$ acts around the lateral force lever $n_{\tau,t}$ offset behind the wheel centres (Figs 3.92 and 3.98 and Equations 3.32 and 3.32a):

$$M_{S,y} = F_{y,W,f,t} \cos \sigma \, n_{\tau,t} \tag{3.34}$$

Fig. 3.97 The lateral forces acting on the centres of tyre contact of the front wheels must be resolved in the direction of the steering axis and vertical to it, shown here for the wheel on the outside of the bend. $F_{y,W,f,o}$ cos σ then has a righting effect and $F_{y,W,f,o}$ sin σ strengthens the vertical force $F_{z,W,f,o}$.

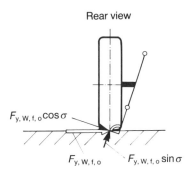

Rear view

$F_{y,W,f,o}\cos\sigma$

$F_{y,W,f,o}$ $F_{y,W,f,o}\sin\sigma$

Section 2.8.4. describes the coefficient of friction $\mu_{y,W}$.

If the axis of rotation is vertical in the side view, Case 2 ($\tau = 0$, Fig. 3.89), the formulas remain unchanged; only $r_{\tau,W} + r_{\tau,T}$ appears for $n_{\tau,t}$, i.e. the path around which the wheel centre is located behind the steering axis, together with the tyre caster.

If the vehicle has negative caster (Case 3, Fig. 3.90), the lateral force could cause the wheels to steer into the bend if this were not counterbalanced by caster $r_{\tau,T}$ and by the tractive force F_a which also has a righting effect on front-wheel drive vehicles (Figs 3.91a and 3.100). In the case of negative caster, only $r_{\tau,T} - r_{\tau,k}$ needs to be inserted into the equation.

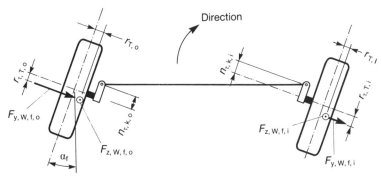

Direction

Fig. 3.98 On wheels which roll at an angle of α_f, the lateral cornering forces $F_{y,W,f}$ act behind the wheel centres, offset by the tyre caster $r_{\tau,T}$ and push the centres of tyre contact (and therefore also the vertical forces $F_{z,W,f}$, Fig. 3.91a) to the bend centre by the trail r_T. The marked forces and paths are of different sizes on the outside (o) and inside (i) of the bend:

$$F_{z,W,f,o} = F_{z,W} + \Delta F_{z,W} \text{ and } F_{z,W,f,i} = F_{z,W} - \Delta F_{z,W}$$
$$F_{y,W,f,o} = \mu_{y,W} F_{z,W,f,o} \text{ and } F_{y,W,f,i} = \mu_{y,W} F_{z,W,i}$$

The kinematic caster $r_{\tau,k}$ has been ignored here and the steering axis is shown vertically for reasons of simplification.

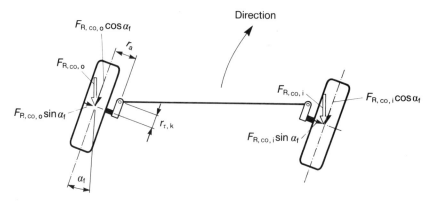

Fig. 3.99 The rolling resistance forces $F_{R,co,o}$ and $F_{R,co,i}$, which have increased on a bend due to the tyre slip, must be resolved by the angle α_f; the component $F_{R,co} \cos \alpha_f$ then appears in the wheel centre with the lever r_a. The greater α_f, and the longer the caster trail $r_{\tau,k}$, the stronger the self-righting due to $F_{R,co} \sin \alpha_f$.

For reasons of clarity, the tyre caster $r_{\tau,T}$ and lateral offset r_T (Fig. 3.91a) have been ignored here and the steering axis has been shown vertically.

The increased rolling resistance force $F_{R,co,f}$ during cornering on the outside and the inside is:

$$F_{R,co,o} = k_{R,co} \, F_{z,W,f,o}$$
$$F_{R,co,i} = k_{R,co} \, F_{z,W,f,i}$$

and seen together this must be resolved into the two components:

$$F_{R,co} \cos \alpha_f$$
$$F_{R,co} \sin \alpha_f$$

In the case of positive caster, Case 1, the last component has a righting effect on both wheels (Fig. 3.99 and Equations 3.32 and 3.32a):

$$M_{S,x_1} = k_{R,co} \, F_{V,f} \sin \sigma \sin \alpha_f \, n_{\tau,t} \left[\text{or} \, (r_{\tau,W} + r_{\tau,T}) \right] \tag{3.35}$$

The rolling resistance force is (in accordance with the transfer of wheel forces $\Delta F_{z,W,f}$ during cornering, Fig. 1.2) greater on the outside of the bend than on the inside, so that the difference in vertical force $\Delta F_{z,W,f}$, together with $\cos \alpha_f$, can be a factor.

$$F_{z,W,f,o} - F_{z,W,f,i} = 2 \, \Delta F_{z,W,f} \tag{3.35a}$$

$$M_{S,x_2} = k_{R,co} \, 2 \, \Delta F_{z,W,f} \cos \alpha_f \cos \tau \, (r_a - 2 \, r_T) \tag{3.35b}$$

This deals with the longitudinal forces that act at the wheel centres – shifted

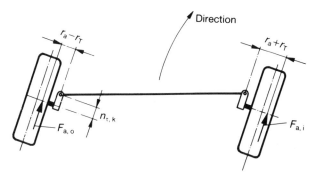

Fig. 3.100 At $r_a + r_T$, the tractive force $F_{a,i}$ on the inside of the bend has a larger lever than that on the outside of the bend $F_{a,o}$ at r_a-r_T; the steering axis is shown vertically, for reasons of simplification.

to the middle of the bend (Figs 3.86 and 3.91a); it is possible to calculate the coefficient of rolling resistance $k_{R,co}$ required using Equations 2.4a to 2.4c.

The previous figures also refer to the tractive forces F_a on front-wheel drive vehicles (related to one wheel). These must be resolved first in the wheel centre by the angle τ (Fig. 3.87) and considered offset by r_T, in the rolling direction of the wheel. Provided that the differential spreads the moment equally to both front wheels when the wheels are turned as the wheel load changes $\pm\Delta F_{z,w}$, the tractive force component $F_A \cos \tau$ (relating to the entire axle) would cause the following moments (Fig. 3.100):

$$M_{S,A} = -F_a \cos \tau(r_a - r_T) + F_a \cos \tau \, (r_a + r_T) = F_a \cos \tau \, 2 \, r_T$$

$$M_{S,A} = F_A \cos \tau \, r_T \tag{3.36}$$

The size of the force F_A depends either on the coefficient of friction $\mu_{x,w}$ ($F_A = \mu_{x,w} \, F_{V,f}$, see Equation 2.5) or on the drive torque (see Equation 6.36); the lateral offset length r_T is contained in the caption to Fig. 3.91a (see also Section 2.10.3.4).

3.10.4 Kingpin inclination, camber and caster alteration on steering input

Due to the spatial movement of the steering axis (onto which the vertical force $F_{z,w}$ must be shifted, see Figs 3.81 and 3.82) the righting moment for one wheel can only be calculated precisely if the kingpin inclination, when the wheels are turned, is taken into account. If, in the zero position, the steering axis is inclined exclusively by the angle σ_0, i.e. there is either no caster or this has been achieved by shifting the wheel centre (Fig. 3.89), then it is easy to determine the kingpin inclination angle $\sigma_{o \text{ or } i}$ which becomes smaller in both input directions:

outside of bend: $\tan \sigma_o = \tan \sigma_0 \cos \delta_o$ (3.37)
inside of bend: $\tan \sigma_i = \tan \sigma_0 \cos \delta_i$ (3.37a)

As shown in Fig. 3.80, kingpin inclination and camber are directly related, i.e. if either one changes the other one must change too. This means that the camber values $\epsilon_{W,o \text{ or } i}$, adopted by the wheel on the outside and the one on the inside of the bend when the wheels are turned, can be determined simultaneously:

$$\epsilon_{W,o} = (\sigma_0 + \epsilon_{W,0}) - \sigma_o \text{ and } \epsilon_{W,i} = (\sigma_0 + \epsilon_{W,0}) - \sigma_i \quad (3.38)$$

σ_0 and $\epsilon_{W,0}$ are the angles prevailing when the wheels point straight ahead in the design load or the particular load position (this applies equally to τ_0). If the steering axis is also inclined by the positive caster angle τ_0, the two auxiliary angles σ' and δ' must first be calculated using σ_0 and τ_0;

$$\tan \delta' = \frac{\tan \tau_0}{\tan \sigma_0} \quad \text{and} \quad \tan \sigma' = \frac{\tan \tau_0}{\sin \delta'} \quad (3.39)$$

They can then be used to determine directly the angles $\delta_{o, \text{ or } i}$ on the wheels on the outside and inside of the bend:

outside of bend: $\tan \sigma_o = \tan \sigma' \cos (\delta' - \delta_o)$ (3.39a)
inside of bend: $\tan \sigma_i = \tan \sigma' \cos (\delta' - \delta_i)$ (3.39b)

Equation 3.38 again applies to the camber $\epsilon_{W,o}$. Using a passenger car with the following axle settings as an example:

$\epsilon_{W,0} = 15'$, $\sigma_0 = 9°53'$ and $\tau_0 = 10°4'$

This gives $\delta' = 45.54°$ and $\sigma' = 13.97°$ and, where δ_o and $\delta_i = 20°$, the values are as follows:

$\sigma_o = 12°39'$, $\epsilon_{W,o} = -2°31'$, $\sigma_i = 5°53'$ and $\epsilon_{W,i} = +4°15'$

The wheel on the outside of the bend therefore goes into negative camber on this vehicle, while the inner one goes into positive camber. This is not the same as the case of a front-wheel drive vehicle with the following axle settings (Fig. 3.101):

$\epsilon_{W,0} = +40'$, $\sigma_0 = 12°25'$ and $\tau_0 = +36'$

Owing to the front-wheel drive and the manual non-power-assisted steering, the vehicle has a minor caster and therefore positive camber on the turned front wheel on the outside of the bend.

 Mercedes Benz designed their passenger car with a non-driven strut damper front suspension to have negative caster offset $-r_{\tau,W}$ (Figs 1.25 and 3.91) and a large angle τ (Figs 1.25 and 3.91). Figure 3.101a shows the success

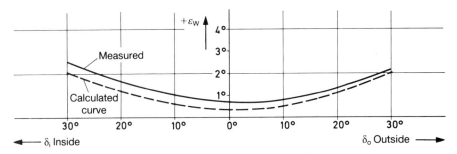

Fig. 3.101 Camber alteration measured and calculated as a function of the steering angle on a front-wheel drive vehicle. Due to the large kingpin inclination $\sigma_0 = 12°25'$, the wheels on both the inside and outside of the bend go into positive camber.

The values measured are higher than those calculated because the camber of the vehicle tested was in the plus tolerance. The calculation was made on the basis of the manufacturer's information ($\epsilon_w = 20'$ and $\tau = 0°$) and this accounts for the slightly different inclination of the curves.

of this design: the wheel on the outside of the bend goes into severe negative camber and the one on the inside of the bend goes favourably into positive camber.

For demonstration purposes, the camber alteration based on $\epsilon_{w,0} = 0°$, $\sigma_0 = 6°$ and various caster angles were calculated (Fig. 3.102); a larger kingpin inclination would only have resulted in a higher curvature for all curves. It

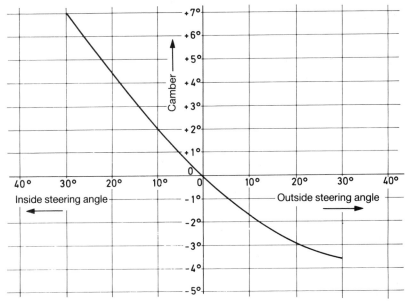

Fig. 3.101a Camber alteration measured on a Mercedes as a function of the steering angle. The axle settings in the design position were: $\epsilon_w = 0°$, $\sigma = 14°40'$, $\tau = 10°10'$, $r_s = -14$ mm and the negative caster offset $r_{\tau,w} = -28$ mm.

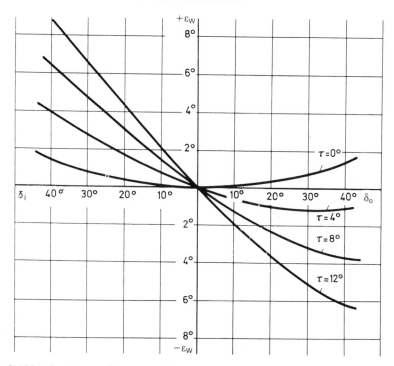

Fig. 3.102 Camber angle $\epsilon_{W,o}$ and $\epsilon_{W,i}$, as a function of the steering angle δ_a (outside of bend) and δ_i (inside of bend). The influence of the various caster angles τ can be clearly seen. Given was: $\sigma_0 = 6°$ and $\epsilon_{W,0} = 0°$

can be clearly seen how an increase in the angle τ_0 improves the lateral grip properties of the entire front axle as the wheel on the outside of the bend goes into more negative camber and the one on the inside into positive camber. Trail and caster angle alter in exactly the same way as kingpin inclination and camber alteration when the wheels are turned. In a passenger vehicle with rear-wheel drive for example, $r_{\tau,k}$ is 6.5 mm when the wheels are pointing straight ahead. The trail increases on the inside of the bend when the wheel is turned, with a decrease on the outer wheel (Fig. 3.102a). Negative caster occurs as $\delta_o \approx 8°$, and at $\delta_o = 30°$ it amounts to $r_{\tau,k} \approx -30$ mm, which would lead to the outer wheel turning into the bend under lateral force if there were no caster.

The caster angle alteration can be calculated just as simply as that of the kingpin inclination:

outside of the bend: $\tan \tau_o = \tan \sigma' \sin (\delta' - \delta_o)$ (3.40)

inside of the bend: $\tan \tau_i = \tan \sigma' \sin (\delta' - \delta_i)$ (3.40a)

If the vehicle has $\tau_0 \approx 0°$, only the kingpin inclination angle σ_0 plays a role thereby simplifying the formulas as follows:

outside of the bend: $\tan \tau_o = -\tan \sigma_0 \sin \delta_o$ $\qquad$ (3.41)

inside of the bend: $\tan \tau_i = +\tan \sigma_0 \sin \delta_i$ $\qquad$ (3.41a)

The equation shows that negative caster can occur on the wheel on the outside of the bend even with a small steering input; this is clearly demonstrated in Figs 3.103 to 3.104a, which show a comparison of curves calculated with various τ_0 and σ_0 angles and also one measured curve.

3.10.5 Caster alteration on front wheel travel

If there are two people seated in the front of a vehicle, the body jounces almost parallel and the caster hardly changes. However, if two or three people are seated in the back, or the boot at the back of the vehicle is loaded, it is a very different story. The rear axle springing complies more strongly than that of the front axle and the body's position, which was almost parallel to the ground, alters by $\Delta\theta = 1°$ to $2\frac{1}{2}°$ (Fig. 3.105). The caster angle increases by the same amount $\Delta\tau$; something which designers should bear in mind when specifying axle settings. The increase in the caster angle under load is probably the main reason why the steering is heavier on a fully laden vehicle even though (as shown in Fig. 5.5 and Fig. 5.6) the front axle load reduces. An alteration in caster has its disadvantages, as this in turn causes the self-right-

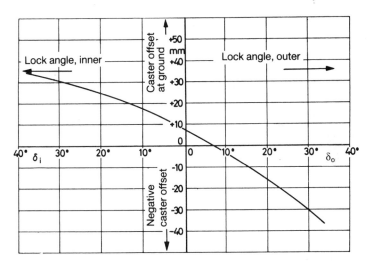

Fig. 3.102a   The length of a caster trail $r_{\tau,k}$ at the ground alters based on the steering input, shown using the example of a standard passenger car and the axle settings:

$\epsilon_W = +20'$, $\sigma = 11°5'$, $\tau = 8°20'$
$r_{\tau,W} = -32.5$ mm and $r_S = 56$ mm

The large kingpin inclination angle results in the deviation of the curve from the horizontal.

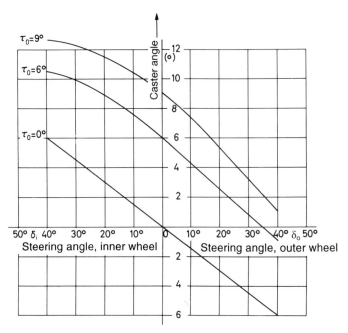

Fig. 3.103 Caster angles calculated as a function of $\sigma = 9°$ and $\tau_0 = 0°$, 6° and 9°. The smaller the τ_0 value in the normal position, the faster negative caster occurs on the wheel on the outside of the bend.

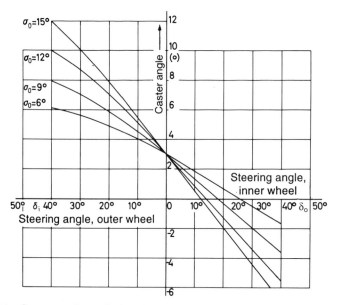

Fig. 3.104 Caster angles calculated as a function of the steering input with $\tau = 3°$ and $\sigma_0 = 6°$, 9°, 12° and 15°. The larger the kingpin inclination, the sooner the wheel on the outside of the bend goes into negative caster.

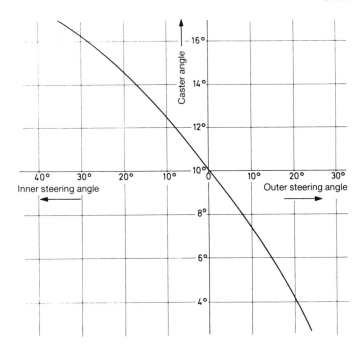

Fig. 3.104a Caster alteration measured as a function of the steering angle on the wheels of a Mercedes. The kingpin inclination angle of $\sigma = 14°40'$ is the determining factor for the severe curvature of the curve and the caster angle $\tau = 10°10'$ for the angle position.

ing torque to alter, but it is unavoidable if the brake dive on the front axle is to be kept within limits by means of vehicle pitch poles (see Section 6.3.2).

On double wishbone suspensions, the axes of rotation 1 and 2 of the two suspension control arms are usually parallel to one another (Fig. 3.106); in the standard configuration of the McPherson strut and strut damper there is a

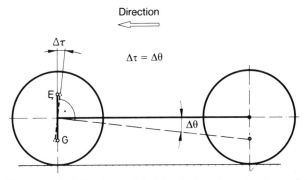

Fig. 3.105 When loaded, the body tail sinks further than the front; the caster angle $\Delta_{\tau 0}$ increases by its angle alteration $\Delta\theta$ (see also Fig. 6.15).

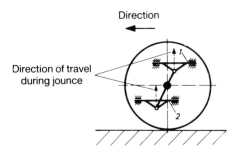

Fig. 3.106 On most double wishbone suspensions, the axes of rotation 1 and 2 are parallel to one another; in such cases, caster does not change when the wheels compress and rebound.

right angle between the centre line of the damping part and suspension control arm (Fig. 3.107). In such cases – regardless of the position of the compressed or rebounded wheel – the caster is retained. This is not the case where there are different angles between the suspension control arm axes of

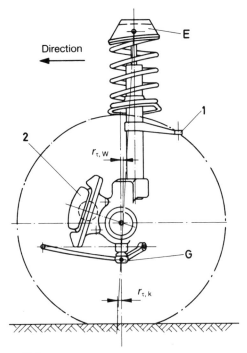

Fig. 3.107 If the line $\overline{EG}$ and control arm axis form a right angle on the McPherson strut and strut damper there is no caster alteration. Point G moves vertical to the suspension control arm axis when the wheels bottom out, i.e. parallel to the line $\overline{EG}$. The axle shown has a negative caster offset $-r_{\tau,w}$ and the lower link G shifted forwards. The line $\overline{EG}$ gives a small caster angle and the trail $r_{\tau,k}$.

The steering arm 1 is positioned high up and inclined backwards; the disc brake calliper 2 is at the front, giving the disadvantage of a higher wheel bearing load during braking.

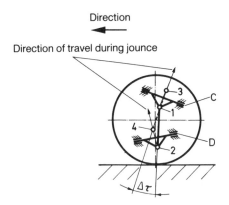

Fig. 3.108   To create a pitch pole on the front axle (see Fig. 3.120) on double wish-bone suspensions, the axes of rotation C and D must be inclined against one another. The disadvantage of this is that when the wheels compress, point 1 moves to point 3 and point 2 to 4, increasing the caster angle by Δτ, equivalent to twisting the steering knuckle by this angle.

rotation (Fig. 3.108), or the damper centre and suspension control arm (Fig. 3.109).

When the front wheel compresses in Fig. 3.108, the upper ball joint 1 of the steering knuckle moves backwards and the lower one forwards, resulting in an increase in caster. Rebounding has the opposite effect; the caster (if in the normal position) decreases and may even become negative. In the case of McPherson struts and strut dampers (Fig. 3.109), point 2 moves to 4, parallel to the axes of rotation, and compresses the damping element, which is fixed in point 1. This shortens and there is rotation by the angle Δτ.

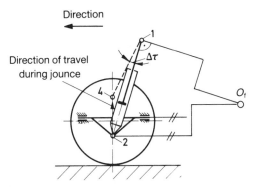

Fig. 3.109   When the McPherson strut or strut damper compresses, point 2 moves to 4 and the caster angle increases by Δτ. The intersection of a parallel to the suspension control arm axis of rotation (drawn through point 2) and a vertical on the damper centre line in point 1 gives the vehicle pitch pole O_f. The steering knuckle fixed to the damping part also rotates by this angle.

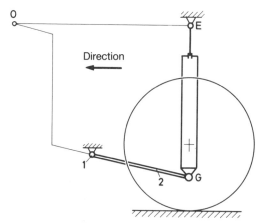

Fig. 3.110 So as not to reduce the ground clearance of the front axle and the front overhang (Fig. 1.18), the back 1 of the anti-roll bar must be raised. The arms 2 which support the lower transverse links in the longitudinal direction therefore drop backwards. The result is a vehicle pitch pole O_f in front of the axle, which causes the front end to be pulled further down during braking and adverse lift of the front end when a front-wheel drive vehicle moves off (Figs 4.0 and 3.111).

In the case of a rear wheel suspension, this position of the vehicle pitch pole O_r would be favourable.

As can be seen in Figs 3.120 and 3.121, a pitch pole O_f, which lies behind the front axle, would be more readily achieved with a double wishbone suspension than with McPherson struts and strut dampers (Fig. 3.121a). If the anti-roll bar is located in front of the axle and used for the longitudinal wheel control, its back must be raised to provide enough ground clearance (Fig. 3.110). This will cause the pitch pole to lie in front of the axle and will also draw the front end down when the brakes are applied. Figure 3.11 shows the kinematic caster alteration measured on three passenger cars with spring dampers or McPherson axles. The curve shape clearly shows whether there is an 'anti-dive' or a 'pro-dive' mechanism.

From a design point of view, the alteration angle $\Delta\tau = f(s)$ can easily be determined by drawing verticals to the suspension control arm axes of rotation C and D through the centres 1 and 2 of the wheel joints, as shown in Fig. 3.108. Fixed paths must be marked off on one of the two verticals and, using a compass with the path 1–2 the corresponding point on the other determined. The angle $\Delta\tau$ of the connecting line 3–4 to the initial position 1–2 is the caster alteration. In the case of McPherson struts and strut dampers (Fig. 3.109) the upper point 1 is fixed in the wheel house, so that the distance 1–2 shortens when the spring compresses (path 1–4) and lengthens when it rebounds.

Figure 3.108 shows that when the vehicle is designed with a pitch pole, the steering knuckle rotates by the angle $\Delta\tau$ – clockwise at the front (Figs 3.109 and 3.111) and anticlockwise at the rear axle (Fig. 3.111a); this is demon-

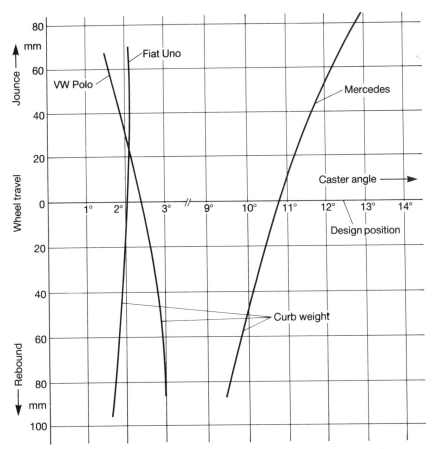

Fig. 3.111 Typical caster alteration curves for McPherson struts and strut dampers measured on three front axles. The strut damper of the Mercedes has a large caster angle that increases even further when the springs compress, i.e. a positive anti-dive mechanism. There is no such mechanism on the suspension strut of the Fiat Uno (the almost vertical curve shape indicates this) and the McPherson suspension on the VW Polo up until production year 94 had a pro-dive mechanism (Fig. 4.0).

The front end is further drawn down when the vehicle brakes; this phenomenon becomes more pronounced the more it dips. The reasons are the vertical position of the suspension strut and the high location of the anti-roll bar back; the pitch pole is therefore far in front of the axle: Figs 3.107, 3.110 and 4.0 give details. On the Mercedes, the lateral force lever $n_{\tau,k}$ on the compressing wheel on the outside of the bend increases; this means that a kind of speed-dependent lateral force understeering occurs.

strated by the shape of the curves in the above figures. This wheel travel-dependent rotation causes a changing relative speed between stator and rotor in the wheel sensors of all wheel slip control systems, which adversely affects the response speed of the ABS and ASR traction control.

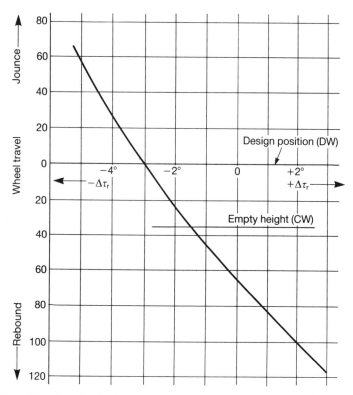

Fig. 3.111a Alteration $\Delta\tau_r$ in the theoretical negative caster angle, measured as a function of the compression and rebound travel on the rear axle of a Mercedes. The company specifies the trail as $r_{\tau,k} = -15$ mm; in the design position this would correspond to an angle of $\tau_r \sim -3°$. This increases as the springs compress and decreases or may even go into positive caster, as they rebound. The inclined position of the curve indicates high pitch poles that move further upwards when the wheels rebound and therefore progressively reduce the brake dive. Furthermore, the negative caster trail increases on the compressing outer wheel during cornering, resulting in favourable lateral force understeering, increasing with speed.

3.10.6 Wheel travel-dependent rotation of the rear steering knuckle

On the multi-link independent rear suspensions fitted in Mercedes models, five rods are used to control the steering knuckle with four of them providing lateral force reaction support. The extensions of the two upper rods intersect in pole E and those of the lower rods in G. The lines connecting the two poles give the theoretical steering axis $\overline{EG}$ (Fig. 3.111a). A negative caster was set (Figs 3.111b and 3.90) to obtain lateral force understeering (Fig. 3.58).

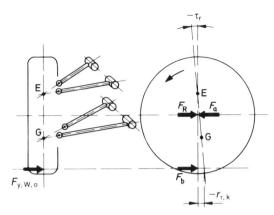

Fig. 3.111b If, on a multi-link rear suspension, there are four bars supporting the lateral forces, when viewed from the rear, their extensions meet in the points E and G. When they are connected in the side view, the result can be the theoretically negative caster angle $-\tau_r$ and the caster trail $-r_{\tau,k}$ on the ground.

Where the brake is on the outside, the braking force F_b should be regarded as acting at the centre of tyre contact. The rolling resistance force F_R and the tractive force F_a have to be shifted into the wheel centre.

The tyre caster $r_{\tau,T}$, which reduces this, must also be considered. The lateral force lever is then (Fig. 3.92, see also Equation 3.32):

$$n_{\tau,t} = r_{\tau,T} \cos \tau_r - n_{\tau,k} \tag{3.41b}$$

If anti-dive behaviour is desired, the pitch pole O_r must lie in front of the rear axle, as shown in Figs 3.110 and 3.119.

A parallel development is the multi-link rear axle, which is becoming more and more popular. This contains a trailing link that forms one piece with the steering knuckle, with a pivot in front of the axle centre which, simultaneously, represents the pitch pole O_r (Figs 1.0, 1.45 and 1.59). The kinematic movement of the wheel carrier corresponds to that of the longitudinal link suspension (Figs 3.123 and 6.17).

3.10.7 Resolution of the vertical wheel force on caster

If the steering axis $\overline{EG}$ on a double wishbone suspension is angled by the caster angle τ, the lower ball joint lies in front of the wheel centre and the upper one behind it. If the spring is supported on the lower suspension control arm, its force $F_{G,z}$ may be the same size as the vertical wheel force less the weight of the axle side (Fig. 3.112, Equation 5.2), but the moment $M_z = F_{G,z} (f - e)$ occurs, causing the forces $F_{E,x}$ and $F_{G,x}$. The elasticity present causes the caster angle to reduce. If the spring were on the top, it would increase.

Where there is caster (Case 1), the vertical force component $F_{z,W} \cos \sigma$,

Side view

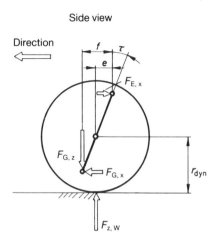

Fig. 3.112 If the spring is supported on the lower suspension control arm and if the front axle has caster, the supporting ball joint will be in front of the wheel centre. Forces $F_{z,W}$ and $F_{G,z}$ form a moment, which generates the reaction forces $-F_{E,x}$ and $+F_{G,x}$ in the direction of the suspension control arm axes of rotation. In the example these are assumed to be parallel to the ground.

shown in Fig. 3.81, would be further resolved by the angle τ, i.e. in $F_{z,W} \cos \sigma \cos \tau$ and $F_{z,W} \cos \sigma \sin \tau$ (Fig. 3.113). The last component tensions the wheels via the lever r_n at the front (Fig. 3.113a). If the caster angles τ on the left and right wheels are different, the same will apply to the tensioning forces, i.e. the vehicle could deviate from the direction of travel if the steering wheel were let go, and would pull to one side when held (Fig. 3.114 and Equation 3.41c). A 2° difference means that there is a 30 N to 40 N higher tie rod force on the side with the greater angle τ.

If the caster is achieved by relocating the wheel centre to the back (Case 2), the component $F_{z,W} \sin \sigma$ pushes the wheels together at the front via the force lever $r_{\tau,W}$ (Figs 3.115 and 3.116), i.e. even here the parallel position of the left and right steering axes to one another plays a role.

In addition, equal kingpin inclination angles are required on both wheels,

Direction

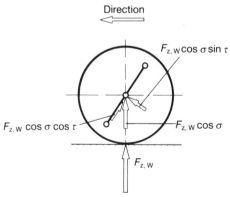

Fig. 3.113 If the steering axis is at the caster angle τ in the side view, the vertical force component $F_{z,W} \cos \sigma$ calculated in the rear view in Fig. 3.81 must be further resolved.

Fig. 3.113a The forces $F_{z,W} \cos \sigma \sin \tau$ push the front wheels together at the front via the levers r_n (i.e. into toe-in) when the vehicle is in a stationary position or moving in a straight line, and generate the forces F_T in the tie rods. The caster angles τ left and right should therefore only deviate slightly from one another (see Equation 3.42a and Fig. 3.80).

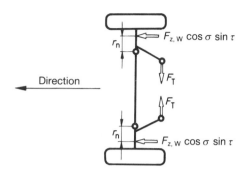

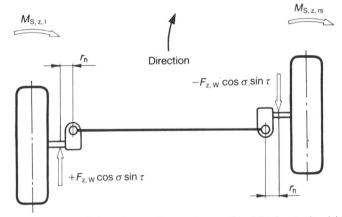

Fig. 3.114 Caster on the left and negative caster on the right front wheel (or caster angles τ of different sizes) cause the vehicle to pull to the right when travelling in a straight line. This is caused by opposed moments.

$$M_{S,z,\tau} = \pm F_{z,W} \cos \sigma \sin \tau \, r_n \tag{3.41c}$$

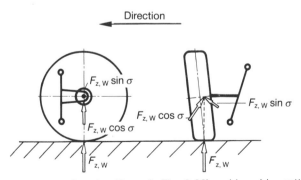

Fig. 3.115 In the case of caster (Case 2, Fig. 3.89), achieved by setting the wheel centre back, the vertical force component $F_{z,W} \sin \sigma$ comes behind the steering axis.

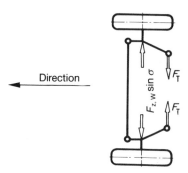

Fig. 3.116 Left and right vertical force components $F_{z,W} \sin \sigma$ push the front wheels into toe-in when the vehicle is stationary or moving in a straight line and put the tie rods under stress (forces F_T). The camber angle ϵ_W (and therefore also the king-pin inclination angle σ) should be largely the same left and right (see Equations 3.4b and 3.80).

and because these are generally directly related to the camber (see Section 3.9), only a small camber deviation between left and right front wheels is permissible (see Equation 3.4c).

Where the σ angles are different, the length of the vertical force lever $r_n = (r_S + r_{dyn} \tan \sigma) \cos \sigma$ (Equations 3.46 and 3.21a), which appears in all formulas, changes and, with caster as in Case 2 above, the vertical force $F_{z,W} \sin \sigma$ is no longer the same on the right and the left. Both instances cause the steering wheel to pull to one side. The negative offset n_τ shown in Fig. 3.91 – together with the angle τ – requires a more in-depth look at the correlations (Fig. 3.117). The vertical force $F_{z,W}$ on the wheel axes resolved in the direction of the kingpin inclination, gives $F_{z,W} \cos \sigma$ and $F_{z,W} \sin \sigma$. The first component

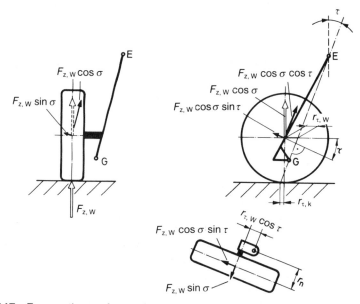

Fig. 3.117 Force ratios on front axles with negative caster offset $-r_{\tau,W}$. The opposed moments $F_{z,W} r_{\tau,W} \sin \sigma \cos \tau$ and $F_{z,W} r_n \cos \sigma \sin \tau$ can cancel one another out.

must be further divided up in the side view into $F_{z,\text{W}} \cos \sigma \sin \tau$ and $F_{z,\text{W}} \cos \sigma \cos \tau$. As can be seen in the top view, when the vehicle is moving in a straight line, there are two opposing moments on each wheel (which can cancel each other out). These are:

$$M_{\text{s},\tau} = F_{z,\text{W}} \left(r_{\text{n}} \cos \sigma \sin \tau - r_{\tau,\text{W}} \sin \sigma \cos \tau\right) \tag{3.42}$$

3.10.8 Settings and tolerances

The caster value of the empty vehicle should appear on the drawing and in workshop manuals. Optical measurement is also carried out in this load condition, as specified in DIN 70 020.

Where there is no caster offset, in order to ensure favourable steering righting, passenger cars of a standard design have caster angles of around 4° to 8°. However, in the case of a designed offset $-r_{\tau,\text{W}}$, the values can rise to $\tau = 8°$ to 11°. The type of steering system is also a factor here. If it is power-assisted (see also Sections 4.2.5 and 4.3.3), the steering moment must also right the parts in the hydraulics. In such cases a greater caster angle is preferable. If no power steering is available, lower angles have generally to be designed to limit the steering effort, especially during parking manoeuvres. Front-wheel drive vehicles are set to $\tau = 1°$ to 4°. The righting moment, which is strengthened by the tractive forces $M_{\text{T,f}}$ (Fig. 3.91a), means caster values are not absolutely necessary.

In addition to the absolute value, a tolerance is required, which is usually around ±30' but can be as much as ±1°30' to make manufacturing more cost-effective. The additional requirement (as in the case of camber, see Equation 3.4c) that there should be no greater difference than 30' between left and right wheels is necessary to prevent the vehicle pulling to one side (Fig. 3.114). The details given on the drawing would then be:

$$\tau = 4° \pm 30' \text{ maximum difference between left and right } 30' \tag{3.42a}$$

3.10.9 Measuring the caster, kingpin inclination, camber and toe-in alteration

3.10.9.1 Measurement conditions   In repair workshop manuals, the axle settings (with a few exceptions) relate to the vehicle when empty, and when checking only the manufacturer's specified values, this condition should be assumed. To eliminate the influencing friction in the suspension parts, the vehicle should be briefly settled by hand on both axles before measurement begins. The initial basis for all alterations resulting from the wheels compressing and rebounding is the design position. The vehicle is loaded with three people (each weighing around 68 kg, see Section 5.3.4), and the static settings are determined in this condition. Even load distribution is important because otherwise the body can tilt and therefore take on different camber on the left and the right. It is therefore essential that the third person should sit in the middle of the rear seat. The compression travel between the empty condition and the design position should be taken from the wheel house arches so that the vehicle can later be drawn down as far as possible against a fixed resistance for the static measurements.

3.10.9.2 Measuring the camber angle   A spirit level or electronic measuring device can be used to measure the static camber angle precisely if the zero position of the device corresponds to the wheel centre plane. If the wheel is turned slowly, the device holder can be aligned.

3.10.9.3 Measuring the caster angle   Determining the static angle can (regardless of the measurement condition) require a steering angle input of $\delta = \pm 20°$. The greater the angle τ the more the body sinks down over the wheel on the outside of the bend and is accordingly pushed up over the wheel on the inside of the bend (Fig. 3.118). The body tilts slightly, and the positive camber on the side of the vehicle on the outside of the bend consequently

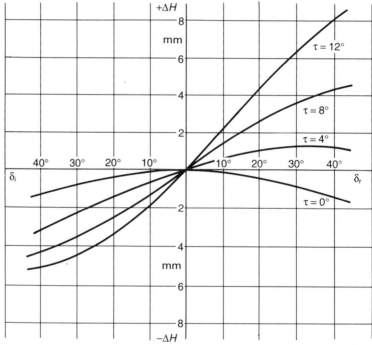

Fig. 3.118 Lift heights ΔH calculated for the wheel on the outside and the one on the inside of the bend as a function of the steering angle with the settings $\sigma_0 = 6°$, $r_s = +25$ mm and various caster angles. Where $\tau = 0°$, the centre of tyre contact of both wheels theoretically moves below ground (ΔH becomes negative), which is equivalent to lifting the body. The larger is τ, the more the body is raised on the inside of the bend ($-\Delta H$ at δ_i), but drops on the outside of bend. When kingpin inclination and caster are measured, these relationships must be taken into account (see Section 3.10.9.3). In the case of $r_s = 0$, straight lines rather than sets of curves are produced and when the kingpin offset on the ground is negative, the curves bend in the other direction.

increases, and that on the inside of the bend reduces. The associated decrease in the kingpin inclination on the outside (and increase on the inside) of the bend can lead to a measurement error, if the body is not braced against a fixed resistance to obtain the necessary horizontal position during the measurement process.

3.10.9.4 Measuring the caster alteration To avoid a pitch angle distorting the measurement (Fig. 3.105), the body should be drawn down parallel (or pushed up parallel). Only the alteration values $\pm\Delta\tau$, are ascertained and these must be deducted from or added to the initial data in the design position. The simplest way of doing this is to determine the rotation of the wheel with a measuring device. It is important that the brakes be locked. The floating plates (on which the wheels stand) flex longitudinally and laterally. No other measurement method can be used on the rear axle; Figs 3.111 and 3.111a show curves recorded in this way.

3.10.9.5 Measuring the kingpin angle Once the static caster angle has been determined in the empty condition, a mean value between left and right should be calculated to eliminate the angle τ_0 calculated in this manner, by raising the tail end (or lowering it in the case of negative caster) and thereby obtaining steering axes that are vertical from the side view. To measure the angle σ, steering inputs (where possible up to $\delta \pm 20°$) are necessary, and the kingpin inclination angle is determined via the three-dimensional movement of the wheel centre plane. The modification values should be the same for left and right inputs. Figure 3.102 indicates the correlations clearly. If the vehicle has the wheelbase l, the necessary lift height Δh_τ in the middle of the rear axle would be:

$$\Delta h_\tau = l \sin \tau_0 \qquad (3.43)$$

In the raised position, the caster must then be zero, but it is always worthwhile checking.

3.10.9.6 Checking kingpin inclination and camber As shown in Fig. 3.80 the sum of camber and kingpin inclination ($\epsilon_W + \sigma$) left and right should be the same. If the deviation exceeds 30', this may be a measurement error, the result of an accident, or an assembly inaccuracy on McPherson struts and strut dampers (Fig. 3.80a).

3.10.9.7 Measuring kingpin inclination and camber alteration The two are identical and pure alteration values can easily be determined. ($\pm \Delta\epsilon_W = \pm \Delta\sigma$, see Figs 3.41 and 3.42). Only spirit levels or electronic measuring devices need to be fixed to the wheels and corrected by the caster angle, which changes as the vehicle is drawn up or down parallel with the brakes locked. The values should then be added to or subtracted from the data determined in the design position. Figures 3.9 and 3.40 indicate curves measured in this way.

3.10.9.8 Measuring the toe-in alteration   The static toe-in angle $\Delta_{f\,or\,r}$ (at the front or back, see Equation 3.8) can be determined nowadays with opto-electronic measuring devices. The alteration values for the front and rear axle should then be recorded as a function of the wheel travel s_1 and s_2 (separately for the left and right wheel) and added to the basic values. The wheel position is measured relative to the body, so it is sensible to work with optical devices and to fix the scale (or the mirror) to the body itself. Lateral movements of the vehicle, when it is raised or pulled down, could otherwise lead to errors when reading off the figures.

3.11 Anti-dive and anti-squat mechanism

3.11.1 Concept description

The anti-dive mechanism reduces the amount by which the front end of the vehicle dips or the tail rises when the brakes are applied. It can – in the case of brakes which are outside in the wheels – only be achieved if there are pitch poles O_f and O_r between the axles; at the front, at the rear, or on both axles (Fig. 3.119).

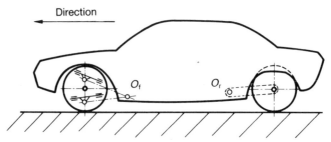

Fig. 3.119   The pitch axis is obtained by linking pitch poles front and rear. If these are available with O_f (at the front) and O_r (at the rear), the body is supported at this point in the longitudinal direction when the brakes are applied, assuming the brakes are on the outside of the wheels.

The anti-squat mechanism reduces the amount by which the tail sinks on rear-wheel drive vehicles or the front end lifts (on front-wheel drive vehicles). It acts during acceleration and only on the driven axle. On independent wheel suspensions it is important for the pole to be higher than the wheel centre of the driven axle (as can be seen in Figs 3.121 and 3.124) or, on a rigid axle, the differential is located in the axle housing (Figs 1.12 and 1.27).

The anti-dive and anti-squat angle is also a consideration here; ϵ and κ are entered in Fig. 3.124; the greater these can be the better is the pitch equalization.

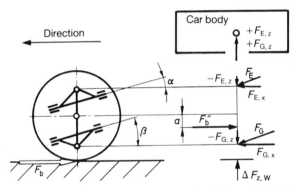

Fig. 3.119a If the front brake is on the inside on the differential, brake dive can be compensated by disposing the suspension control arms in the same direction, but at an angle. The braking force must be regarded as being under the wheel centre by $a = r_a \sin \sigma$ (see Equation 3.28b). When it compresses, the wheel moves forward. The diagonal springing angle is: $\kappa = (\alpha + \beta)/2$.

3.11.2 Vehicle pitch pole front

Left and right suspensions are generally identical so the poles determined by the momentary position of the suspension control arms are in the same position on both sides, which leads to the so-called pitch axes. If these are at infinity (i.e. for practical purposes they do not exist, Fig. 3.106) the longitudinal forces are concentrated in the wheel centre, which applies if the brake is located on the inside (on the differential). Here the brake dive can be countered by the two double wishbones being set at an angle in the same direction (Fig. 3.119a).

As can be seen from the illustration, the brake force operating as F''_b, shifted from the wheel centre vertical to the steering axis (shown in Fig. 3.86 for the rolling resistance), causes the reaction forces $F_{E,x}$ and $F_{G,x}$ in the suspension control arms, which (due to the angled position) cause the vertical component $-F_{E,z} = F_{E,x} \tan \alpha$ and $-F_{G,z} = F_{G,x} \tan \beta$. The sum of forces in one effective direction must be zero, i.e. $+F_{E,z}$ and $F_{G,z}$ work against the vehicle front end jounce. Two suspension control arms, which are placed at an angle in this manner, have the advantage of no caster change but the disadvantage that they move forward during jounce (in other words in the direction of the obstacle force). The Citroën GSA had this type of suspension control arm configuration and therefore an almost 100% anti-dive system (see Section 5.2.4).

Where the brake is on the outside (as in Figs 3.119 and 6.16), it is also necessary for the suspension control arms to be at an angle to achieve a pitch axis and, therefore, reaction forces in the vertical direction. However, the two suspension control arms must be inclined against one another. The right-hand side of Fig. 3.120 shows the statics with the (compared with Fig. 3.119a) significantly increased components $F_{G,z}$, caused by the higher force $F_{G,x} = F_b + F_{E,x}$ in the case of outside brakes (in the case of inside brakes $F_{G,x} = F_b - F_{E,x}$).

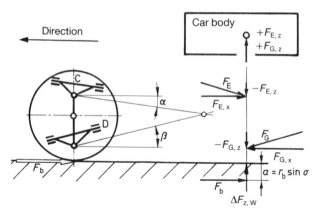

Fig. 3.120 To reduce brake dive when the brakes are on the outside, the suspension control arms must be inclined against one another. The forces $F_{E,x}$ and $F_{G,x}$ must be calculated assuming the braking force F'_b below ground by $a = r_b \sin \sigma$ (or in the case of negative kingpin offset, above ground by the same amount, Equation 3.27). The components acting against front end dip are then $+F_{E,z}$ and $+F_{G,z}$; in the case of $-r_S$, all forces are smaller. In the case of caster $F'_b = F_b \cos \tau$ (Fig. 3.88).

Meanwhile, all front-wheel drive vehicles coming on to the market have a negative kingpin offset. The prerequisite for the counter-steering effect, which can be achieved in this way (Fig. 6.12), is a brake inside the wheel. By angling the lower suspension control arm, on a double wishbone suspension it is possible to reduce both the brake dive and the squat. The brake force F''_b acting now on the steering axis by the amount a above ground causes the component $F_{G,z}$ supporting the body (Fig. 3.121) and – as shown on the left –

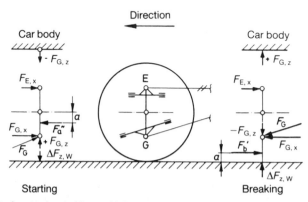

Fig. 3.121 In front-wheel drive vehicles, both the lifting of the vehicle as it moves off and front-end brake dive can be reduced by disposing the lower suspension links only at an angle, if (as is usually the case) the brake is in the wheel. In the case of a negative kingpin offset, F'_b acts on the steering axis by the amount a above ground (see Equation 3.27).

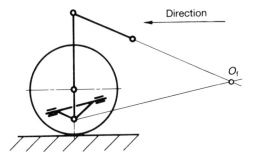

Fig. 3.121a To determine the vehicle pitch pole O_f on the longitudinal transverse axle the upper suspension control arm must be lengthened and a parallel must be drawn to the suspension control arm axis of rotation through the ball joint centre. When the front end jounces, O_f moves towards the wheel, this being the equivalent of a progressive anti-dive mechanism.

the drive-off force F''_a acting below the wheel centre causing the force $-F_{G,z}$ pulling downwards. The upper suspension control arm is horizontal. Its job can also be done by a vertically positioned McPherson strut or strut damper. In this type of suspension there is an anti-dive and anti-squat mechanism.

The pitch axis on double wishbones can be shown on a drawing using parallels to the suspension control arm axes of rotation C and D, drawn through the centres of the ball joints E and G (Fig. 3.120). The McPherson strut and strut damper require a vertical to be set up on the direction of movement of the damper in point E, the intersection of which with the suspension control arm parallel passing through G, gives the point O_f (Fig. 3.109), or an extension of the tension rods or anti-roll bars absorbing the longitudinal forces (Fig. 3.110).

On the trailing link axle, in order to get the pole O_f, the upper control arm must be extended and again a parallel to the axis of rotation must be drawn through the lower wheel joint (Fig. 3.121a). When the front end of the vehicle jounces, the upper suspension control arm moves to a greater angle of inclination and the pole O moves closer to the wheel. This means a progressively increasing anti-dive system, which also applies if, in the case of a double wishbone suspension, the braking forces are absorbed upwards through a trailing or semi-trailing link (Figs 1.23a and 3.25) or by the legs of the anti-roll bar.

3.11.3 Pitch poles rear

The requirement for a reduction in the brake dive demands a pitch axis that is close to the wheel and as high as possible, however, both of these result in a severe caster change on the front axle. Here – particularly where the vehicle is fitted with ABS (see the end of Section 3.10.5) – a compromise must be struck between both criteria. On the rear axle, the picture is different. The poles O_r can be positioned close in front of the axle whereby the length of the

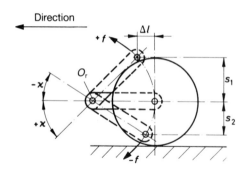

Fig. 3.122 Longitudinal links on a rear axle have the advantage of the favourable pitch pole O_r. The suspension control arm should be as short as possible, however, with the required spring travel s_1 and s_2, no angular deflections $\pm\kappa$, which are too large, may arise. These would lead to significant diagonal springing $\pm f$; the driver will hardly be aware of the change in wheel base associated with this.

suspension control arms and the ABS performance objectives represent the limits.

Too short a suspension control arm gives unfavourably large rotation angles $\pm\kappa$ (Fig. 3.122) to achieve the desired spring travel s_1 and s_2. The wheel base change associated with the pitch axis and Δl should not affect the handling properties. As proof, we can look at the earlier Renault models on which there were different wheel bases left and right.

The trailing link suspension and the compound crank axle (Figs 1.1, 1.8 and 1.42) have the best position of the pitch pole among the wheel suspensions commonly used as rear axles. These lie in the centre of the suspension control arm axis of rotation and the force $-F_{O,z}$, which draws the rear end down during braking, is in accordance with Fig. 3.123:

$$F_{O,z} = F_b\, g/d \tag{3.44}$$

i.e. the higher the height g and the shorter the distance d can be, the stronger the effect.

Figure 3.123 also applies to all 'multi-link axles' (Figs 1.0, 1.45 and 1.59) as well as rigid axles carried on two trailing links:

- the torsion crank axle (Fig. 1.44)
- the drawbar or A-bracket axle (Fig. 1.43a)
- the off-road vehicle axle shown in Fig. 1.27.

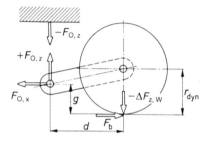

Fig. 3.123 On trailing link and multi-link suspensions with axes of rotation parallel to the ground, the mounting point on the body is also the pitch pole. The higher the pole lies (path g) and the closer it is to the wheel (path d), the more the force $-F_{O,z}$ pulls the tail end down during braking.

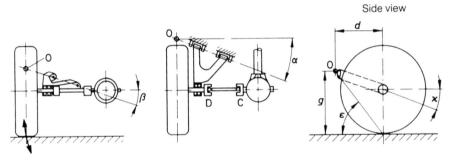

Side view

Fig. 3.124 On the semi-trailing link suspension the point at which the extension of the axis of rotation goes through the plane of the wheel centre gives the pitch pole O. The brake reaction support angle ϵ can be calculated from the existing paths:

$$\tan \epsilon = g/d \tag{3.45}$$

The same applies to the anti-squat or diagonal springing angle κ (see Section 5.4.4 and Fig. 3.15):

$$\tan \kappa = (g - r_{dyn})/d \tag{3.46}$$

except that here the integer is important. In the case of $+\kappa$ (as shown) when the vehicle accelerates, the squatting tail is pushed upwards and in the case of $-\kappa$, it is pulled further down.

In the case of the semi-trailing link axle, the top view must first be drawn to ascertain the pitch pole (Figs 3.124 and 3.29). Using the angle α, the distance d (pitch pole O to wheel centre) is determined and then, in the rear view, the height g of point O is also determined. The side view then shows the actual position.

If a watt linkage or a pair of control arms per side is used to link the rigid rear axle, the centre lines of the suspension arms must be extended and made to intersect to obtain O_r (Fig. 3.125). The shorter upper suspension control arms ensure that the pitch pole moves favourably towards the axle when the vehicle is loaded (in other words the tail sinks at points E and G) and therefore the anti-dive mechanism is reinforced.

Fig. 3.125 If a rigid rear axle is controlled by two trailing link pairs, its extensions give the pitch pole O_r. When the vehicle is laden, points E and G on the body side move down, i.e. O_r moves towards the wheel in a favourable manner.

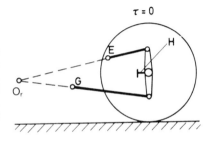

If, as can be seen in Fig. 1.27, the differential is contained in the axle housing, the moments coming from the engine and driving the wheels are vertical to one another (Fig. 1.12); because the forces on the axle housing are jointly supported, the anti-squat mechanism is, at the same time, supported at the pitch pole.

4

Steering

The steering system is type-approved on all new passenger cars and vans coming on to the market; it is governed by the following EC directives.

70/311/EEC   91/662/EEC
74/297/EEC   92/62/EEC

Figure 4.0 shows the complete steering system of a front-wheel drive passenger vehicle.

4.1 Steering system

4.1.1 Requirements

On passenger cars, the driver must select the steering wheel angle to keep deviation from the desired course low. However, there is no definite functional relationship between the turning angle of the steering wheel made by the driver and the change in driving direction, because the correlation of

- turns of the steering wheel
- alteration of steer angle at the front wheels
- development of lateral tyre forces
- alteration of driving direction

is not linear (Fig. 4.1) because of the elastic compliance in the components of the chassis. To move a vehicle, the driver must continually adjust the relationship between turning the steering wheel and the alteration in the direction of travel. To do so, the driver will monitor a wealth of information, going far beyond the visual perceptive faculty (visible deviation from desired direction). These factors would include for example, the roll inclination of the

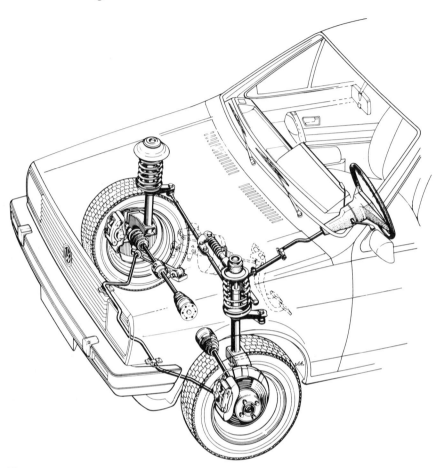

Fig. 4.0 Damper strut front axle of a VW Polo with 'short gear steering', long tie rods and a 'sliding clutch' on the steering tube; the end of the tube is stuck onto the pinion gear and fixed with a clamp. The steering arms, which consist of two half shells and point backwards, are welded to the damper strut outer tube. An 'additional weight' sits on the longer right drive shaft to damp vibrations. The anti-roll bar carries the lower control arm. To give acceptable ground clearance, the back of it was designed to be higher than the fixing points on the control arms. The pitch pole is therefore in front of the axle and the vehicle's front end is drawn downwards when the brakes are applied (Figs 3.110 and 3.111).

body, the feeling of being held steady in the seat (transverse acceleration) and the self-righting torque the driver will feel through the steering wheel.

It is therefore the job of the steering system to convert the steering wheel angle into as clear a relationship as possible to the steering angle of the wheels and to convey feedback about the vehicle's state of movement back to the steering wheel. This passes on the actuating moment applied by the dri-

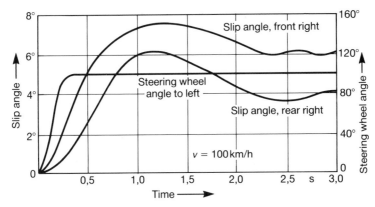

Fig. 4.1 Delayed, easily manageable response of the right front wheel when the steering wheel is turned by 100° in 0.2 s, known as step steering input. A slip angle of $\alpha_f \sim 7°$ on both front tyres is generated in this test. The smaller angle α_r on the rear axle, which later increases, is also entered. Throughout the measurement period it is smaller than α_r (x-axis), i.e. the model studied by Mercedes Benz understeers and is therefore easy to handle.

ver, via the steering column to the steering gear 1 (Fig. 4.2) which converts it into pulling forces on one side and pushing forces on the other, these being transferred to the steering arms 3 via the tie rods 2. These are fixed on both sides to the steering knuckles and cause a turning movement until the required steering angle has been reached. Rotation is around the steering axis $\overline{EG}$ (Fig. 3.80), also called kingpin inclination, pivot or steering rotation axis (Fig. 1.1a).

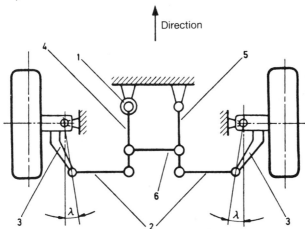

Fig. 4.2 Synchronous steering square on the front axle of a left-hand drive passenger car or light van; on the right-hand drive vehicle, the steering gear is on the other side. The pitman, idler and steering arm rotate in the same direction. The tie rods are fixed to these arms.

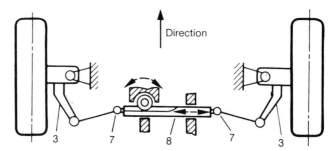

Fig. 4.3 Rack and pinion steering with the steering triangle behind the front axle. The spigots of the inner tie rod joints 7 are fixed to the ends of the steering rack 8 and the outside ones to the steering arms 3 (see also Figs 1.24 and 1.37).

4.1.2 Steering system on independent wheel suspensions

If the steering gear is of a type employing a rotational movement, i.e. the axes of the meshing parts (screw shaft 4 and nut 5, Fig. 4.12) are at an angle of 90° to one another, on independent wheel suspensions, the insides of the tie rods are connected on one side to the pitman arm 4 of the gear and the other to the idler arm 5 (Fig. 4.2). As shown in Figs 4.10 and 4.21 to 4.23, parts 4 and 5 are connected by the intermediate rod 6. In the case of steering gears, which operate using a shift movement (rack and pinion steering), it is most economical to fix the inner tie rod joints 7 to the ends of the steering rack 8 (Fig. 4.3).

4.1.3 Steering system on rigid axles

Rack and pinion steering systems are not suitable for steering the wheels on rigid front axles, therefore only steering gears with a rotational movement are used. The intermediate lever 5 sits on the steering knuckle (Fig. 4.4). The intermediate rod 6 links the steering knuckle and the pitman arm 4. When the wheels are turned to the left, the rod is subject to tension and turns both wheels simultaneously, whilst when they are turned to the right, part 6 is sub-

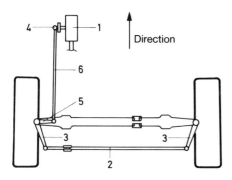

Fig. 4.4 On rigid axles, apart from the two steering arms 3, only the tie rod 2, the idler arm 5 and the drag link 6 are needed to steer the wheels. If leaf springs are used to carry the axle, they must be aligned precisely in the longitudinal direction, and lie vertical to the lever 5 when the vehicle is moving in a straight line.

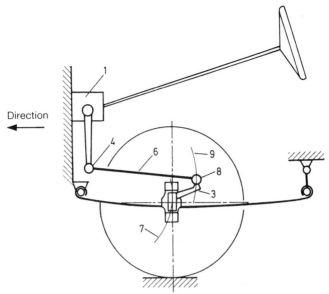

Fig. 4.5 Side view of a rigid front axle showing the movement directions 9 and 7 of the drag link and axle housing. The path of point 7 is determined by the front half of the leaf spring and can be calculated on a spring-balance by measuring the change in length when a load is added to and removed from the spring.

ject to compression. A one-piece tie rod connects the wheels together via the steering arm.

However, on front axles with leaf springs, the pitman arm joint 4, which sits on the steering gear 1, must be disposed in such a manner that when the axle bottoms out, the lower joint 8 describes the same arc 9 as the centre of the front axle housing (Figs 4.5 and 1.22). The arc 9 must be similar to the curved path 7, otherwise there is a danger of the wheels experiencing a parallel toe-in alteration when the axle bottoms out, i.e. both being turned in the same direction (Fig. 4.5a). If a rigid axle is laterally controlled by a panhard rod, the steering rod must be parallel to it. Its construction is similar to that of the intermediate rod of the steering linkage shown in Fig. 4.11; length adjustment and ball joints on both sides are necessary.

4.2 Rack and pinion steering

4.2.1 Advantages and disadvantages

This steering gear with a shift movement is used not only on small and medium-sized passenger cars, but also on heavier and faster vehicles, such as for example the Audi A8, Mercedes E-Class, Ford Scorpio, Porsche 928 GTS, etc., plus almost all new light van designs with independent front wheel

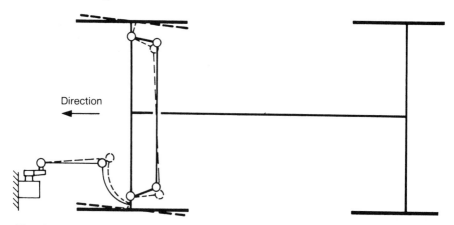

Fig. 4.5a If the movement curve 7 of the axle housing and curve 9 of the rear steering rod joint do not match when the body bottoms out, the wheels can turn and therefore an unintended self-steering effect occurs.

suspension. The advantages over manual recirculating ball steering systems are (see also Section 4.3.1):

- simple construction
- economical and uncomplicated to manufacture
- easy to operate due to good degree of efficiency
- contact between steering rack and pinion is free of play and even internal damping is maintained (Fig. 4.8)
- tie rods can be joined directly to the steering rack
- minimal steering elasticity (Fig. 3.77a)
- compact (the reason why this type of steering is fitted in all European and Japanese front-wheel drive vehicles)
- the idler arm (including bearing) and the intermediate rod are no longer needed
- easy to limit steering rack travel and therefore the steering angle.

The main disadvantages are:

- greater sensitivity to impacts
- greater stress in the case of tie rod angular forces
- any disturbance of the steering wheel is easier to feel
- tie rod length sometimes too short where it is connected at the ends of the rack (side take-off design, Fig. 3.52a)
- size of the steering angle dependent on steering rack travel
- this sometimes requires short steering arms 3 (Fig. 4.3) resulting in higher forces in the entire steering system
- decrease in steering ratio over the steer angle (Fig. 3.75) associated with heavy steering during parking if the vehicle does not have power-assisted steering
- cannot be used on rigid axles.

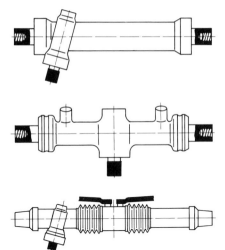

Fig. 4.6 The three most common types of rack and pinion steering on left-hand drive passenger cars; right-hand drive vehicles have the pinion gear on the other side on the top and bottom configurations (shown in Fig. 4.24). The pinion gear can also be positioned in the centre to obtain longer steering rod travel.

4.2.2 Configurations

There are four different configurations of this type of steering gear (Fig. 4.6):

Type 1   Pinion gear located outside the vehicle centre (on the left on left-hand drive and on the right on right-hand drive) and tie rod joints screwed into the sides of the steering rack (side take-off).

Type 2   Pinion gear in vehicle centre and tie rods taken off at the sides.

Type 3   Pinion gear to the side and centre take-off, i.e. the tie rods are fixed in the vehicle centre to the steering rack.

Type 4   'Short steering' with off-centre pinion gear and both tie rods fixed to one side of the steering rack.

Types 1 and 3 are the solutions generally used, whilst Type 2 can be found in some Porsche vehicles and Type 4 is preferred by Audi and VW (Figs 4.0 and 1.40).

4.2.3 Steering gear, manual with side tie rod take-off

Type 1 is the simplest solution, requiring least space; the tie rod joints are fixed to the sides of the steering rack (Fig. 4.7), and neither when the wheels are turned, nor when they bottom out does a moment occur that seeks to turn the steering rack around its centre line. It is also possible to align the pinion shaft pointing to the steering tube (Figs 4.9a, 4.14, 4.16b and 1.40) making it easy to connect the two parts together. Using an intermediate shaft

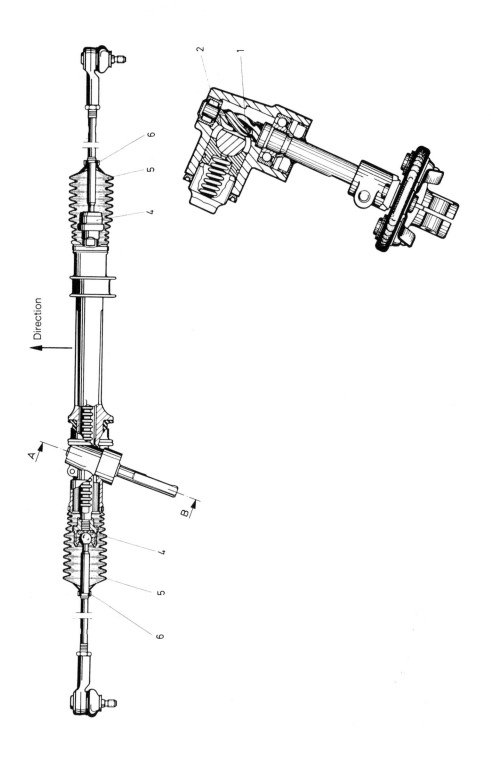

Direction

A

B

1

2

4

5

6

4

5

6

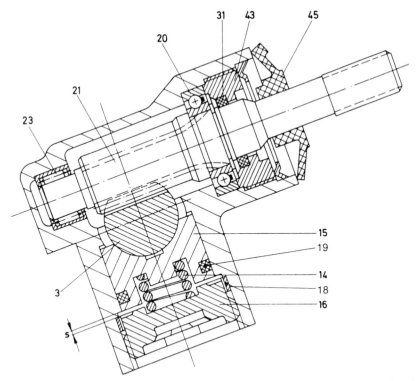

Fig. 4.8 ZF, rack and pinion steering, section through pinion gear, bearing and rod guide. The distance ring 18 is used for setting the play, and the closing screw 16 is tightened against it. The O-ring 19 provides the damping function and prevents rattling noises.

with two joints (Figs 4.16, 1.31 and 1.34) enables the steering column to bend at this point in an accident. In this event the entire steering gear is turned when viewed from the side (i.e. around the *y*-axis).

Figure 4.8 is a section showing how, on all rack and pinion steering systems, not only can the play between the steering rack and the pinion gear be easily eliminated, but it also adjusts automatically to give the desired damping. The pinion gear 21 is carried by the grooved ball bearing 20; this also absorbs any axial forces. Ingress of dirt and dust are prevented by the seal 31 in a threaded ring 43 and the rubber cap 45. The lower end of the pinion gear is supported in the needle bearing 23.

Fig. 4.7 Rack and pinion steering on the Vauxhall Corsa. The tie rod axial joints 4 bolted to the side of the steering rack and the sealing gaiters 5 can be seen clearly. To stop them from being carried along when the toe-in is set (which is done by rotating the middle part of the rod) it is necessary to loosen the clamps 6.

The pinion 1 has been given a 'helical cut', due to the generous ratio, and is carried from below by the needle bearing 2. The bearing housing has been given a cover plate to facilitate assembly and prevent dirt ingress.

In a left-hand drive passenger car or light van, the right-hand side of the steering rack 3 is carried by the plastic bearing shell shown in the figure below and the left-hand side by the guide 15, which presses the steering rack against the pinion gear. On a right-hand drive vehicle this arrangement is reversed. The half-round outline of the guide 15 does not allow radial movement of the steering rack. To stop it from moving off from the pinion gear, when subject to high steering wheel moments (which would lead to reduced tooth contact), the underside of the guide 15 is designed as a buffer; when it has moved a distance of $s \leqslant 0.12$ mm it comes into contact with the screw plug 16.

Depending on the size of the steering system, the coil spring 14 has an initial tension force of 0.6 kN to 1.0 kN, which is necessary to ensure continuous contact between steering rack and pinion gear and compensate for any machining imprecision, which might occur when the toothing is being manufactured or the steering rack broached or the pinion gear milled or rolled. The surface of the two parts should have a Rockwell hardness of at least 55 HRC; the parts are not generally post-ground due to the existence of a balance for the play. Induction-hardenable and annealed steels such as Cf 53, 41 Cr 4, in accordance with the German standard DIN 17 200, are suitable materials for the steering rack and case-hardened steels, such as for example 20 MnCr 5, 20 MoCr 4, in accordance with DIN 17 210, are suitable for the pinion gear. Sealing the steering rack by means of gaiters to the side (Fig. 4.7) makes it possible to lubricate them with grease permanently, and lubrication must be provided through a temperature range of $-40°C$ to $+80°C$. It is important to note that if one of the gaiters is damaged, the lubricant can escape, leading to the steering becoming heavier and, in the worst case, even locking. Gaiters should therefore be checked at every service inspection. They are also checked at the German TÜV (Technischer Überwachungs Verein) annual vehicle inspection.

4.2.4 Steering gear, manual with centre tie rod take-off

As shown in Figs 1.40 and 4.0, in the case of McPherson struts and strut dampers, the tie rods must be taken off from the centre if the steering gear has to be located fairly high up. In such cases the inner joints are fixed in the centre of the vehicle to the steering rack itself, or to an isolator that is connected to it. The designer must ensure that the steering rack cannot twist when subject to the moments that arise. When the wheels rebound and compress, the tie rods are moved to be at an angle, something which also happens when the wheels are steered. The effective distance a between the eye-type joints of the tie rods and the steering rack centre line, shown in Fig. 4.9, gives a lever, via which the steering could be twisted. Two guide pieces which slide in a groove in the casing stop this from happening. However, the need to match the fit for the bearing of the steering rack and the guide groove can lead to other problems. If they are too tight, the steering will be heavy, whilst if they are too loose, there is a risk of rattling noises when the vehicle is in motion.

4.2.5 Rack and pinion hydraulic power steering

The advantages over recirculating ball steering (as listed in Section 4.2.1) can be transferred directly from the manual gears. Cost advantage and low weight are also likely to be an important factor in any decision here. As for the main disadvantage; the sensitivity to road conditions, and therefore the influence of torsional impacts and vibrations coming through to the steering wheel are considerably less compared with the configurations that are not power assisted, due to the hydraulic self-damping. This is also likely to be the reason why an additional steering damper is not necessary on most vehicles with hydraulic power-assisted steering, whilst equivalent vehicles with manual steering may require it (see Section 4.5).

Manual steering systems serve as a basis for power-assisted steering, with the advantage that, regardless of the power assistance, the mechanical link between steering wheel and wheel with all components is retained (Figs 4.0 and 4.15). Only the steering gear is replaced by a more expensive configuration fitted with a steering valve and a working cylinder. The ratio i_S (Fig. 3.75) is mostly configured more directly on hydraulic steering because it does not need to be kept large to avoid high steering forces.

Figure 4.9a shows the components that are added for power-assisted steering, using the example of a configuration fitted by Vauxhall in the Cavalier. The hydraulic fluid necessary for steering assistance is carried from the vane pump 1 by the high pressure line 2 directly to the steering valve 10 in the pinion gear housing of the steering gear 9. Depending on the direction in which the steering wheel is being turned and the corresponding countervailing force at the wheels, it is distributed to the right or left cylinder line (positions 12 and 11). Both lead to the working cylinder 13 which, in the configuration shown, is housed in the right-hand part of the rack housing. The force which arises here supports – via a piston located on the steering rack – the shifting movement and therefore the steering movement of the wheels introduced by the tie rods. If the steering wheel is in its neutral position, the oil flows from the steering valve 10 via the return line 5 directly back into the oil tank and from here via the low pressure line 6 to the pump 1.

If the steering system is designed with side tie rod take-off, instead of centre take-off, the hydraulic system does not alter. Only the working cylinder in the steering gear is modified to meet the design conditions.

4.3 Recirculating ball steering

4.3.1 Advantages and disadvantages

Steering gears with a rotating movement are difficult to house in front-wheel drive passenger cars and, in a standard design vehicle with independent wheel suspension, also require the idler arm 5 and a further intermediate rod (position 6 in Fig. 4.2 and Figs 4.10 and 1.3) to connect them to the pitman

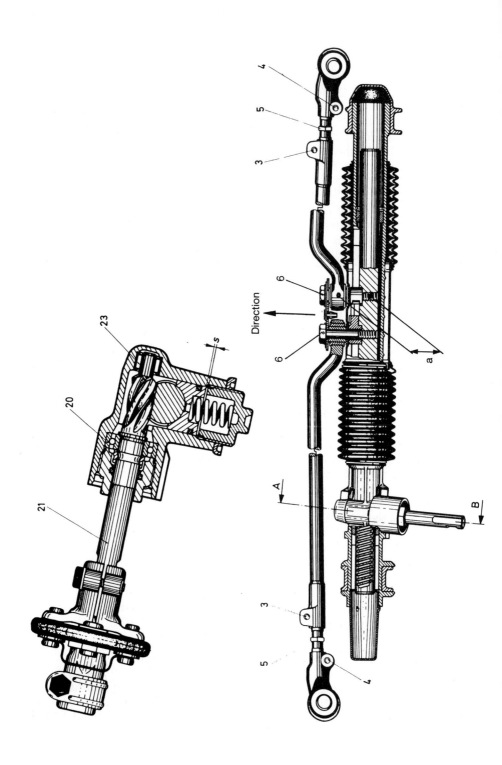

Direction

arm 4; the tie rods are adjustable and have pre-lubricated ball joints on both sides (Figs 4.11 and 4.11a).

This type of steering system is more complicated on the whole in passenger cars with independently suspended front wheels and is therefore more expensive than rack and pinion steering systems; however, it sometimes has greater steering elasticity, which reduces the responsiveness and steering feel in the on-centre range (see Section 3.7.4). Comparing the two types of configuration (without power-assisted steering) indicates a series of advantages:

- can be used on rigid axles (Figs 4.4 and 1.22)
- able to transfer high forces
- large wheel input angle possible; the steering gear shaft has a rotation range up to ±45°, which can be further increased by the steering ratio, and therefore
- it is possible to use long steering arms, which results in
- only low load to the pitman and intermediate arms in the event of tie rod diagonal forces occurring.

It is also possible

- to design tie rods of any length desired, and to have steering kinematics which allow an increase in the overall steering ratio i_S with increasing steering angles. The operating forces necessary to park the vehicle are reduced in such cases (see Section 3.7.3).

4.3.2 Steering gear, manual

The input screw shaft 4 (Fig. 4.12) has a round thread in which ball bearings run, which carry the steering nut 5 with them when the steering wheel is rotated. The balls which come out of the thread at the top or the bottom

---

Fig. 4.9 Top view of the rack and pinion steering of the front-wheel drive Vauxhall Belmont and Cavalier; the steering arms on the McPherson strut point backwards and the steering gear is located relatively high. For this reason the tie rods have to be jointed in the middle and (in order not to come into contact with the gear housing when the wheels are turned) have to be bent. The guide in the groove of the housing prevents the steering rack from twisting. On the inside, both tie rods have the eye-type joint shown in Fig. 5.30; the distance a to the steering rack centre, which causes a bending moment, and a torque (when the wheels jounce and rebound) is also shown. The two bolts 6 gripping into the steering rack are secured (see also Fig. 4.9a). Once the screws 3 and 4 have been loosened, toe-in to the left and right can be set by turning the connecting part 5.

The steering gear has two fixing points on the dashpanel, which are a long way apart and which absorb lateral force moments with minimal flexing. As also shown in Fig. 4.8, the pinion is carried by a ball and a needle bearing (positions 20 and 23) and is also pressed onto the steering rack by a helical spring. The illustration shows the possible path s of the rack guide. Fig.s 4.31 to 4.33 show the reason for the length of the tie rods on McPherson struts and strut dampers.

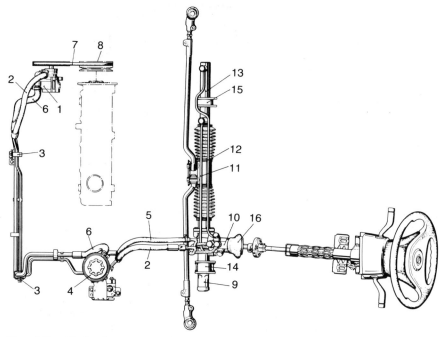

Fig. 4.9a Hydraulic power-assisted steering system of the Vauxhall Cavalier. The individual components are:

  1 vane pump, driven by the vehicle engine
  2 high-pressure line from the pump to the steering valve
  3 noise-insulating line mounting
  4 fluid reservoir with fine filter insert
  5 return flow line, from the steering valve to the reservoir
  6 suction line, between tank and pump
  7 V-belt
  8 belt pulley of the vehicle engine
  9 steering gear, fixed to vehicle dashpanel
 10 steering valve, rotary principle
11/12 left and right cylinder line
 13 pressure cylinder
14/15 bracket for fixing steering gear
 16 dashpanel seal.

(depending on the direction of rotation) are returned through the tube 6. The nut has teeth on one side which mesh with the toothed segment 7 and therefore with the steering output shaft 8. When viewed from the side, the slightly angular arrangement of the gearing can be seen top right. This is necessary for the alignment bolt 1 to overcome the play when the wheels are pointing straight ahead. If play occurs in the angular ball bearings 2 and 3, the lock nut must be loosened and the sealing housing cover re-tightened.

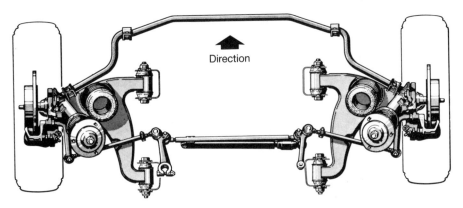

Fig. 4.10 Top view of the strut damper front axle on a Mercedes vehicle. The inter-mediate rod and the tie rods are fixed side by side on the pitman and idler arms and one grips from the top and the other from the bottom into the two levers; the steering square is opposed. The steering damper is supported on the one side at the interme-diate rod and on the other side on the suspension subframe.

The anti-roll bar is linked to the lower wishbone type control arms whose inner bear-ings take large rubber bushings. The defined springing stiffness of these bearings, together with the inclined position of the tie rods (when viewed from the top) means that when the vehicle corners, there is a reduction in the steering input, i.e. elasticity steering, tending towards understeering (Fig. 3.63). The strut dampers are screwed to the steering knuckles; the negative kingpin offset is $r_S = -14$ mm. Fig. 1.25 contains further details.

Only a few standard design larger saloons can be found on the road with manual recirculating ball steering. For reasons of comfort, newer passenger cars of this type have hydraulic power-assisted steering. The same applies to commercial vehicles; only a few light vans are still fitted with manual configu-rations as standard and even these are available with power-assisted steering as an option.

4.3.3 Recirculating ball hydraulic power steering

The basis for the steering gear is the manual configuration (Fig. 4.12). However, it differs in that the working piston and the control valve are in a larger but common housing 1. The internal diameter functions as a working cylinder (Fig. 4.13). The working piston 2 also has toothing similar to the steering nut of the manual, which meshes into the counter-profile of the toothed segment on the steering output shaft 4.

In front of and behind the piston are the two pressure chambers D_1 and D_2 that are sealed against one another with piston rings and provide the support-ing force, based on the level of the pressures $p_{hyd,\,1}$ or $p_{hyd,\,2}$ and the size of the piston diameter D_{Pi}.

$$F_{Pi} = p_{hyd,1\ or\ 2}\, A_{Pi} = p_{hyd,1\ or\ 2}\,\frac{\pi D_{Pi}^2}{4} \tag{4.1}$$

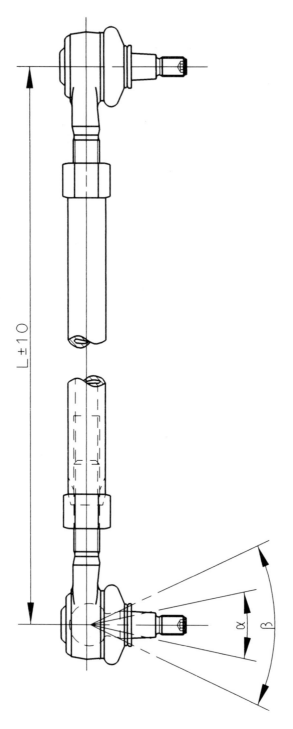

Fig. 4.11 Configuration of an adjustable tie rod with pre-lubricated joints and buckle-proof central tube, the interior of which has a right-hand thread on one side and a left-hand thread on the other. It can usually be continuously adjusted by ±10 mm. When toe-in has been set, the length on the right and left tie rod may differ, resulting in unequal steering inputs and different size turning circles; for this reason, the central tube should be turned the same amount on the left and right wheel.

The configuration shown in the illustration is used on rigid front axles and as a drag link (illustration: Lemförder Metallwaren)

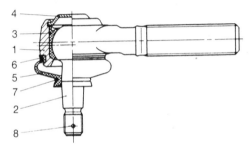

Fig. 4.11a Lemförder Metallwaren pre-lubricated tie rod joint, used on passenger cars and light vans. The joint housing 1 has a fine thread on the shaft (M14 × 1.5 to M22 × 1.5) and is made of annealed steel C35V, DIN 17 200 (German standard, V = quenched and tempered), whilst surface-hardenable steel 41Cr4V, DIN 17 200 is used for the ball pivot 2.

The actual bearing element – the one-part snap-on shell 3 made from polyacetal (e.g. DELRIN, made by Dupont) snap on shell 3 – surrounds the ball; the rolled-in panel cover 4 ensures a dirt and waterproof seal. The polyurethane or rubber sealing gaiter 5 is held against the housing by the tension ring 6. The gaiter has a bead at the bottom (which the second tension ring 7 presses against the spigot) and a sealing lip, which comes into contact with the steering arm.

The ball pivot 2 has the normal 1:10 taper and a split pin hole (position 8). If there is a slit or a hexagonal socket (with which the spigot can be held to stop it twisting), a self-locking nut can be used instead of a slotted castle nut and split pin.

Based on this design, and a significant advantage for matching the steering valve shown at the top of the illustration, is the fact that the effective surfaces included in the calculation are of the same size on both sides of the piston, since the pressure $p_{hyd, 2}$ also acts behind the steering screw 5 in the interior of the piston. Both pressure chambers are connected with the radial groove 13 or 14 of the valve housing 5 via holes within the steering housing and, depending on the direction in which the steering wheel is rotated, are subjected to hydraulic pressure.

4.4 Steering column

In accordance with the German standard DIN 70 023 'nomenclature of vehicle components', the steering column consists of the jacket tube (also known as the outer tube or protective sleeve), which is fixed to the body, and the steering shaft, also called the steering tube. This is only mounted in bearings at the top (or top and bottom, positions 9 and 10 in Fig. 4.15) and transfers the steering wheel moment M_H (Fig. 3.77a) to the steering gear.

A cardan elastic joint (part 10 in Fig. 4.14) can be used to compensate for small angular deviations. This also keeps impacts away from the steering

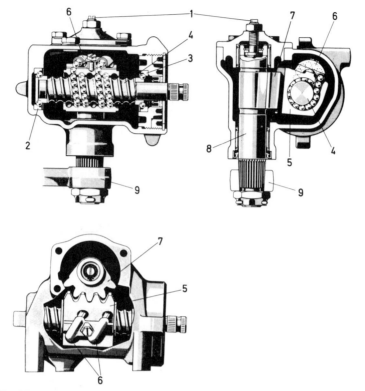

Fig. 4.12 Mercedes Benz recirculating ball steering suitable for passenger cars and light vans; today, apart from in a few exceptional cases, this is only fitted as a hydraulic power-assisted version.

The pitman arm 9 is mounted onto the tapered toothed profile with a slotted castle nut 11.

wheel and, at the same time, performs a noise insulation function on hydraulic power-assisted steering. If the steering column does not align with the extension of the pinion gear axis (or the steering screw), an intermediate shaft with two universal joints is necessary (part 6 in Fig. 4.15).

The steering tube should be torsionally stiff to keep the steering elasticity low but be able to flex in a defined manner in the longitudinal direction with the jacket tube to limit steering wheel intrusion into the vehicle in the case of a head-on crash and so as not to impede the effectiveness of the airbag. Three types of steering tube configuration meet these requirements with vehicle-specific deformation paths on passenger cars:

- steering tubes with flexible corrugated tube part (Fig. 4.14)
- collapsible (telescopic) steering tubes (Figs 4.16 and 4.16a)
- detachable steering tubes (Figs 4.16b and 4.0).

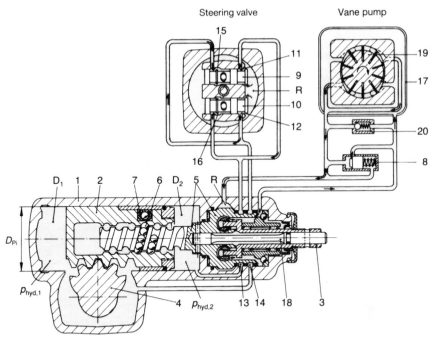

Fig. 4.13 Illustration of the principles of the ZF recirculating ball steering in the neutral position (vehicle travelling in a straight line). The steering valve, the working piston and the mechanical gear sit in a common housing. The two valve pistons of the steering valve have been turned out of their operating plane to make the diagram easier to see. The individual parts are:

     1 gear housing
     2 piston with steering nut
     3 steering tube connection
     4 steering shaft with toothed segment
     5 steering screw with valve body
     6 balls
     7 recirculation tube
     8 fluid flow limitation valve
  9/10 valve piston
 11/12 inlet groove
 13/14 radial groove
 15/16 return groove
    17 fluid reservoir
    18 torsion bar
    19 hydraulic pump
    20 pressure-limiting valve.

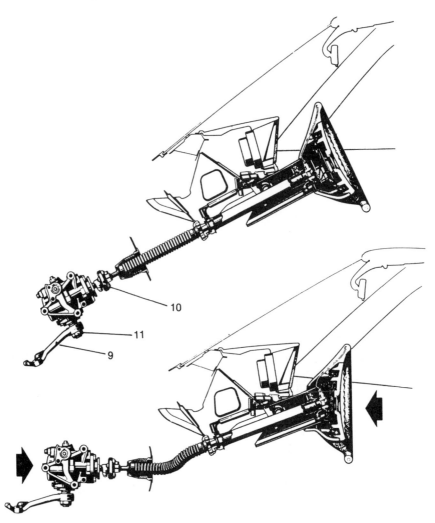

Fig. 4.14 Mercedes Benz safety steering tube and pot-shaped steering wheel; it is fixed to the recirculating ball steering gear with an 'elastic joint'.

The bottom illustration shows the corrugated tube bending out in a head-on crash. The illustration also shows the energy-absorbing deformation of the steering wheel and the flexibility of the steering gear mounting.

To increase ride and seating comfort, most automobile manufacturers offer an adjustable steering column, either as standard or as an option. The position of the steering wheel can then be altered backwards and forwards as well as up and down (positions 1 and 2 in Fig. 4.16c). As can be seen in the illustrations, electrical adjustment is also possible.

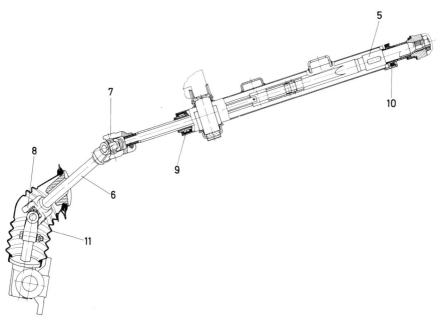

Fig. 4.15 Steering column of the VW Golf and Vento. The collapsible steering tube shown below is carried from the bottom by the needle bearing 9 and through the top by the ball bearing 10 in the jacket tube; the spigot of the steering lock grips into part 5. The almost vertical pinion gear of the rack and pinion steering is linked to the inclined steering tube via the intermediate shaft 6 with the universal joints 7 and 8. The dashpanel is sealed by a gaiter between this and the steering gear (illustration: Lemförder Metallwaren).

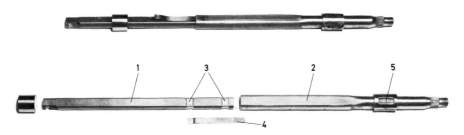

Fig. 4.16 Telescopic collapsible steering tubes consist of a lower part 1, which is flattened on the outside, and a hollow part 2, which is flattened on the inside. The two will be stuck together; the two plastic bushes 3 ensure that the arrangement does not rattle and that the required shear-off force in the longitudinal direction is met. The tab 4 fixed to part 1 ensures the flow of electric current when the horn is operated. The spigot of the steering wheel lock engages with the welded-on half shells 5 (illustration: Lemförder Metallwaren).

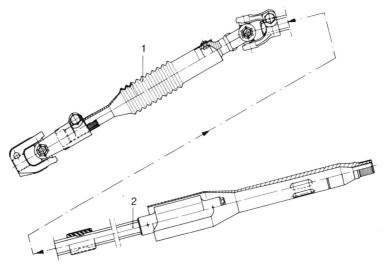

Fig. 4.16a Volvo steering column. Both the corrugated tube 1 in the intermediate shaft and the collapsible steering tube meet the safety requirements. To save weight, the universal joints are made of aluminium alloy Al Mg Si 1 F31, DIN 1725 (German standard) (illustration: Lemförder Metallwaren).

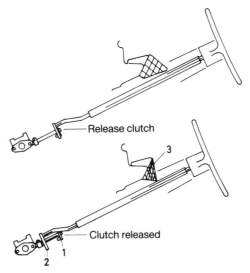

Fig. 4.16b 'Release clutch' used by VW on steering columns. A half-round plate sits on the short shaft that is linked to the steering pinion gear, and carries the two pins 1 which point downwards. They grip into the two holes of the clutch 2 sitting on the steering tube from the top. The jacket tube is connected to the dashboard via a deformable bracket. As shown in a head-on crash, this part 3 flexes and the pins 1 slide out of part 2.

Fig. 4.16c Electrically adjustable steering column manufactured by Lemförder Metallwaren. The electric motor 3 turns a ball nut via the gears 4 and this engages with the grooves 5 of the steering tube and shifts it (position 6) in the longitudinal direction (position 1). To change the height of the steering wheel (position 2), the same unit tips around the pivot 8 by means of the rod 7.

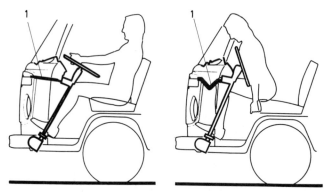

Fig. 4.16d   The VW light van LT has an almost vertical steering column. In a head-on crash, first the steering wheel rim gives and then the retaining strut 1, which is designed so that a given force is needed to make it bend in.

On light vans which have a steering gear in front of the front axle, the steering column is almost vertical (Figs 1.3 and 1.22). In a head-on crash the jacket tube bracket 1 and the steering wheel skeleton must flex (Fig. 4.16d)

4.5 Steering damper

Steering dampers absorb jolts and torsional vibrations from the steering wheel and prevent the steering wheel over-shooting (also known as free-control) on front-wheel drive vehicles – something which can happen when the driver pulls the steering wheel abruptly. The dampers therefore increase ride comfort and driving safety, mainly on manual steering gears. The setting, which generally operates evenly across the whole stroke range, allows sufficiently light steerability but stops uncontrollable wheel vibrations where the front wheels are subjected to uneven lateral and longitudinal vibrational disturbances; in this event the damper generates appropriate forces according to the high piston speeds involved.

The dampers are fitted horizontally. As shown in Fig. 1.40, on rack and pinion steering, one side of the damper is fixed to the steering rack via an eye or pin type joint and the other to the steering housing. On recirculating ball steering systems, the pitman arm on independent wheel suspensions or the intermediate rod can be used as a pivot point (Figs 4.10 and 1.23a) or the tie rod on rigid axles. As shown in Fig. 4.4, this is parallel to the axle housing. Section 5.6.5 describes how the non-pressurized monotube damper works.

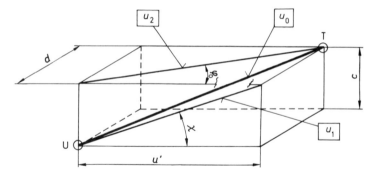

Fig. 4.17 On independent wheel suspensions, the tie rod $\overline{UT}$ is spatially inclined. The path u' (i.e. the lateral distance of points U and T from one another) or the angle κ must be determined when viewed from the rear. From the top view, the distance d or the angle φ_0 is more important; the projected lengths which appear in both views are u_1 and u_2. The true tie rod length is then:

$$u_0 = (u'^2 + c^2 + d^2)^{\frac{1}{2}}$$

4.6 Steering kinematics

4.6.1 Influence of the steering gear

Calculating the true tie rod length u_0 (Fig. 4.17) and the steering arm angle λ (Fig. 4.2) creates some difficulties in the case of independent wheel suspensions. The position of the steering column influences the position of the steering gear by the type of rotational movement. If this deviates from the horizontal by the angle ω (Fig. 4.18), a steering gear shaft, which is also inclined by the angle ω, becomes necessary. The inner tie rod joint T which sits on the pitman arm, is carried through a three-dimensional arc influenced by this angle ω when the wheels are turned. However, the outer joint U on the steering knuckle and the steering axis is inclined inwards by the kingpin inclination angle σ and is usually inclined backwards by the caster angle τ (shown in Fig. 4.18). This joint therefore moves on a completely different three-dimensional arc (Figs 3.4, 3.6 and 3.8). The designer's job is to calculate the steering arm angle λ (and possibly also the angle o of the pitman arm, Fig. 4.22) in such a manner that when the wheels are turned, the specified desired curve is produced as much as possible.

Figure 3.71 shows two curves that are desirable on passenger cars with an initially almost horizontal shape ($\Delta\delta \pm 30'$) and a subsequent rise in the curve to nearly half the nominal value when the wheels are fully turned. The more highly loaded wheel on the outside of the bend can even be turned further in than the inner wheel (and not just parallel to it, $\Delta\delta < 0$); due to the higher slip angle that then has been forced upon it, the tyre is able to transfer higher lateral forces. When the wheels are fully turned, the actual curve should, never-

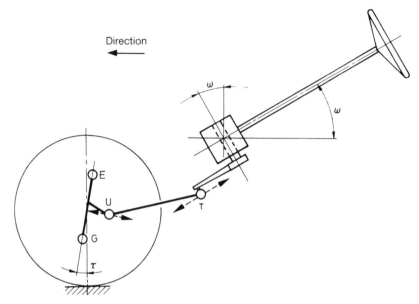

Fig. 4.18 The central points of the tie rod joints (T on the inside and U on the outside) change their position relative to one another, based on the wheel travel on independent wheel suspensions. The reasons for this are the different directions of movement of pitman arm and steering arm. The former depends on the inclined position of the steering gear (angle ω) and that of the point U from the inclination of the steering axis EG, i.e. the kingpin inclination σ and the caster angle τ.

theless, remain below the nominal curve to achieve a smaller turning circle (see Equation 3.14).

The steering angle $\Delta\delta_o$ of the wheel on the outside of the bend depends on the angle of the one on the inside of the bend δ_i via the steering difference angle $\Delta\delta$:

$$\Delta\delta = \delta_i - \delta_o \text{ (axis of the ordinate, Fig. 3.71)}$$

4.6.2 Steering square and steering triangle

The main influences on $\Delta\delta$ are the steering arm angle λ, the inclined position of the tie rod when viewed from the top (angle φ_o, Fig. 4.17) and the angle O of the pitman and idler arms on steering gears with a rotational movement. The tie rod position is determined by where the steering gear can be packaged. The amount of space available is prescribed and limited and the designer is unlikely to be able to change it by more than a little. The actual task consists of determining the angles λ and O by means of drawing or cal-

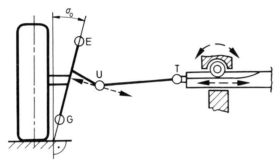

Fig. 4.19 When viewed from the rear, the inner tie rod joint T on rack and pinion steering moves parallel to the ground, whilst the outer tie rod joint U moves on an arc running vertical to the steering axis $\overline{EG}$. Any caster angle τ must also be considered.

culation. Both also depend on the bearing elasticities, which are not always known precisely.

The configuration of the steering kinematics on rack and pinion steering is comparatively simple; here, it is only necessary to transfer a straight line lateral shift movement into the three-dimensional movement of the steering knuckle (Fig. 4.19). However, the extension of the tie rod $\overline{UT}$ must point to pole P (Fig. 4.20); this is necessary on all individual wheel suspensions for determining the body roll centre R_f and is therefore known (see Sections 3.4.3 and 4.6.3).

On steering gears with a rotational movement, the steering square can be either in front or behind the axle and can be opposed or synchronous; Figs 4.2 and 4.21 to 4.23 show four different configurations.

From a kinematic point of view, rack and pinion steering systems have a steering triangle that can either be in front of or behind the axle or even across it. Figures 4.3 and 4.24 to 4.26 show the individual options for left and

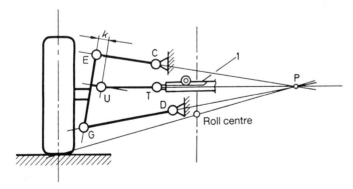

Fig. 4.20 Path and movement points necessary for determining the tie rod length and position. The position of the tie rods is given by the connecting line $\overline{UP}$ (to the pole). The illustration also shows the roll centre R.

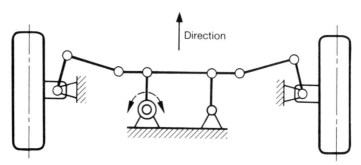

Fig. 4.21 Synchronous steering square with steering arms pointing forwards. The inner joints are fixed to the sides of the intermediate rod.

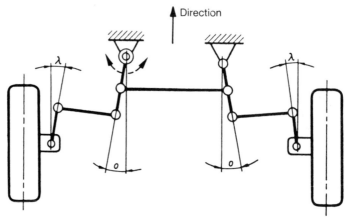

Fig. 4.22 Opposed steering square located in front of the wheel centre. Steering arm and pitman arm rotate in opposite directions towards one another, similar to meshing gears. The tie rods are fixed directly to pitman and idler arms. For kinematic reasons, these can have the pre-angle o (see also Fig. 1.3).

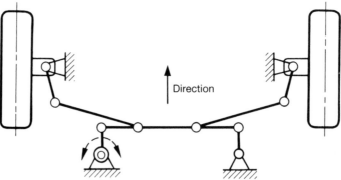

Fig. 4.23 Opposed steering square located behind the wheel centre. The inner tie rod joints can be fixed to the middle part of the intermediate rod or directly to the pitman and idler arm. (See Figs 4.10 and 1.23a.)

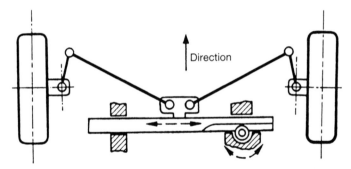

Fig. 4.24 The rack and pinion steering is behind the wheel centre and the steering arms point forward (shown for a right-hand drive vehicle). For kinematic reasons, the inner tie rod joints are fixed to a central outrigger – known as a central take-off; this type of solution (also shown in Figs 1.40 and 4.0) is necessary on McPherson and strut damper front axles with a high-location steering system.

right-hand drive vehicles and also where the pinion gear must be located – above or below the steering rack – to make the wheels turn in the direction in which the steering wheel is turned. The steering arms (negative angles λ) which point outwards, shown in Fig. 4.26, allow longer tie rods; something which is useful when the inner joints are pivoted on the ends of the steering rack (Fig. 3.52a).

The significantly simpler steering kinematics on rigid axles are shown in Figs 4.4 to 4.5a.

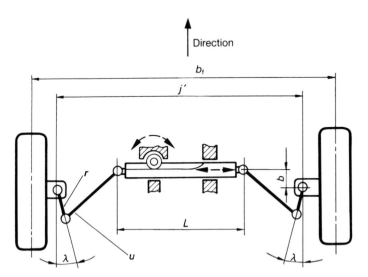

Fig. 4.25 The steering is in front of, and the steering triangle behind, the wheel centre and the inner joints are fixed to the ends of the steering rack.

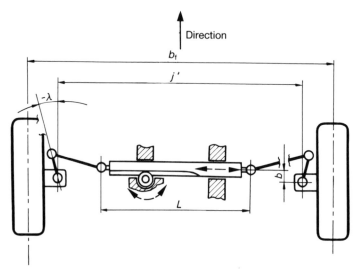

Fig. 4.26 Where rack and pinion steering and steering triangle, are shifted in front of the wheel centre, for kinematic reasons the steering arms must point outwards, making longer tie rods possible (see also Fig. 1.24).

4.6.3 Tie rod length and position

When the wheels compress and rebound, there should be no, or only a certain determined, toe-in alteration; both depend primarily on the tie rods being the correct length and on their position. The individual figures in Section 3.6 show the results of incorrect toe-in and the possibility of achieving a roll–steer effect on the front wheels and counter-steering during braking. The elasticity in the steering system (Figs 3.77a and 3.77b) or that in the bearings of the steering control arms, is also a contributory factor.

4.6.3.1 On double wishbone suspensions   There are two ways of determining the central point T of the inner tie rod joint as a function of the assumed position U of the outer joint, the template and the pole procedure. Both methods consider one side of the front axle when viewed from the rear (here the left side, Fig. 4.20). The projected length u' of the tie rod shown in Fig. 4.7 and the angle κ, which determines its position, must be calculated. This must match the line connecting the outer joint U with pole P, which is also needed for calculating the roll centre (see Section 3.4.3).

Initially, the position of the outer tie rod joint U is unknown when viewed from the rear; to obtain an approximation of this point, the height of the steering gear must be specified (Fig. 4.20). The angle λ is assumed so that, together with the known steering arm length r, the path required for configuring it

$$k = r \sin \lambda$$

can be calculated (for r and λ see Fig. 4.25).

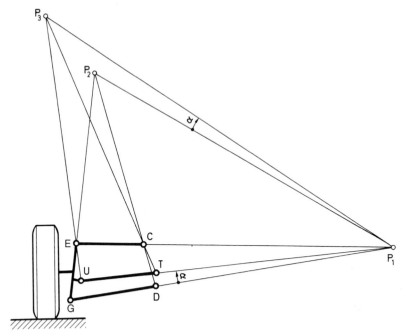

Fig. 4.27 Double wishbone suspension with steering arm pointing inwards. The tie rod is above the lower control arm.

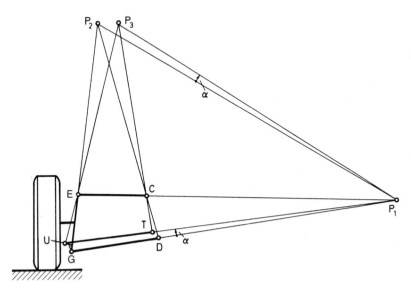

Fig. 4.28 In the case of a steering gear located in front of the wheel centre, the centre of the tie rod joint U lies outside the steering axis $\overline{EG}$.

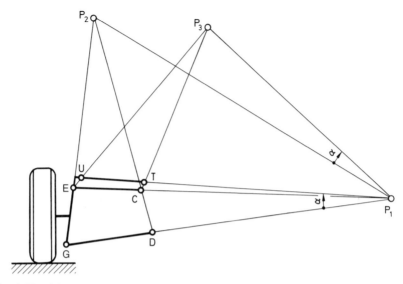

Fig. 4.29 A high-location steering gear can involve a tie rod above the upper control arm. The steering arm points backwards and towards the inside in the example.

The templates that are used for finding point T by drawing have already been described in Section 3.3 and can be seen in Figs 3.4 to 3.8. All figures contain point U and the curve of its movement. It only remains to find point T on the connecting line $\overline{UP}$. T would be the centre point of the arc which best covers the path of point U.

It is likely to be simpler and more precise to determine the point T graphically using the poles. First, as shown in Figs 4.27 and 3.19 to 3.21b, the pole P (marked here as P_1) must be calculated so that it can be connected to U. The extension of the paths $\overline{EG}$ and $\overline{DC}$ gives the pole P_2, which is also required and from which a line to P_1 must be drawn. If the path $\overline{UP_1}$ is above $\overline{GD}$, the angle α enclosed by the two must be moved up to $\overline{P_1P_2}$; if $\overline{UP_1}$ were to lie below it, the line would have to be moved down. A line drawn from P_1 at the angle α must be made to intersect with the extension of the connecting path $\overline{UE}$ to give the tie rod pole P_3. To calculate the point T sought – i.e. the centre of the inner joint – P_3 is connected to C and extended.

The path k (i.e. the distance of point U from the steering axis $\overline{EG}$, Fig. 4.20 and Equation 4.1a) is the determining factor for the position of pole P_3 in the lateral direction. Figure 4.28 shows the case of point U, which lies left of the path $\overline{EG}$; something that is only possible where the steering gear is located in front of the axle (Fig. 4.26). Pole P_3 moves to the right, resulting in an inner link moving further away from the centre of the vehicle. This is beneficial if it is to be fixed to the end of the steering rod.

A tie rod that is located above the upper suspension control arm (Fig. 4.29) causes a large angle α and a pole P_3 that is shifted a long way to the right. Where the control arms are parallel to one another (Figs 4.30 and 3.20), P_1 is

Fig. 4.30 Suspension control arms, which are parallel to one another in the design position of the vehicle, have to have a tie rod in the same position.

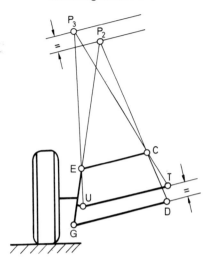

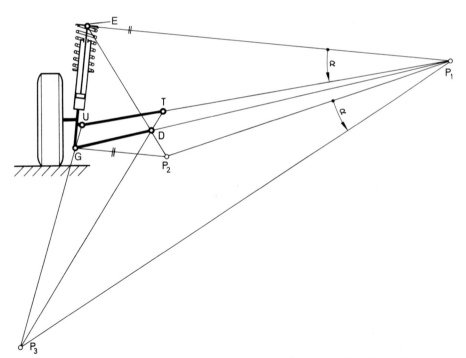

Fig. 4.31 On the McPherson strut or strut damper, the tie rod is above the lower control arm; the steering arms point inwards with the result that the outer joint U lies more to the vehicle centre.

at ∞. In such cases, a line parallel to the path $\overline{GD}$ must be drawn through U and, at the same distance, a further one drawn through the pole P_2. The intersection of this second parallel with the extension of the path $\overline{UE}$ gives P_3, which must be linked to C to obtain T.

4.6.3.2 In the case of McPherson struts and strut dampers   When the vehicle is fitted with McPherson struts or strut dampers – due to the alteration in distance between E and G when the wheels compress and rebound – point T is determined by a different method. To obtain point P_1, a vertical to the centre line of the shock absorber is drawn in the upper mounting point E and made to intersect with the extension of the suspension control arm $\overline{GD}$ (Figs 4.31 and 3.22); P_1 linked with U gives the position of the tie rod. A line parallel to $\overline{EP_1}$ must be drawn through G; the intersection with the extension of $\overline{ED}$ then gives the second pole P_2. The angle α, included by the paths $\overline{EP_1}$ and $\overline{UP_1}$, must be entered downwards to the connection $\overline{P_1P_2}$ to obtain P_3 as the intersection of this line with the extension of the path $\overline{UG}$. The extension of the connecting line $\overline{P_3D}$ then gives the central point T of the inner tie rod joint on $\overline{UP_1}$.

If, in the case of λ = 0°, point U is on the steering axis $\overline{EG}$, which dominates the rotation movement (Figs 4.32 and 3.23), P_3 is on the extension of this path. The determining factor for the position of P_1 is the direction of the shift in the damping part of the McPherson strut; for this reason, the vertical

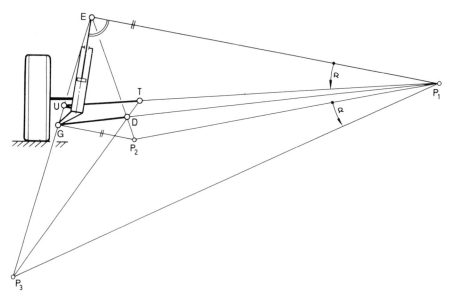

Fig. 4.32   On a McPherson strut with the joint G shifted to the wheel, the outer one, U of the tie rod, can lie in the plane of the steering axis (i.e. on the connecting line $\overline{EG}$) when viewed from the rear. Extending the path $\overline{UG}$ is crititical for determining the pole P_3, whilst the direction of movement of the damper, i.e. a vertical on the piston rod in point E must be the starting point for calculating P_1.

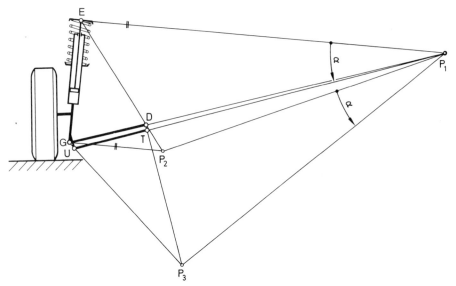

Fig. 4.33 The tie rod can also lie under the control arm when the steering arm points inwards.

in point E must be created on its centre-line (not on the steering axis $\overline{EG}$). The important thing in this calculation is the position of point U, i.e. the extension of the connecting line $\overline{UG}$ downwards. U is shown on the steering axis $\overline{EG}$ simply for reasons of presentation.

A low mounted tie rod causes the pole P_3 to move to the right (Fig. 4.33) and this then causes a shorter rod. This situation is favourable if the inner joint needs to sit on the ends of the steering rack. The figures clearly show that the higher U is situated, the longer the tie rods must be, i.e. a centre take-off becomes necessary on rack and pinion steering (Figs 1.40, 4.0, 4.9 and 4.24).

4.6.3.3 On longitudinal transverse axles   On longitudinal wishbone axles the upper point E moves in a straight line vertical to the steering axis $\overline{CF}$ and the lower point G on an arc around D (Figs 4.34 and 3.25). To obtain P_1, a parallel to $\overline{CF}$ must therefore be drawn through E and made to intersect with the control arm extension $\overline{GD}$. A parallel to $\overline{EP_1}$ laid through point D gives the pole D_2 on the connecting line $\overline{EG}$. The angle α enclosed by the paths $\overline{EP_1}$ and $\overline{UP_1}$ must be drawn downwards to the connecting line $\overline{P_1P_2}$ to obtain the pole P_3 as the intersection with the extension of the path $\overline{UG}$. P_3 linked with D then gives the centre T of the inner tie rod joint.

4.6.3.4 Reaction on the steering arm angle λ   Figures 4.27 to 4.34 indicate that shifting the outer joint U to the side results in a slight alteration in the distance $\overline{UT}$. However, this shift is necessary if the angle λ has to be reduced

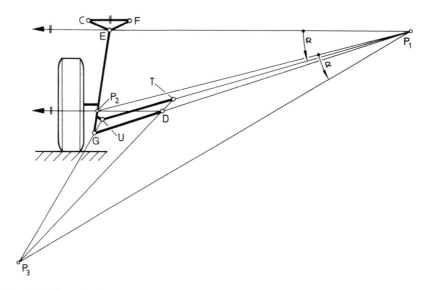

Fig. 4.34   Longitudinal transverse axle with the tie rod located above the lower control arm and the steering arm pointing inwards.

or increased with a given steering arm length *r*. The projected length u' of the tie rod, and therefore also its overall length u_0 (Fig. 4.17), changes when viewed from the rear. However, the latter is one of the determining factors for the aspects relating to the steering angles δ_i (inside) and δ_o (outside), i.e. for the actual steering curve sought (Fig. 3.71). It is, therefore, likely to be essential to check the position of point T with the tie rod, which has become longer or shorter.

5

Springing

5.1 Comfort requirements

Springing and damping on a vehicle are mainly responsible for the

- ride comfort

and play an important part in the

- handling (Fig. 5.0) and
- the tendency of the body to roll and pitch.

Other important influences on the handling are the kinematic changes of the wheels as they compress and rebound and the elastokinematics. Details are given in Chapter 3.

The ride comfort experienced by the vehicle occupants is attributable to a combination of the acceleration and the mechanical vibration acting upon them. The critical frequency range is 1 Hz to 80 Hz. It is sensible to subdivide this into two ranges to which different comfort terms are allocated. These are the

- springing or ride comfort, which lies below $n = 240$ min^{-1}, i.e. 4 Hz
- wheel comfort or road harshness ($f > 4$ Hz).

The split is sensible because the two frequency ranges are experienced differently by the human body and become important if individual parts of the suspension, such as springs, shock absorbers, suspension link bearings etc., are to be evaluated for their influence on comfort.

The springing balance (which expresses how well the front and rear axles are matched to one another) should also be taken into consideration. If a vehicle does not pitch when it goes over bumps in the ground, but instead moves up and down in parallel, it has a good springing balance. To provide

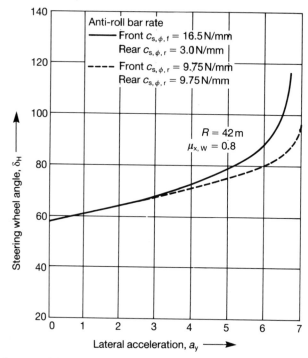

Fig. 5.0 Influence of the anti-roll bar rate on the steering angle, measured whilst the vehicle is steady-state driving on a circuit (R = 42 m) on a standard design passenger car with $m_{V,t}$ = 1544 kg. The understeering can be reinforced, or a incipient tendency to oversteer reduced by strengthening the front anti-roll bar and/or adding a thinner rear one. On front-wheel drive vehicles a more highly stabilized rear axle is usually necessary.

In the case of low lateral acceleration, and therefore also on wet or slippery roads, the anti-roll bar rate has no effect on the self-steering behaviour.

an objective evaluation of the comfort, measurement devices are used which, based on the VDI (Verein Deutscher Ingenieure) directive 2057, evaluate the vibration that occurs (vibration stroke, speed and acceleration) dependent on frequency, in accordance with existing knowledge of how the human body experiences it. The measurement result is then available as a numerical value, the so-called K-value. Low values indicate good ride comfort, whilst high values refer to poor ride comfort.

5.1.1 Springing comfort

This comfort range is mainly influenced by the acceleration acting on the upper part of the human body and lies in a frequency range of 1 to 4 Hz. With a given vehicle body mass m_{Bo} (see Equation 6.5), the configuration of the rate of the body springing and the body's resonant frequency are critical

Fig. 5.1 When a wheel rebounds by the path s_2, the wheel load reduces by the amount $\Delta F_{z,W}$. The level of the residual force which still ensures wheel grip $F'_{z,W} = F_{z,W} - \Delta F_{z,W}$ depends mainly on the springing hardness, or rather the rate $c_{f\,or\,r}$.

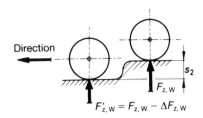

variables. In accordance with the substitute system for the sprung-mounted car body shown in Fig. 5.3 (single mass vibrator) the mass portion $m_{Bo,f\,or\,r}$ exhibits free undamped vibration at the natural frequency in accordance with Equation 5.4, and the corresponding vibration rate in accordance with Equation 5.4a.

The softer the springing, i.e. the lower the springing rate $c_{f\,or\,r}$ of the body (front or rear), the lower the natural frequency for a specified body mass and, accordingly, the higher the ride comfort. Unfortunately, at the same time the roll pitch increases on bends (it must be reduced by anti-roll bars, see Section 5.5.4 and Fig. 5.0), as does the tendency to pitch when the brakes are applied (see Sections 3.11 and 6.3). Vibration values of $n_{f\,or\,r} = 60$ min^{-1} (where $f = 1$ Hz) are desirable, but cannot necessarily be easily achieved (see Section 5.2).

Another advantage of soft springing would be the improvement in the absorbency of bumps and the wheel grip. If, for example, a front wheel loaded at $F_{z,W} = 3000$ N drops into a 60 mm deep pothole (Fig. 5.1), with soft linear springing at the rate $c_f = 15$ N mm^{-1} the residual force at the bottom of the pothole is

$$F'_{z,W,f} = F_{z,W,f} - c_f s_2 = 3000 - 15 \times 60 = 2100 \text{ N} \tag{5.0}$$

whilst with sporty hard springing at $c_f = 30$ N mm^{-1}, it would be only $F'_{z,W,f} = 1200$ N. The greater residual force equates with better road holding. The same can be said of a vehicle travelling over a 40 mm high bump (Fig. 5.2). With hard springing, the force transferred from the axle to the body as an impact, ignoring the damping and time influence, would be $\Delta F_{z,W} = 1200$ N; soft springing only transfers 600 N and therefore generates lower wheel load fluctuation.

The disadvantage (as already mentioned) is the greater body roll pitch on bends and the concomitant lower ability of the wheels to transfer lateral

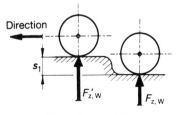

Fig. 5.2 When the wheel compresses by the path s_1 the wheel load increases by $\Delta F_{z,W}$. The size of the increase in force in the body depends mainly on the springing hardness, i.e. the rate $c_{f\,or\,r}$.

forces (see Section 5.4.3 and Equation 2.16). As shown in Fig. 1.2, the wheels incline with the body on independent wheel suspensions. The wheel on the outside of the bend, which absorbs most of the lateral forces, loses negative camber (or goes into positive camber), resulting in the need for a larger tyre slip angle (see Section 2.8.5.5). The springing comfort, and associated with it also the handling, depends not only on the weight of the vehicle and the body springing rate, but also on other variables and the interaction of the individual components:

- the load distribution (see Section 5.3.6)
- the type of mounting and design of the springs (see Section 5.3)
- the anti-roll bars (Fig. 5.0 and Section 5.5.4)
- the torsional rate of the rubber bushings (Figs 5.2a, 3.6 and 3.65a to 3.67)
- the shock absorbers and their mountings (see Section 5.6.7)
- the weight of the axles (see Section 6.1.3)
- the type of engine and gearbox mounting
- the wheelbase (see Section 3.2)
- the tread width (see Section 3.3) and, to a very large extent
- the tyres (see Section 2.4).

5.1.2 Running wheel comfort

Even new-looking road surfaces have almost invisible slight irregularities and bumps, which are transferred to the body as high frequency acceleration and jolts (4 to 80 Hz). The vehicle occupants feel them in the underbody of the vehicle, in the seat cushion, and the driver also feels them in the steering wheel. They determine the wheel comfort and the concomitant road harshness.

The cause of this is in the, often limited, vibration insulation between the suspension parts and the body, i.e. the suspension links, suspension subframe and McPherson strut mount, plus the friction in the suspension control arm bearings (Fig. 5.2a), the wheel joints (Fig. 1.23) and in the shock absorbers or spring dampers (Fig. 5.36).

On McPherson struts and strut dampers the friction in the piston rod guide generated by transverse forces can be the cause (see Figs 1.4, 1.6 and 3.23). The springing does not respond as well and today's ever-wider (and therefore harder tyres) no longer absorb the bumps, which are transferred directly to the body.

The aspects related to this can easily be explained using the hysteresis of a springing curve (Fig. 5.2b). The friction force is 200 N per wheel in the central range, i.e. starting from the centre line it is:

$$F_{fr} = \pm 100 \text{ N}$$

The rate of body springing at the front should be $c_f = 15 \text{ N mm}^{-1}$ and the height of a bump $s_1 = 6 \text{ mm}$ results in a spring force of:

$$\Delta F_{z,W,f} = c_f s_1 = 15 \times 6 = 90 \text{ N} \qquad (5.0a)$$

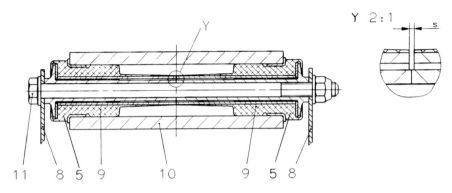

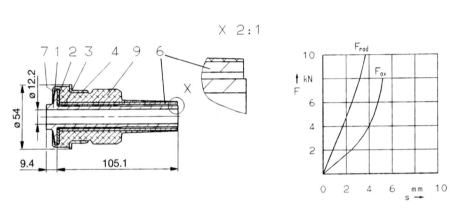

Fig. 5.2a The mounting of the upper control arm of the double wishbone front axle on the Mercedes C class, manufactured by Lemförder Metallwaren. The inner tubes 1 within the two brackets 8 on the wheel house panel are fixed using the hexagonal bolt 11. Rubber parts (position 9) are vulcanized onto the intermediate tubes 6, which are pressed into the suspension control arms. Flanges 5 on both sides absorb the axial forces $F_{a,x}$. The compliance in this direction and the low compliance in the radial direction (F_{rad}) are indicated in the diagram. To keep the friction moment $M_{fr} = 1$ N m there are PTFE coated guide bushes 3 between the tubes 1 and 6 and the discs 2 between the lateral flanges 5. The lips 7 provide the seal to the maintenance-free mountings. The smaller the moment M_{fr} can be, the more favourable the ride comfort and the absorbency of the springs become.

The outer tubes 6 are slightly shorter than the inner ones (position 1), between them is the clearance s, which evens out installation tolerances and provides the longitudinal mobility to take the radial tyre rolling hardness (see Section 3.6.5.2). In the case of high (axial) braking forces, depending on the compliance of the rubber flange 5, the outer tubes 6 butt up against one another and ensure the necessary longitudinal stiffness.

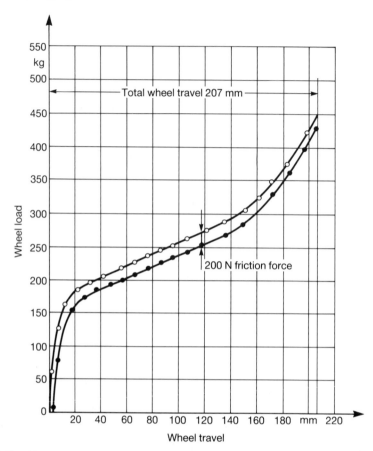

Fig. 5.2b Hysteresis of the curve of front wheel springing shown in Fig. 5.4a; the line distance indicates the friction force in the suspension parts, i.e. the self-damping. This is 200 N in total, i.e. $F_{fr} = \pm100$ N (taking the mean value as nominal).

Since $F_f > \Delta F_{z,w,f}$, in this instance the soft springing would not absorb the bump and the suspension would transfer the force onto the body (see also Section 2.5).

However, if the spring rate is $c_f = 30$ N mm^{-1}, the force would be absorbed by the spring. The problem here is reversed, as shown on the basis of Figs 5.1 and 5.2.

Soft springing creates greater difficulties in achieving the desired running wheel comfort in terms of road harshness than harder springing, particularly on front-wheel drive vehicles.

There is also the longitudinal vibration caused by the steel belts of the radial tyre, particularly on rough cobbles. Section 2.2.2 contains details and Section 3.6.5.2 explains how this vibration can be kept away from the body.

The design complexity is likely to be greater on driven wheels than on non-driven ones.

5.1.3 Preventing 'front end shake'

'Front end shake' is a term used to describe short, hard jolts (in the vertical direction) in the body floor and the front end of the vehicle which, particularly on front-wheel drive vehicles, are triggered by movement of the engine on the rubber parts of the engine mountings and are in a frequency range of around 8 to 12 Hz. This vibration does not occur continuously, but whenever the engine mounting, the frequency of which is often very close to that of the suspension, begins to resonate. The softer these bearings can be, the less engine noise and vibration will be transmitted to the vehicle interior, but the higher will be the tendency to 'front end shake'. Conversely, hard mountings reduce front end shake, but more engine noise is transferred to the vehicle interior. To solve this conflict of aims, hydraulically dampened engine mountings, so-called hydro-mounts, are used and these have a lower static spring rate and, in the event of resonance, generate far higher damping than is possible with normal elastomer mountings.

5.2   Masses, vibration and spring rates

For determining the vibration rates, $n_{f\,or\,r}$ (front or rear) of the body and the spring rate $c_{f\,or\,r}$ the front axle load $m_{V,f,pl}$ (or $m_{V,max,f}$) and the rear axle load $m_{V,r,pl}$ (or $m_{V,max,\,r}$) must be known in the design position (see Section 5.3.4, index pl = partly loaded) and for a permissible gross vehicle weight $m_{V,t,max}$. Mostly, in the case of the maximum payload the permissible rear axle load $m_{V,max,r}$ is fully utilized, whilst, in this instance, the resulting front axle load $m_{V,f,lo}$ (index lo = loaded) needs to be calculated from the maximum gross vehicle weight $m_{V,t,max}$ (see Equation 5.9):

$$m_{V,f,lo} = m_{V,t,max} - m_{V,max,r} \; [\text{kg}] \tag{5.1}$$

The mass proportions $m_{1,Bo,f}$ and $m_{1,Bo,r}$ of the body, which each load one axle side, front and rear, can be calculated using the axle load and the weight $m_{u,f}$ and $m_{u,r}$ of the front and rear axles (unsprung masses) based on both axle sides (see Section 6.1.3).

$$m_{1,Bo,f} = \frac{m_{V,f} - m_{u,f}}{2} \tag{5.2}$$

$$m_{1,Bo,r} = \frac{m_{V,r} - m_{u,r}}{2} \tag{5.3}$$

The axle masses comprise the weights of the wheels and wheel carriers. The latter can be the steering knuckles or, in the case of rigid axles, the axle housing including the differential. There is also the proportional (sometimes half) weight of the parts, which flexibly connect the actual axle with the body or frame. This includes:

- suspension control arms
- tie rods
- axle shafts
- leaf or coil springs
- shock absorbers
- anti-roll bar arms
- panhard rod etc.

The other half of the weight is accounted for by the body. Torsion bars are in the underbody, so their weight forms part of the sprung mass.

Section 6.1.3 contains all details and Equation 6.4c contained therein makes it possible to determine the approximate weight of an axle based on its design.

The spring rate $c_{f \, or \, r}$ (Fig. 5.4a) is required for calculating the spring itself and the configuration of the suspension. This should appear in N mm^{-1} on drawings and as a measurement value, whilst in all calculations the unit is N m^{-1}. If this stipulation is not complied with, there is a risk of minor calculation errors, unless these have been recognized when a dimension equation

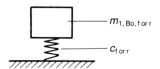

Fig. 5.3 On the simple vibration system, the level of the body frequency $n_{f \, or \, r}$ (front or rear) depends only on the weight or mass proportion $m_{1, Bo, f \, or \, r}$ of the body over a front or rear axle side and the spring rate $c_{f \, or \, r}$, which on linear springing is a quotient of force and travel: $c_{f \, or \, r} = F/s$. On progressive springing, the change in force ΔF over a minimum travel range Δs plays a part $c_{f \, or \, r} = \Delta F/\Delta s$ (see Fig. 5.7).

was done. With the international units the equation for the angular frequency ω is as follows (Fig. 5.3):

$$\omega = \left(\frac{c}{m} \right)^{\frac{1}{2}} \left[\left(\frac{N}{m \, kg} \right)^{\frac{1}{2}} \right]$$

The conversion 1 N = 1 $\dfrac{kg \, m}{s^2}$ results in: (5.4)

$$\left[\frac{kg \, m}{s^2 \, m \, kg} \right]^{\frac{1}{2}} = s^{-1} \quad and \quad [1 \, s^{-1} = 1 \, Hz]$$

To obtain the vibration rates $n_{\text{f or r}}$ (per minute) used in the springing layout, the angular frequency needs to be multiplied by:

$$60/2\pi = 9.55 \, [\text{s min}^{-1}]$$

Related to the body, if the damping, the influence of the mountings and the tyre were ignored, the equation (with indexes) would then be:

$$n_{\text{f or r}} = 9.55 \left(\frac{c_{\text{f or r}}}{m_{1,\text{Bo,f or r}}} \right)^{\frac{1}{2}} [\text{min}^{-1}] \tag{5.4a}$$

The calculation of the vibration rate $n_{\text{u,f or r}}$ of one axle side (front or rear) includes half the axle weight

$$m_{1\text{u,f or r}} = m_{\text{u,f or r}}/2 \tag{5.5}$$

in kilograms and the tyre spring rate $c_{\text{T,f or r}}$ in N m^{-1}. Figures 2.18 and 2.18a show statically measured values which increase during driving (see also Section 2.2.8). The factor k_{T} includes the springing hardening of around 1% per 30 km h^{-1} (see also Section 2.2.8):

$$k_{\text{T}} \approx 1.04 \text{ at } 120 \text{ km h}^{-1} \tag{5.5a}$$

The equation for the axle vibration rate is then (Fig. 5.4).

$$n_{\text{u,f or r}} = 9.55 \left(\frac{k_{\text{T}} \cdot c_{\text{T,f or r}} + c_{\text{f or r}}}{m_{1,\text{u,f or r}}} \right)^{\frac{1}{2}} [\text{min}^{-1}] \tag{5.6}$$

In the case of passenger cars with steel springs the body vibration rate should be:

Front $n_{\text{f}} = 60 \text{ min}^{-1}$ to 80 min^{-1}
Rear $n_{\text{r}} = 70 \text{ min}^{-1}$ to 90 min^{-1}

For reasons of comfort, n_{f} should be approximately 60 min^{-1}, which is extremely difficult to achieve on the front axles of small to medium-sized passenger cars and can only be achieved at the back if the vehicle is fitted with an automatic level control. The load difference between the loading condi-

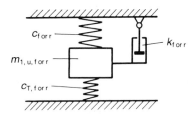

Fig. 5.4 The level of the wheel vibration rate $n_{\text{w,f or r}}$ is a function of the axle mass $m_{1,\text{u,f or r}}$, of the body springing rate $c_{\text{f or r}}$, the tyre springing rate $c_{\text{T,f or r}}$ and the damping $k_{\text{f or r}}$. The driving speed also has an influence (shown in Fig. 2.18a).

tions for one person and fully laden (Figs 5.9 and 5.10) makes it difficult to design the springing on the rear axle to be soft.

There are further limitations on the front axle. The fact that the engine hood is low, both for aesthetic reasons and because of the requirement for low air drag values, limits the space available for the springs, particularly in the case of McPherson struts. So as not to exceed the material stresses, soft springs are longer and their block length is therefore larger; harder springing does not have this disadvantage (Fig. 5.8). However, this reduces the comfort despite the fact strut dampers do allow longer spring travel (Figs 1.25 and 5.7).

The spring rate $c_{\text{f or r}}$ can be calculated on the basis of a specified vibration level $n_{\text{f or r}}$ using the transformed equation 5.4a:

$$c_{\text{f or r}} = 0.011 \, n^2_{\text{f or r}} \, m_{1,\text{Bo,f or r}} \qquad (5.7)$$

The frequency figure in min^{-1} and the mass in kg must be inserted.

The front wheel springing of a front-wheel drive vehicle can be used as an example, whereby the specified loads correspond to the lower limit, i.e. with one person in the vehicle:

Front axle load	$m_{\text{V,f}} = 455$ kg
Rate of the unsprung mass	$m_{\text{u,f}} = 55$ kg
Specified frequency of sprung mass	$n_{\text{f}} = 60 \, \text{min}^{-1}$

In accordance with Equations 5.2 and 5.7:

$m_{1,\text{Bo,f}} = (475 - 55)/2 = 210$ kg, and
$c_{\text{f}} = 0.011 \times 60^2 \times 210 = 8316 \, \text{N m}^{-1}; c_{\text{f}} = 8.3 \, \text{N mm}^{-1}$

Figure 5.4a shows the springing curve with the calculated rate (and associated long paths). The design or zero position (i.e. when there are three people each weighing 68 kg in the vehicle, see Section 5.34), is entered as a further point of reference and the weighed wheel load as a function of the wheel travel is shown. This load is observed in the centre of tyre contact.

In the reverse situation, the springing rate can be calculated from an existing springing curve as a function of the various loading conditions. In the case of a curve that is linear over the middle range (as shown in Fig. 5.4a), it only needs to be extended over the whole spring travel for it to be possible to read the load difference at the end point (here 338.5 kg and 164 kg). This, multiplied by 9.81 and divided by the total travel ($s_{\text{t}} = 207$ mm), gives the spring rate. The formula in the caption to the figure indicates the forces in kN.

In the case of a progressive curve, a tangent must be drawn to the curve for the loading condition to be observed and for it to be possible to read the difference values of loads and paths from it. Figure 5.7 shows an example relating to the design position. The vibration rate can then be calculated from spring rate, axle load and estimated axle weight. This is usually more precise than settling because most vehicles have McPherson struts, strut dampers or

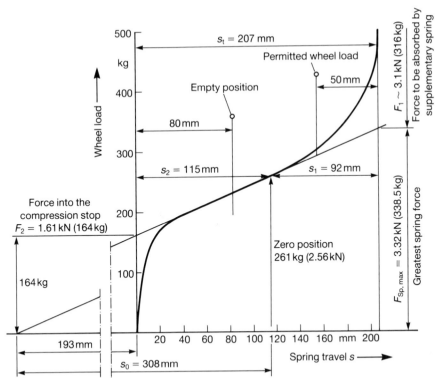

Fig. 5.4a Curve of the front wheel springing of a Renault model, the wheel load (in kg) is entered as a function of the wheel travel (in mm). The soft springing shown requires stops; if the compression stops were missing (Fig. 5.33), the front wheel could rebound from the zero position (the vehicle occupied with three people each weighing 68 kg) by $s_0 = 308$ mm. Where there is no supplementary spring (Figs 5.13 and 5.35), at $F_{Sp,max} = 3.32$ kN the axle would make a hard contact. The residual forces to be absorbed by the spring travel limiters are entered in kN. The progressivity achieved by the supplementary spring can be seen clearly. If the stops are in the shock absorber (Fig. 5.19a), the compliance of the suspension parts also appears in the curve. The rate of the body springing is:

$$C_{f,pl} = \frac{\Delta F_f}{\Delta s_f} = \frac{3.32\ 2\ 1.61}{0.0207}$$

$$C_{f,pl} = 8.26\ \text{kN m}^{-1} = 8.3\ \text{N mm}^{-1}$$

spring dampers and the inherent friction in these parts means a correct result is unlikely.

5.3 Weights and axle loads

Without knowledge of the weight in the empty or loaded condition and dis-tribution of the load to the two axles, springing on a passenger car can neither be designed nor evaluated. The variables of weight and load laid down in German standards relate, in accordance with DIN 70 020, page 2, to the mass (in kilograms or tons) of the vehicle occupants, the transportable items or goods and the vehicle itself.

The following information and details relate to vehicles of class M_1 in accordance with the directive 71/320/EEC of the European Union. These vehicles must be suitable for carrying passengers but may not, apart from the driver's seat, have more than eight seats. They must have at least four (or three) wheels and a total mass of $m_{V,t,max}$ which does not exceed 1 t (tonforce) fully laden.

5.3.1   Curb weight and vehicle mass

The actual weighed curb weight $m_{V,ul}$ of the vehicle is essentially determined by:

- the weight of the body with interior trim and the fuel tank
- the engine and gearbox weight with all necessary accessories, such as starter motor, generator, exhaust system, etc.
- the weight of the chassis
- the optional equipment such as automatic gearbox, air-conditioning system, sun roof, etc. (see Equation 5.8a).

In accordance with the German standard DIN 70 020 the curb weight also includes:

- the charged battery
- lubricant, coolant and brake fluid
- the standard tool set
- a fuel tank at least 90% full.

However, Section 42 of the German Straßen-Verkehrs-Zulassungs-Ordnung (StVZO), the regulations for vehicle approval, demands a full tank.

There are also the pieces of equipment, such as jack, spare wheel, etc. to be carried in the vehicle and, in most countries, the triangular safety reflector and first aid kit. The international recommendation ISO/R 1170 contains further details.

5.3.1.1 Curb weight, manufacturer's data Since the curb weight information required by law allows a tolerance of ±5% – which means a weight range of 110 kg on a 1100 kg vehicle – vehicle manufacturers try to set the curb weight $m_{V,ul,o}$, shown in the vehicle identification card, such that it is both as low as possible (since this governs the balance weight class, which itself is important for the vehicle's fuel consumption and emission rating) and yet still covers as many model versions as possible to keep the technical expenditure (for example, in type approval) low. These aims lead to the optional and supplementary equipment sometimes being ignored. It is therefore not easy for the vehicle's registered keeper to calculate the actual permissible luggage, roof and trailer load (see Section 5.3.3).

5.3.1.2 Mass of driveable vehicle   From 1/1/1996, all new models in the European Union, and from 1/1/1998, all newly registered vehicles of class M_1 must be tested in accordance with EU directive 92/21/EEC.

This specifies that all vehicle manufacturers must quote the mass $m_{V,dr}$ in the driveable condition, i.e. including the weight of the vehicle driver at $m_p = 68$ kg and the luggage mass at $m_b = 7$ kg must be included. Until now, in Germany, this approval condition was only specified for vans and commercial vehicles (class N in the EU Directive 71/320/EEC, see Section 5.3.6.3).

5.3.1.3 Mass of the driveable vehicle when towing a trailer   If the vehicle is intended for towing a trailer, the weight m_{Th} of the towing device and the permissible tongue load ΔM_{Th} under static conditions must be added to the mass $m_{V,dr}$ (Fig. 5.4b and Section 5.3.3.4). The permissible rear axle load must generally be based on this.

5.3.2 Permissible gross vehicle weight and mass

This is specified by the vehicle manufacturer taking into consideration the minimum load – which corresponds to the nominal payload m_t in accordance with ISO 2416 (see Equation 5.7c) – required by law, based on the number of seats provided.

5.3.3 Permissible payload

The permissible payload $m_{t,pe}$ of a passenger car is the load that the driveable vehicle can carry without exceeding the permissible gross vehicle weight. It therefore results from the difference between the permissible gross vehicle weight $m_{V,t,max}$ and the actual curb weight $m_{V,ul}$:

$$m_{t,pe} = m_{V,t,max} - m_{V,ul} \tag{5.7a}$$

Vehicle manufacturers generally specify the payload higher than the regulations demand. This is reflected in a larger permissible gross vehicle weight. The calculation takes into account the component and material stress to be guaranteed, the tyre and wheel bearing load capacity and the loss of braking capacity and handling usually associated with a higher load. The distribution of the goods being transported and the spring travel limitation also play a part in this loss of handling (see Section 5.3.6 and 5.5.3).

There is also the risk of the permissible rear axle load $m_{V,max,r}$ being exceeded with a full boot and it is possible that the front axle then might lift. This is bound to lead to reduced steerability. On a front-axle drive vehicle, traction and climbing capacity are reduced (see Sections 1.1.7 and 6.4). The

Calculation of useful load

Minimum payload for 5-seater	275 kg
Options	70 kg
Trailer support weight	75 kg
Required minimum payload for basic models	520 kg
Manufacturer's determined payload	**530 kg**

Calculation of permissible gross vehicle weight

Curb weight of car (without options)

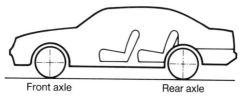

Front axle   Rear axle

Curb weight =   665 kg   +   485 kg   =   **1150 kg**

Calculation of permissible rear axle load

(Distribution of payload to people and baggage)

Without drawbar and trailer support weight

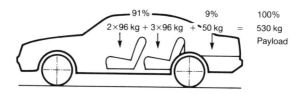

91% — 9%   100%
2×96 kg + 3×96 kg + 50 kg = 530 kg
Payload

Resulting permissible rear axle load   **880 kg**

With drawbar and trailer support weight

(Payload reduced by weight of drawbar and trailer support weight)

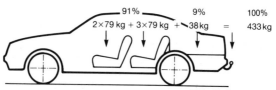

91% — 9%   100%
2×79 kg + 3×79 kg + 38 kg = 433 kg

75 kg Trailer support weight
22 kg Weight of support bar

Resulting permissible rear axle load   **940 kg**

EU directive 92/21 EEC therefore specifies that the front axle load $m_{V,f}$ may not be less than 30% of the vehicle total weight $m_{V,t}$, i.e.

$$m_{V,f} \geqslant 0.3\, m_{V,t} \tag{5.7b}$$

5.3.3.1 To ISO 2416   This standard specifies the minimum payload for passenger cars, i.e. the nominal payload m_t. This depends on the number n of seats provided by the vehicle manufacturer and the passengers' luggage or on the number n_0 of the occupied seats and the luggage mass m_{tr} of the goods then transportable.

To determine the number n, a weight of $m_p = 68$ kg for each person – including clothing – must be added, plus a luggage mass of $m_b = 7$ kg per person. The nominal payload m_t must be:

$$m_t \geqslant (m_p + m_b)\, n \tag{5.7c}$$

The greatest value – i.e. the luggage mass actually transportable m_{tr} – is then:

$$m_{tr} = m_t - m_p \times n_0 \text{ or} \tag{5.7d}$$

$$m_{tr} = m_{V,t,max} - m_{V,ul} - m_p \times n_0 \tag{5.8}$$

Experience has shown that the actual or weighed curb weight $m_{V,ul}$ exceeds the manufacturer's stated curb weight $m_{V,ul,o}$ by the weight of the optional equipment Δm_V found in the vehicle

$$m_{V,ul} = m_{V,ul,0} + \Delta m_V \tag{5.8a}$$

A five-seater passenger car with a permissible payload of $m_{t,pe} = 400$ kg and 20 kg optional equipment can be used as an example:

$$m_{tr} = m_{t,pe} - \Delta m_V - m_p \times n \tag{5.8b}$$
$$m_{tr} = 400 - 20 - 68 \times 5 = 40 \text{ kg}$$

The transportable luggage mass m_{tr} is therefore above the minimum value:

$$m_b = 7 \times 5 = 35 \text{ kg}$$

Fig. 5.4b   The directive 92/21/EEC, due to enter into force in 1996, specifies that when the permissible axle loads are determined, 90.7% (i.e. around 91%) of the permissible payload must be on the seats and 9.3% (or 9%) evenly distributed in the boot. The front seats should be in the furthest back steering (or sitting) position. The medium-size passenger car used as an example has front-wheel drive and, when empty, an axle load distribution 58%/42%.

Without a towing device, the permissible axle load is 880 kg. If the towing device and a tongue load are considered, it is 940 kg.

5.3.3.2 Nominal payload   It is the manufacturer who specifies the payload – and therefore also the permissible gross vehicle weight – taking into consideration the expected use of the vehicle (saloon, estate car, sports coupé, etc.) whilst complying with the legally required nominal payload m_t, i.e. based on the number n of seats provided. In accordance with Equation 5.7c m_t will be for

two people 136 kg + 14 kg luggage = 150 kg
three people 204 kg + 21 kg luggage = 225 kg
four people 272 kg + 28 kg luggage = 300 kg
five people 340 kg + 35 kg luggage = 375 kg
etc.

This means that for a nominal payload of m_t =375 kg a saloon will still be legally approved as a five-seater. The precondition is that the other requirements are met, e.g. in respect of seat belt anchoring.

If five people, each weighing 75 kg, get into a five-seater passenger car, the permissible payload of which, at 375 kg, is at the lower limit, this already gives a figure of 375 kg. If the vehicle has retrofitted accessories not included in the weight calculation or optional equipment Δm_V beyond the normal amount (see Equation 5.8a), the vehicle is already overloaded and it would not be possible to carry any luggage. If, without being aware of the situation, the driver nevertheless puts items of luggage into the boot, the vehicle will exceed the permissible total weight and probably also the permissible rear axle load. If the resulting deterioration in handling or the now insufficient tyre pressure leads to an accident, the driver would be regarded under law in Germany as responsible for the overload. Legal decisions back this up.

5.3.3.3 To EU directive 92/21/EEC   Unlike in Section 5.3.1.2, EU directive 92/21/EEC does not assume the driveable condition $m_{V,dr}$, but rather the curb weight $m_{V,ul}$. The permissible number of occupants must then be calculated in accordance with Equation 5.7c.

5.3.3.4 When towing a trailer   When the vehicle is towing a trailer, EU directive 92/21/EEC specifies that the weight m_{Tr} of the towing device and the greatest tongue load Δm_{Tr} allowed by the manufacturer must be included in the calculation of the necessary nominal payload (Section 5.3.1.3). A five-seater passenger car would then be permissible for the following nominal payload (Fig. 5.4b).

Minimum value for five people	m_p = 375 kg
Weight of optional equipment including towing device (assumed)	$\Delta m_V + m_{Th}$ = 70 kg
Tongue load when towing a trailer	Δm_{Tr} = 75 kg
Required nominal payload	m_t = 520 kg

If the payload were under 520 kg, the relationships would be different, e.g.:

Nominal payload	420 kg
Optional equipment	−30 kg
Towing device	−15 kg
Tongue load	−75 kg
Minimum value	300 kg

In accordance with Equation 5.7c, the vehicle would just count as a four-seater and so the number of permissible seats would have to be altered in the vehicle identification papers.

5.3.4 Design weight

The design weight determines the design position of the vehicle, also known as the normal or zero position. Under the specified payload, starting from the empty condition, the body compresses front and rear and the result is a particular position *vis-à-vis* the ground. ISO/IS 2958 'Road vehicles: Exterior protection for passenger cars' therefore specifies the design position as follows.

Number of seats	Distribution
2 and 3	two people each weighing 68 kg on the front seats
4 and 5	two people on the front seats and one person on the rear seat
6 and 7	two people on the front and rear seats

Luggage is ignored. The vehicle should be shown with this number of passengers on the drawing board. When vehicle manufacturers are exchanging vehicle dimensions, the design weight is always specified for determining the design position. The German Directive VDA 239-01 (Verband der Automobilindustrie – Automobile Industry Federation) covers all aspects relating to this field.

5.3.5 Permissible axle loads

5.3.5.1 In accordance with Section 34 of the German Straßen-Verkehrs-Zulassungs-Ordnung (StVZ0) The permissible axle loads front and rear are specified by the vehicle manufacturer. Several points on which the axle loads have a direct effect must be considered:

- component strength of the body and wheel suspension or axles
- load capacity and therefore minimum size of the tyres
- configuration of the brake and brake force distribution
- springing and damping.

The permissible axle loads are included in the ABE (Allgemeine Betriebserlaubnis or General Operating Approval) in type-testing in Germany or, in the case of the approval of an individual vehicle in accordance with Section 21 of the StVZ0, are included in the report of an officially approved expert. The values are indicated on the type plate.

To date, for passenger cars, this specification has not been governed by any particular legal regulations, with the result that only the nominal payload m_t (Equation 5.7c) in accordance with the number of seats approved had to be considered and that the sum of permissible axle loads front $m_{V,max,f}$ and rear $m_{V,max,r}$ has to be greater than, or at least equal to, the permissible gross vehicle weight (see also Equation 5.1):

$$m_{V,max,f} + m_{V,max,r} \geq m_{V,t,max} \tag{5.9}$$

To be able to match the payload to the load compartment in the vehicle better, the gross vehicle weight is usually kept larger than the permissible total value $m_{V,t,max}$ (see Fig. 5.6).

On drive tests and in vehicle behaviour simulations (see Sections 6.3 and 6.4), the least favourable loading condition, i.e. the permissible rear axle load $m_{V,max,r}$ must be assumed. The front axle load $m_{V,f,lo}$ which arises, is then usually below the permissible $m_{V,max,f}$ (Equation 5.1).

The vehicle manufacturer is given the option of the residual compression paths, i.e. there are no regulations on how far a fully laden axle may compress the springs. If this is less than $s_{Re} = 50$ mm, the desired springing effect would be compromised. Furthermore, the body can barely go any further down on the outside of the bend when cornering, so its centre of gravity rises and the cornering behaviour changes and tends to 'oversteer', as a result of which situations can arise which are beyond the competence of the driver (Figs 5.10 and 5.11).

5.3.5.2 In accordance with EU Directive 92/21/EEC   Directive 92/91/EEC (see Section 5.3.1.2) makes the loading of the vehicle and therefore the axle loads subject to stricter regulations. The permissible payload $m_{t,pe}$ (see Equations 5.7a and 5.8a) to be calculated from the difference between the permissible gross vehicle weight $m_{V,t,max}$ and the actual curb weight $m_{V,ul}$ must be divided up as a percentage into flat rate mass:

91% (to be precise 90.7%) is then allocated to the seats and 9% (or 9.3%) evenly distributed throughout the boot (Fig. 5.4b and Section 5.3.6).

The manufacturer must certify the resulting axle loads as permissible values.

5.3.5.3 When towing a trailer   If the vehicle has a towing device, a reduced loading by its component weight must be assumed and, furthermore, the tongue load in the centre of the coupling point must also be included (see Section 5.3.1.3). The remaining payload is then set at 100% and (as described in Section 5.3.5.2) distributed 91% to the seats and 9% to the boot.

The permissible rear axle load (Fig. 5.4b) is then greater. Two options then result.

- The manufacturer specifies the higher axle load for all vehicles, which means that the vehicle components listed above must be designed with this in mind, with the disadvantage that harder springs reduce the comfort and, under certain circumstances, tyres, axle parts and wheel bearings with a higher load capacity may become necessary.

- The manufacturer specifies two separate axle loads with and without a trailer towing device; the manufacturer must then ensure that the requirements listed under section 5.3.5.1 are met.

Shock absorbers with an automatic level control system or supplementary springs (Figs 5.13 and 5.36) can balance the springing.

5.3.6 Axle load distribution

The springing of a vehicle, irrespective of whether it is a passenger car, commercial vehicle or trailer, can only be designed if the axle load distribution has previously been calculated or determined by weighing. The important thing is how many kilograms of payload (and not what percentage) will be on the respective axle, and whether the permissible axle load is fully utilized or exceeded.

The permissible roof load is between 50 kg and 100 kg; it can be taken from the service manual of the respective vehicle.

5.3.6.1 On passenger cars with a non-variable boot volume   Figure 1.21 shows the axle load distribution as a percentage. Where the axle load weight is known, once the weight of the people has been added, the axle loads in the various loading conditions can be calculated.

Section 5.3.3 describes calculation of the permissible axle load which gives the axle load distribution. In industry, and at the TÜV, this is determined with weights placed on the seats at the hip centre H, i.e. the centre of gravity of a person. The position of this point is laid down internationally in the standards SAE-J 826a, ISO 6549 and in DIN 33408.

The adjustable front (and, where applicable, also rear) seats must be moved into the end position for calculating the load distribution and, in accordance with ISO 2416, the weight of the occupants arranged in such a manner that their H-points act 100 mm in front of the respective H-point of the seats. Where the rear seats are not adjustable, the distance is only 50 mm. However, EU Directive 92/21/EEC specifies exclusively the furthest back steering or sitting position and no shifting forwards of the H-points (see Section 5.3.5.2). Both cases are therefore a purely theoretical determination of the load distribution, which ignores whether the vehicle can be steered and operated at all with the sitting position set.

The permissible payload $m_{t,pe}$ calculated in accordance with Equations 5.7a and 5.8a has to be distributed in accordance with Fig. 5.4b in the ratio 91% to 9%, and the 9% luggage mass must be put into the centre of the boot. The standard design passenger car shown in Table 5.5 at $m_{t,pe}$ = 427 kg, an

Table. 5.5 Axle load distribution determined on a standard medium-size passenger car by means of weighing. The vehicle was fitted with an electric sun roof. This and further special features meant it weighed 1173 kg empty (instead of 1100, as specified by the manufacturer).

Manufacturer's details	Number of seats	5		Permissible axle load	
	Curb weight	1100 kg		Front	750 kg
	Payload	500 kg		Rear	850 kg
	Permissible gross vehicle weight	1600 kg		Sum	1600 kg

State of loading	Load	Weight of vehicle	Axle load		Axle load distribution	
			front	rear	front	rear
	(kg)	(kg)	(kg)	(kg)	(%)	(%)
Empty	0	1173	623	550	53.1	46.9
2 passengers	136	1309	692	617	52.8	47.2
2 passengers in front and 1 passenger in rear	204	1377	705	672	51.2	48.8
4 passengers	272	1445	718	727	49.6	50.4
5 passengers	340	1513	731	782	48.4	51.6
Maximum load	427	1600	721	879	45.1	54.9

assumed real occupant and load weight of $m_{P,r} = 77$ kg and $m_{b,r} = 42$ kg, would have the following loads and axle loads:

State of loading	Vehicle Load weight		Axle Load		Axle load distribution	
			Front	Rear	Front	Rear
	[kg]	[kg]	[kg]	[kg]	[%]	[%]
Empty	0	1173	623	550	53.1	46.9
2 passengers	154	1327	701	626	52.8	47.2
2 passengers in front 1 passenger in rear	231	1404	716	688	51.0	49.0
4 passengers	308	1481	731	750	49.3	50.7
5 passengers	385	1558	746	812	47.9	52.1
Fully laden	427	1600	741	859	46.3	53.7

In this case the permissible rear axle load is only exceeded by 9 kg.

In practice, it would be easier to load the car with people of any weight and calculate the difference values afterwards, and certainly less tiring than loading many individual weights into the car body and boot. In order to work as precisely as possible, the driver (who should weigh around 68 kg and be approximately 1.70 m tall) should adjust the seat into a position which is comfortable for him or her. As a result of the centre of gravity of the passengers, the weight of all the people should not deviate too greatly from this standard mass m_p.

Table 5.5 shows the load distribution of a standard design passenger car whose curb weight is 73 kg heavier than specified by the manufacturer, because of its optional equipment; the permissible payload therefore falls from 500 kg to 427 kg. Although the transportable luggage weight is now only 87 kg, the older weighing method (with occupant values of m_p = 68 kg) has resulted in the rear axle load being exceeded by 29 kg. However, the 185/65 R 15 88 H tyre fitted carries 490 kg at $v \leqslant 190$ km h^{-1} with the specified air pressure for full load of p_T = 2.5 bar (Fig. 2.10 and Equation 2.14), so the overload would affect neither the tyres nor, as shown in Fig. 5.9, the springs.

The axle load distribution at 45%/55% (front to rear) in the fully laden condition is likely to cause a slight deterioration in the driving properties of this standard vehicle, whilst significantly improving the traction.

The situation on a front-wheel drive vehicle also studied in the Laboratory for Chassis Engineering of the University of Cologne shows a different picture (Table 5.6). The axle load distribution of 46%/54% calculated under full passenger load, indicates such a severe load alleviation on the driven front wheels that difficulties will be encountered in wet weather conditions, during uphill driving and when the vehicle is towing a trailer (Fig. 6.22). Passenger weights of 70 kg were used to compensate somewhat for the manufacturer's specified excessively high additional load of 500 kg. When empty, the vehicle weighs 6 kg more than shown on the log book; nevertheless, 144 kg of lug-

Table. 5.6 Axle load distribution determined on a front-wheel drive compact family car by means of weighing. Empty, the vehicle weighed only 6 kg more than quoted. The manufacturer's approved high payload of 500 kg (or here 494 kg) would be extremely difficult to achieve. If it is fully utilized, serious effects on the driving safety cannot be ruled out (Fig. 5.11). The rear axle load can be up to 780 kg which, at the total weight of maximum 1400 kg, would mean a load of 620 kg on the driven front wheels and an unreasonable axle load distribution of 44.2%/55.8% on a front-wheel drive vehicle (Fig. 1.21 and Equation 5.7b).

Manufacturer's details	Number of seats	5		Permissible axle load	
	Curb weight	893 kg		Front	770 kg
	Payload	500 kg		Rear	750 kg
	Permissible gross vehicle weight	1393 kg		Sum	1550 kg

State of loading	Load	Weight of vehicle	Axle load		Distribution of axle load	
			front	rear	front	rear
	(kg)	(kg)	(kg)	(kg)	(%)	(%)
Empty	0	899	548	351	60.9	39.1
2 passengers	140	1039	623	416	60.0	40.0
2 passengers in front and 1 passenger in rear	210	1109	635	474	57.2	42.8
4 passengers	280	1179	647	532	54.8	45.2
5 passengers	350	1249	659	590	52.7	47.3
Maximum load	494	1393	643	750	46.1	53.9

gage weight had to be taken into consideration. If this luggage is, in fact, in the boot, the handling, braking and cornering properties deteriorate (see also Figs 5.8, 5.10, 5.11 and 6.15). The ideal load distribution in accordance with EU Directive 92/21/EEC would certainly give significantly better results.

The 155 R 13 78 S tyres fitted have a load capacity of 410 kg at speeds of up to 160 km h^{-1} with a tyre pressure $p_T = 2.1$ bar. The total of the two wheels (820 kg) is above the permissible rear axle load of 780 kg.

5.3.6.2 On passenger cars with a variable boot volume   On all estate cars, hatch back and fast back saloons (and some notchbacks) the boot volume can be increased by folding the rear seats forwards. In this type of passenger car design, the load distribution must be calculated in accordance with ISO 2416, both for when the vehicle is carrying passengers only and when it has been converted to carry goods. As specified by the vehicle manufacturer, to do this the rear seat cushion must be folded forwards and the seat backs folded down (or seat backs alone folded forwards) or the entire row of seats taken out. One disadvantage can be that, on some vehicles the front seats cannot then be pushed back far enough; the driver seat travel is limited by the seat cushion which has been folded forwards.

The axle loads must be calculated with two people, each weighing 68 kg, on the front seats and the mass of luggage (or goods) determined in accordance with Equation 5.7d. The numerical values of Equation 5.8b (and n_0 for the number of seats occupied) with two people in the vehicle give:

$$m_{tr} = m_{t,pe} - \Delta m_V - \Delta m_p \times n_o$$
$$m_{tr} = 400 - 20 - 68 \times 2 = 244 \text{ kg}$$

This large luggage mass can lead to the rear axle load $m_{V,max,r}$ being exceeded. To avoid this, ISO 2416 allows the weight to be distributed in accordance with the manufacturer's instructions.

Folding the rear seats forward can result in slight axle load changes of the empty and driveable condition (including the driver), or if the rear row of seats is removed, to a lower curb weight and a higher payload.

5.3.6.3 On vans and lorries   Where they have three or more wheels and a total weight exceeding 1 ton, these types of commercial vehicle meet the conditions of class N in the EU Directive 71/320/EC; the weight of 75 kg of the driver here, is therefore included in the kerb weight (see Section 5.3.1.2). Only the load distribution with any mass in the centre of gravity of the cargo area and in the fully laden state needs to be determined, to calculate from this the axle loads at the design weight – calculated on these vehicle types at 85% of the payload and in the fully laden condition.

5.4   Springing curves

5.4.1 Front axle

The springing on the front axle of a passenger or estate car should be soft, to give a high level of comfort to the occupants, making it possible to transport goods without them being shaken around and to give good wheel grip (see Section 5.1.1). At extremely low vibration frequency ($n \approx 30$ min^{-1}) people notice the vibration paths and speeds 80% less than they do on hard springing with frequencies around 100 min^{-1}. However, the softness of the springing is limited by the overall spring travel available

$$s_{f,t} = s_{1,f} + s_{2,f} \tag{5.9a}$$

which comprises the compression and rebound travel of the wheels and should be at least:

$$s_{f,t} \geqslant 160 \text{ mm}$$

It is almost as important that, on the front and rear axles, a residual spring travel of $s_{Re} \geqslant 50$ mm is specified to keep the body centre of gravity from rising too much when the vehicle is cornering (see Fig. 5.11). Measurements on a variety of passenger car models have shown that on comfortable vehicles (fitted with steel springs), frequencies on the front axle are between $n_f = 60$ min^{-1} and 70 min^{-1}, with a total travel path (from stop to stop) of 200 mm; Fig. 5.4a shows a springing curve of this type.

In automotive engineering, presentation of the paths on the x-axis and the wheel loads on the y-axis has become the norm. To make it possible to read path differences and the associated load changes on each wheel easily, it is necessary for them to be entered in a sufficiently large scale, at least 1:1 for the x-axis and 100 kg $\approx$ 40 mm for the y-axis.

In Fig. 5.4a, the spring rate in the linear range is $c_f = 8.3$ N mm^{-1}, and the wheel would travel a path of $s_0 = 308$ mm as it rebounds – starting from the zero position ($F_{z,W,pl} = 2.56$ kN). The travel can be calculated easily using the units of N and mm.

$$s_0 = \frac{F_{z,W,pl}}{c_f} = \frac{2560}{8.3} = 308 \text{ mm} \tag{5.10}$$

From a ride and handling point of view, such a long travel is unnecessary and cannot be designed in. For this reason, a rebound stop limits the rebound travel s_2 on all vehicles; on passenger cars and light lorries, this component is inside the shock absorber (Figs 5.19a, 5.32 and 5.40) or in the McPherson strut or strut damper. In Fig. 5.4a s_2 is relatively large at 115 mm. The kink in the curve at around $s = 30$ mm indicates the point where the stop comes into operation. Soft springing also demands that the compression travel be limited. If there were no buffers the axle would make a hard contact. The buffer force (or load) in Fig. 5.4a is:

$$F_{\text{Sp,max}} = F_{z,\text{W,pl}} + c_f \times s_1 = 2560 + 8.3 \times 92 = 3324 \text{ N} \qquad (5.10a)$$

$$F_{\text{Sp,max}} = 3.32 \text{ kN (or 338.5 kg)}$$

On roads with pot holes, an impact factor of 2.5 is easily possible, i.e. based on the normal force $F_{z,\text{W,pl}}$ in the zero position, the maximum value $F_{z,\text{W,max}}$ could be:

$$F_{z,\text{W,max}} = 2.5 \times F_{z,\text{W,pl}} = 2.5 \times 2.56 = 6.4 \text{ kN} \qquad (5.10b)$$

The main spring, designed with a spring rate of 8.3 N mm^{-1} absorbs $F_{\text{Sp,max}} = 3.32$ kN, whilst the additional rubber or polyurethane spring absorbs the residual force $F_1 \approx 3.1$ kN. Figures 5.13 and 5.35 show various configurations and characteristic curves; Fig. 5.4a shows where it comes into play after 140 mm spring travel. If the vehicle compresses over a path of 67 mm from the zero position, the spring begins to act in a way that is not noticed by the occupants and then becomes highly progressive. Figure 5.7 shows the curve of a soft-sprung standard passenger car (and Table 5.5 the associated load distribution); the frequency $n_{f,\text{pl}} = 63$ min^{-1} is in the soft range desired and, at $s_t = 196$ mm, there is a large total spring travel. In contrast, the front-wheel drive vehicle shown in Fig. 5.8 has a high frequency (i.e. hard springing) at $n_{f,\text{pl}} = 84$ min^{-1} and, at $s_t = 156$ mm, still reasonable total spring travel. The residual spring travel (54 mm) when there are five people in the vehicle is sufficient but if the, very high, permissible front axle load of 770 kg is fully utilized (Table 5.6), s_{Re} returns to the too low a value of 36 mm.

5.4.2 Rear axle

The springing configuration on the rear axle is more difficult because of the greater loading difference. Furthermore, the residual rebound travel $s_{2,\text{Re}}$ is also included in the observation. The fuel tank is located in front of, behind or over, the axle. If it is only part-full and there is only one person in the vehicle, the axle load corresponds to the empty condition. The road-holding can be compromised if the wheel cannot rebound far enough; there should be:

$$s_{2,\text{Re}} \geqslant 50 \text{ mm}$$

At the front, the permissible axle load can be taken up, at most, by the roof luggage. The amount of the difference between occupancy with one and five people actually utilized is only

$$\Delta m_{\text{V,f}} = 73 \text{ kg}$$

in accordance with Tables 5.5 and 5.6.

The weight of the people sitting on the front seats is distributed approximately equally between the front and rear axle. However, if passengers sit on the rear seat, on average 75% of their weight is carried on the rear axle springing.

Both standard design and front-wheel drive vehicles have the boot at the back. When they are loaded, around 100% of the luggage weight is carried on the rear axle. This is the reason for the significantly higher load difference between the empty and permissible axle weight of

$$\Delta m_{V,r} = 300 \text{ kg or almost } 400 \text{ kg}$$

on the rear axles of the two vehicles studied. The result would be the value $\Delta m_{V,r} = 400$ for each axle side. This would correspond to a wheel force difference of $\Delta F_{z,W,r} = 1962$ N. If we assume linear springing at a rate $c_r = 20$ N mm^{-1}, due to $\Delta F_{z,W,r}$ a path of

$$\Delta s_r = \Delta F_{z,W,r}/c_r = 1962/20 = 98.1 \text{ mm}$$

would be needed. There is also a residual rebound and compression path of 50 mm in each case so that total travel can barely be less than $s_{r,t} = 200$ mm.

Figure 5.9 shows the linear rear-wheel springing of a standard passenger car. In spite of the soft springing at a rate of $c_{r,pl} = 18.9$ N mm^{-1}, there is residual travel of 86 mm or 50 mm. The frequency on a partly laden vehicle (with three people) is $n_{r,pl} = 77$ min^{-1} and, with additional loading, it reduces, increasing the comfort (the spring rate remains constant but the mass increases, see Equation 5.4). This type of favourable configuration is achieved by:

- a large total spring travel ($s_{r,t} = 220$ mm)
- a payload level which only corresponds to 45% of the kerb weight
- a long wheelbase ($l = 2665$ mm)
- a boot that does not protrude too far at the back.

The disadvantages are the dropping of the tail when the vehicle is laden and the associated pitch angle ϑ (Fig. 3.105), although the danger of dazzling other road-users, which would otherwise be a problem, can be overcome by the headlight height adjustment, which is fitted as standard.

Shortening of the spring travel and a less pronounced squat on the tail can be overcome with progressive springing from the mid-range. Figure 5.10 shows this type of curve, measured on a front-wheel drive vehicle. The frequency $n_{r,pl} = 93$ min^{-1} (where there are three people in the vehicle) points to harder springing. In spite of the very high load difference of 399 kg, which can be seen in Table. 5.6, the axle compresses by only $\Delta s_r = 76$ mm. The possible loading of 500 kg represents 56% of the manufacturer's stated curb weight (893 kg). This unfavourable ratio leads to the severe load alleviation on the driven front wheels, described in Table 5.6 and to the highly loaded

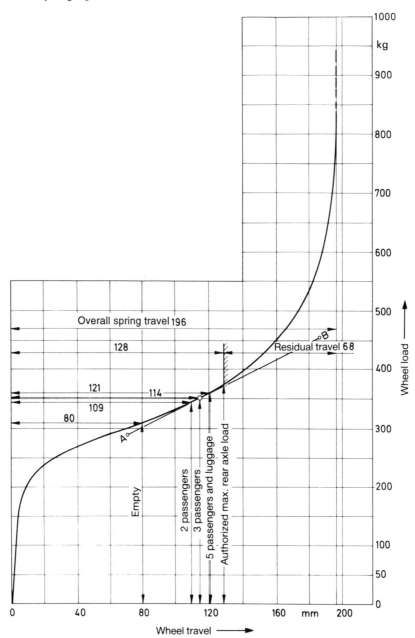

rear axle with a residual spring travel of only $s_{1,Re} = 28$ mm, whilst at $s_{2,Re} = 89$ mm, the rebound travel is large.

The springing curves (of front and rear axles seen together) lead to the assumption that the vehicle initially stood higher by: $\Delta s \approx 20$ mm.

The vehicle having been lowered by the owner (this is assumed due to the unusually large rebound travel) and the high load are likely to be the reasons for the too low residual compression paths front and back.

5.4.3 Springing and cornering behaviour

5.4.3.1 Wheel load change on independent wheel suspensions   As can be seen in Fig. 1.2, the centrifugal force related to the front axle

$$F_{c,Bo,f} = m_{m,f}\, a_y = \mu_{y,w}\, F_{V,f} \tag{5.11}$$

acts at the level of the vehicle centre of gravity. The wheel force change that arises during cornering (outside of the bend $+\Delta F_{z,W}$ and on the inside of the bend $-\Delta F_{z,W}$) can be approximated separately for the two axles. For the rear axle the equation is:

$$\Delta F_{z,W,r} = \mu_{y,w}\, F_{V,r}\, h_S/b_r \tag{5.12}$$

The values of the front-wheel drive vehicle inserted with a permissible axle load and with the centre of gravity height $h_S \approx 530$ mm, the tread width $b_r = 1425$ mm and the lateral coefficient of friction $\mu_{y,W} = 0.7$, give a force of

$$\Delta F_{z,W,r} = 0.7 \times 780 \times 9.81 \times 530/1425 = 1993 \text{ N}$$

The wider the tread width and the lower the centre of gravity, the smaller $\pm\Delta F_{z,W}$ will be.

Fig. 5.7   Soft front wheel springing with long travel and linear coil springs, measured on a medium-size standard passenger car. The progressive characteristic curve is achieved with supplementary spring (see Fig. 1.25); Table 5.5 contains the wheel loads. To be able to determine the spring rate on the design weight (three people each weighing 68 kg), a tangent must be drawn to the progressive curve (path $\overline{AB}$) which is then used to read off two points:

wheel load 450 kg, wheel travel 183 mm wheel load 300 kg, wheel travel 78 mm

The spring rate in the partly laden condition (index pl) is then:

$$c_{f,pl} = \frac{150 \times 9.81}{105} = 14.0 \text{ N mm}^{-1}$$

The axle weight needed to calculate the frequency figure is 59 kg and, in accordance with Equation 5.4, it becomes $n_{f,pl} = 63$ min^{-1}.

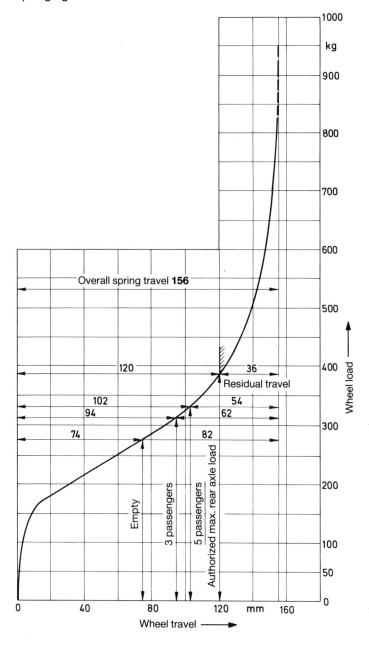

The equation for the front axle is:

$$\Delta F_{z,w,f} = \mu_{y,w}\, F_{V,f}\, h_s / b_f \tag{5.12a}$$

5.4.3.2 Spring travel on independent wheel suspensions The calculated value of 1993 N corresponds to a load change of 203 kg and therefore a wheel load

	on the outside of the bend of	593 kg
and	on the inside of the bend of	187 kg

Assuming a permissible value of 390 kg in Fig. 5.10, this leads to

a compression travel of $\Delta s_{1,r} = 20$ mm and a rebound travel of $\Delta s_{2,r} = 69$ mm

5.4.3.3 Change in the height of the centre of gravity of the body The values inserted into the formula that is valid for the rear axle

$$\Delta h_{Bo,r} = (\Delta s_{2,r} - \Delta s_{1,r})/2 \tag{5.13}$$

give the amount by which the body is pushed upwards above the rear axle (Fig. 6.15):

$$\Delta h_{Bo,r} = (69 - 20)/2 = 24.5 \text{ mm}$$

On the outside of the bend, the body only compresses a little, but on the inside it moves upwards. For this reason, Δs_1 must be deducted from Δs_2 (Fig. 5.11). The higher the centre of gravity R rises, the greater the wheel force change (see Equation 5.12), particularly at the axle with the greater travel difference $\Delta h_{Bo,f \text{ or } r}$. This is usually the rear axle. A tendency to oversteer and the torque steer effect, increase, particularly when the tyre on the outside of the bend is highly compressed, i.e. when its stress goes far beyond the possible load capacity (Figs 2.29a and 2.38).

The difference travel

$$\Delta h_{Bo,f} = (\Delta s_{2,f} - \Delta f_{1,f})/2 \tag{5.13a}$$

Fig. 5.8 Progressive front wheel springing measured on a compact front-wheel drive passenger car. The residual spring travel is high, and at 156 mm the total path is sufficient, which also applies to the residual compression travel of 54 mm when there are five people in the vehicle. Luggage load in the boot would result in the front end rebounding, i.e. it would increase the compression travel. As can be seen in Table 5.6, the manufacturer allows a front axle load of 770 kg, which will be impossible to utilize fully. On the wheel load of 385 kg then possible, the residual compression travel, at 36 mm, is clearly too low. Frequency and rate indicate relatively hard springing; on the design weight it is:

$$c_{f,pl} = 21.8 \text{ N mm}^{-1} \text{ and } n_{f,pl} = 84 \text{ min}^{-1}$$

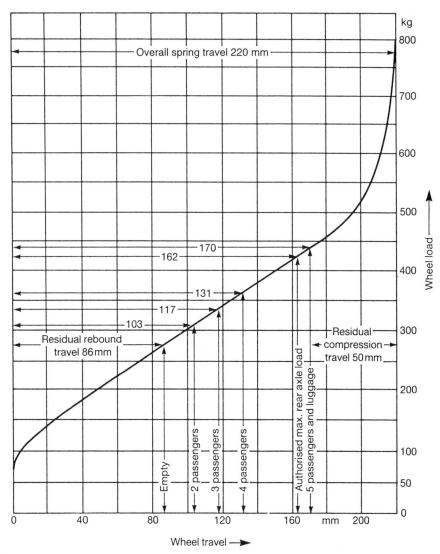

Fig. 5.9 Almost linear and soft rear wheel springing with large travel, measured on a standard passenger car; rebound stop and supplementary springs are located in the shock absorber. Table 5.5 contains the associated wheel loads. With a loading of five people plus luggage (427 kg), the rear wheels still have a residual compression travel of 50 mm. The springing rate (with a design weight, $\dot{m}_{r,pl} = 672$ kg) is $c_{r,pl} = 18.9$ N mm^{-1} and the frequency $n_{r,pl} = 77$ min^{-1}. The manufacturer specifies $m_{u,r} = 91$ kg as the weight of the unsprung mass.

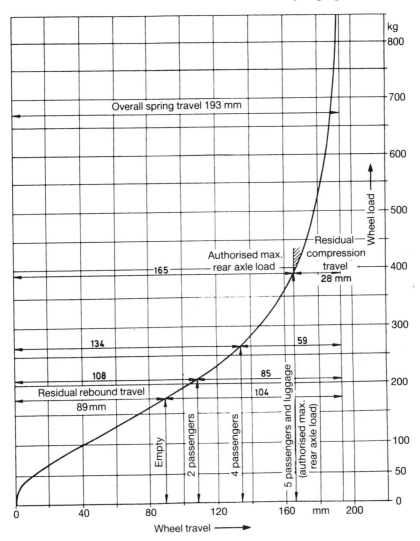

Fig. 5.10 Progressive rear wheel springing, measured on a front-wheel drive vehicle; a poor example in respect of springing design with permissible axle load. Only 28 mm residual spring travel, in association with the very high load of 494 kg, jeopardizes the driving safety (see Fig. 5.11). The associated wheel loads are shown in Table 5.6. With a design weight of $m_{r,pl}$ = 474 kg, the spring rate $c_{r,pl}$ is 20.2 N mm^{-1} and the frequency $n_{r,pl}$ = 93 min^{-1}.

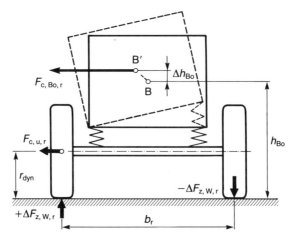

Fig. 5.11   When the vehicle corners, the centrifugal forces $F_{c,Bo,r}$ relating to the body act at its centre of gravity. If the vehicle has too low a residual compression travel, it is not able to compress as much on the outside of the bend as it rebounds on the inside. This means that the body centre of gravity moves up from W to W' by the path Δh_{Bo}, and critical oversteering, which is difficult to control, can be the result.
    Sections 3.4.5 and 5.4.3.5 and Fig. 1.14 contain further details.

by which the body moves upwards in the centre of the front axle, when (as shown in Fig. 5.8) the spring curve is also progressive, can be calculated in the same way.

The distances $l_{Bo,f}$ and $l_{Bo,r}$ from the centre of the front or rear axles should also be taken into consideration when calculating the path Δh_{Bo}, around which the body centre of gravity R changes position (Fig. 6.1, see also Equation 6.2.4):

$$\Delta h_{Bo} = \frac{\Delta h_{Bo,f} l_{Ro,r} + \Delta h_{Bo,r} l_{Ro,f}}{l} \tag{5.14}$$

If the axle loads or the weights are known, they can be inserted into the equation:

$$l_{Bo,r}/l = m_{V,f}/m_{V,t} = F_{V,f}/F_{V,t} \text{ and} \tag{5.14a}$$
$$l_{Bo,f}/l = m_{V,r}/m_{V,t} = F_{V,r}/F_{V,t} \tag{5.14b}$$

5.4.3.4 Body roll angle on independent wheel suspensions   The body roll angle of a torsionally stiff body is the same over the front and rear axles. It can therefore only be determined for the vehicle as a whole, taking into consideration any anti-roll bars fitted and the body roll centres front and rear (see Section 3.4). A two-wheel trailer, which has the springing shown in Fig. 5.10, can therefore be used as an example.

The body roll angle φ_V (Fig. 1.2) can easily be calculated in such cases:

In radians $\varphi_V = \dfrac{(\Delta s_{1,r} + \Delta s_{2,r})}{b_r}$ [rads]

In degrees $\varphi_V = 57.3 \dfrac{(\Delta s_{1,r} + \Delta s_{2,r})}{b_r}$ [°]   (5.15)

If the values of the examples are inserted on the existing progressive springing (see also equation 6.23), the result is:

$$\varphi_V = 57.3 \frac{20 + 69}{1425} = 3.58° = 3° 35'$$

In the case of linear springing over the whole range, compression and rebound travel are equal and the level of the centre of gravity does not change. The larger travel can be easily calculated using Equation 5.10:

$$\Delta s_{1,r} = \Delta s_{2,r} = \Delta F_{z,w,r}/c_{r,pl}$$

If the values of Fig. 5.10 are inserted the result is:

$$\Delta s_r = 1993/20.2 = 99 \text{ mm}$$

This corresponds to a body roll angle of $\varphi_V = 8°$. The example in the calculation should indicate the advantage of the progressive springing.

5.4.3.5 Body roll angle on rigid axles   The springs sit on the axle housing (Fig. 1.13) and the basis of support of the body is the now smaller effective distance b_{Sp}. Furthermore, unlike on all independent wheel suspensions, the rigid axle does not support the tendency of the body to roll (Fig. 1.1b). Therefore the shortened body roll lever arm ($h_{Bo} - h_{Ro,r}$) is included in the equation, and this comprises the level h_{Bo} of the body centre of gravity and the body roll centre height $h_{Ro,r}$ (rear, Fig. 1.4) (see Section 3.4.5 and Fig. 5.11) plus the proportion of weight $F_{Bo,r}$ of the body ($F_{Bo,r} = m_{Bo,r}\, g$, see Equation 6.6b) and the weight $F_{u,r}$ of the axle (see Section 6.1.3).

In accordance with the laws of static, parts that are flexibly connected must be separated. Almost all previous equations are altered by this. The wheel force change (Equation 5.12) becomes smaller,

$$\pm \Delta F_{z,W,r} = \mu_{y,w}\left(F_{Bo,r}\frac{h_{Bo} - h_{Ro,r}}{b_{Sp}} + F_{u,r}\frac{r_{dyn}}{b_r} \right)$$ (5.16)

and the travels $\Delta s_{1,Sp}$ and $\Delta s_{2,Sp}$ calculated from it relate to the springs that are further to the inside. The ratio i_φ is needed to obtain the values related to the centre of tyre contact

$$i_\varphi = b_r/b_{Sp}$$ (5.17)

$$\Delta s_{1,r} = \Delta s_{1,Sp}\, i_\varphi \text{ and } \Delta s_{2,r} = \Delta s_{2,Sp}\, i_\varphi$$ (5.18)

In line with Equation 5.15, the equation for a single-axle vehicle with a rigid axle (for the increased angle φ' in degrees) would then be:

$$\varphi' = 57.3 \frac{(\Delta s_{1,r} + \Delta s_{2,r})}{b_r} \, i_\varphi^2 \qquad (5.18a)$$

The further out the springs can be positioned, the smaller (i.e. more favourable) i_φ becomes; this applies in particular to the drawbar axle (see Fig. 1.43a).

5.4.3.6 Rates on reciprocal springing Apart from slight deviations, the spring rates with parallel and reciprocal springing are equal on all independent wheel suspensions, if we ignore the influence of the anti-roll bar:

$$c_{f\,or\,r} = c_{\varphi,f\,or\,r}$$

The picture is different for rear (or front) rigid axles.

If the springing is parallel, the rate on the centre of tyre contact c_r is equal to that on the spring mounting $c_{Sp,r}$.

However, if the springing is reciprocal, the rigid axle takes on an inclined position (Fig. 1.11).

As Equation 5.18 shows, the differences in travel $\Delta s_{1,r}$ and $\Delta s_{2,r}$ are greater than those at the springs (Δs_{Sp}); however, the changes in force $\Delta F_{z,W,r}$ are smaller:

$$\Delta F_{z,W,r} = \Delta F_{Sp,r} \, b_{Sp}/b_r = \Delta F_{z,W,r}/i_{\varphi,r}$$

With a spring rate $c'_{\varphi,r}$ related to the centre of tyre contact, this yields:

$$c'_{\varphi,r} = \frac{\Delta F_{z,W,r}}{\Delta s_{1,2}} = \frac{\Delta F_{Sp,r}}{i_{\varphi,r} s_{Sp,r} i_{\varphi,r}}$$

$$\qquad (5.19)$$

$$c'_{\varphi,r} = c_{Sp,r}/i_{\varphi,r}^2$$

In the case of reciprocal springing, the elastic parts in the guide joints and struts are tensioned; the actual reciprocal springing $c_{\varphi,r}$ is around 7% higher than the values

$$c_{\varphi,r} = 1.07 \, c'_{\varphi,r} \qquad (5.19a)$$

calculated with Equation 5.19. The equations for a rigid front axle are similar; except the index f appears.

5.4.4 Diagonal springing

Front and rear axle trailing links and longitudinal pairs of links mostly have a vehicle pitch pole $O_{f\,or\,r}$ (per side, see Section 3.11). The wheels then no longer move vertical to the ground when they rebound and compress, but on arcs $\pm f$ around the existing pole (Fig. 3.122). The driving safety is not impeded by the wheels moving out, to the back or the front. When the wheels compress, they move Δl in the direction of the poles (if these are at the height of the wheel centre or below it) and away from the poles as long as the poles are above the wheel centres. When the body compresses, the poles go down with it.

The diagonal springing angle κ on trailing and semi-trailing links is entered in Figs 3.122 and 3.124. This angle also applies to compound crank axles and multi-link axles and to appropriately sprung rigid axles (Figs 1.1, 1.0, 1.27 and 1.44). However, on double wishbones disposed at an angle (Fig. 3.12) it is:

$$\kappa = (\alpha' + \beta')/2 \tag{5.19b}$$

On McPherson struts and strut dampers, the change in the caster angle, i.e. $\Delta\tau$ is a consideration (Fig. 3.109).

5.5 Spring types

Two springs, four stops, two shock absorbers and one anti-roll bar usually control the springing of an axle, the limitation of spring travel and the reduction of body roll inclination for each axle on passenger cars and light commercial vehicles.

5.5.1 Allocation

The springs can be distinguished by media and materials as follows:

- steel springs
- air and gas springs
- composite (leaf) springs
- rubber springs
- springs of polyurethane elastomer.

These last two types are mainly used on passenger car two-wheel trailers or as additional springs parallel to the steel spring. The polyurethane is stressed in compression and the rubber in tension.

Due to the advantage of an almost constant frequency over the load and its ability to hold the body at a constant height level, air springing is used on almost all buses, as well as on an increasing number of commercial vehicles and trailer vehicles (Fig. 1.26). Here, the raising and lowering of the cargo floor are a consideration.

In passenger cars, air springing has not yet replaced the steel springs. This is more expensive, and the increase in comfort compared with vehicles with soft springing and more highly developed damping systems has, to date, been negligible.

The Mercedes 600 had air springing for many years, whilst VW fitted the research passenger car IRVW 3 with air springs, and some Japanese passenger cars in the top price segment were launched with full air springing. Since 1993, the Range Rover, an all-terrain vehicle (Fig. 1.49), has been fitted with air springing. Citroën has installed gas springing – where the forces are transmitted hydraulically via oil pressure, in almost all models since 1953 – as its so-called hydro-pneumatic springing.

5.5.2 Steel springs

The following are manufactured in steel:

- leaf springs
- coil springs
- torsion bars
- anti-roll bars.

5.5.2.1 Leaf springs   Leaf springs are subdivided into longitudinal and transverse leaf springs. Longitudinal leaf springs are used only on rigid axles, more commonly on commercial vehicles and trailers. Figure 5.12 contains a weight comparison between the previously exclusively used multi-layer leaf springs

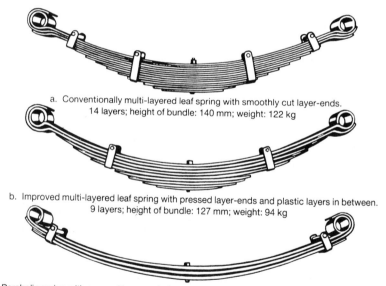

a. Conventionally multi-layered leaf spring with smoothly cut layer-ends.
14 layers; height of bundle: 140 mm; weight: 122 kg

b. Improved multi-layered leaf spring with pressed layer-ends and plastic layers in between.
9 layers; height of bundle: 127 mm; weight: 94 kg

c. Parabolic spring with pressed layer-ends (length approx. 1200 mm) and plastic layers in between.
3 layers; height of bundle: 64 mm; weight: 61 kg

Fig. 5.12 Weight comparison between three different commercial vehicle rear springs with the same data, carried out by Krupp-Brüninghaus; eye distance $L = 1650$ mm, spring rate $c_r = 200$ N mm^{-1} and loadability $F_{Sp} = 33$ kN; however the designs are different.

and modern parabolic springs; Figs 1.10, 1.15 and 1.22 show various designs and also the advantages. For reasons of cost and weight, springs with only a single layer, so-called single-leaf springs, are fitted to an increasing number of passenger cars and light commercial vehicles; Fig. 1.13a shows these on the non-driven rear axle of a van.

Transverse leaf springs, by contrast, can provide the springing on both sides of the axle; they were previously used in independent wheel suspensions of passenger cars.

5.5.2.2 Coil springs Coil springs with a linear curve over the entire wheel travel are used on the front and rear axles of passenger cars (Figs 5.4a and 5.9). If necessary, a certain progression can even be achieved by using conical spring wire and various forming. Figures 1.3, 1.23a, 1.25, 1.28a, 1.43a and 1.63 show springs in their fitted condition, and Fig. 5.10 shows a characteristic curve.

5.5.2.3 Torsion bars Cylindrical torsion bars made of round steel are used to spring the body and as anti-roll bars (see Section 5.5.4). To transfer the springing movement, both ends have warm upset heads, which carry a toothed profile or a square. U-shaped brackets can also be butt-welded and can be very easily fitted to the suspension links. Figures 1.1 and 1.45a show torsion bars in the installed condition.

5.5.3 Stops and supplementary springs

The following are differentiated:

- rebound stops
- compression stops
- supplementary springs.

As shown in Fig. 5.4a, the rebound stops limit the rebound travel of the wheels on soft and medium-hard springing. Apart from a few exceptions, rebound stops are found in shock absorbers or McPherson struts and strut dampers (Figs 5.19a, 5.32, 5.36 and 5.37). In this case, under tensile forces, the elastic attachment parts of the damper and the elastomer, polyurethane or hydraulic rebound stop, all flex (see Figs 5.19a, 5.20a and 5.42).

Compression stops limit the compression travel; they absorb high forces over a short path. The elastic bumpers can also be accommodated in the shock absorber (Fig. 5.32). They can sit within the coil springs (Figs 1.8 and 1.9a) or be fixed on the axle housing (Fig. 1.10) or come into contact with it when the springs bottom out.

In comparison with the relatively flat, hard compression stop, the supplementary or additional springs are much longer, but act softly and, as shown in Figs 5.13, 5.34 and 5.35, have a favourable springing curve and absorb high forces when fully compressed. The parts are made of rubber or polyurethane elastomer. The air bubbles in the elastomer enable the bumpers to be com-

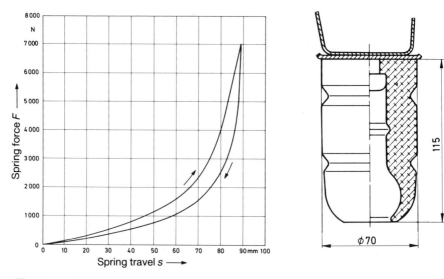

Fig. 5.13 Supplementary spring manufactured by the company Elastogran and fitted by Ford. The part is made of a polyurethane elastomer and remains flexible when cold down to an ambient temperature of –40°C. The 'buckling lip', which can be seen at the lower end ensures a soft contact and a low initial spring rate. The upper end is kept tight against the body within the coil spring.

pressed by 77% where the diameter is increased by only 35%. Like compression stops, they then absorb a force of $F_1 = 7$ kN (Fig. 5.34). Figures 1.13a, 1.24, 1.25, 1.28a and 5.19a show supplementary springs in the installed condition.

Almost any springing curve can be achieved by combining a linear steel spring with a highly progressive supplementary spring (Figs 5.4a and 5.9).

5.5.4 Anti-roll bars

The function of the anti-roll bars is to reduce the body roll inclination during cornering (Figs 1.2 and 5.11) and to influence the cornering behaviour in terms of under or oversteering (Fig. 5.0), i.e. increasing the driving safety. In the case of parallel springing, the back 1 turns (Fig. 5.14) in the bearings L; the anti-roll bar remains inactive. The anti-roll bar rate $c_{S,\varphi}$ on reciprocal springing, which is important for reducing roll inclination, related to both wheels of an axle, depends, for independent wheel suspensions, on the ratio of the wheel joint G to the attachment point T_2 on the suspension link or on rigid axles of distances b_r and b_S (Fig. 1.13). With the rate c_S on the leg ends T_1 of the anti-roll bar, the rate, related to the centre of tyre contact, becomes:

$$c_{S,\varphi} = c_S/i_\varphi^2 \qquad (5.20)$$
and $$i_S = b/a \text{ or } b_r/b_S \qquad (5.21)$$

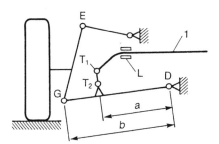

Fig. 5.14 The anti-roll bar is mounted to pivot with its back 1 in the points L. The connection between leg ends T_1 and wishbones (points T_2) is made by an intermediate rod. The ratio $i_S = b/a$ is a long way above one and therefore increases the forces in the suspension links and their mountings.

The closer to the wheel the anti-roll bar operates, the lighter and less expensive it becomes and the lower the forces that occur in all the components. A pendulum-type anti-roll bar, shown in Fig. 1.4 and used only on McPherson struts to date, provides a solution in this direction. The connecting rod 5, whose path is around the same size as that of the wheels, is fixed to the outer tube 1. The caption to Fig. 1.4 describes how it works.

Figures 1.7, 1.9a, 1.26, 1.27, 1.37, 1.39, 1.40 and 1.45a show the configurations of normal anti-roll bars and the various ways in which they are mounted. Apart from the body roll inclination, the cornering behaviour can also be influenced by anti-roll bars. The following rules will apply (see Fig. 5.0).

- A front axle mounted harder anti-roll bar promotes the tendency to understeer and improves the behaviour when changing lanes
- Higher rear axle stabilization means the front-wheel drive vehicle can become more neutral, whereas the rear-wheel drive vehicle oversteers more.

However, the anti-roll bar also has disadvantages. The more the rate $c_{S,\varphi}$ related to the wheels increases, and the more highly the elastic parts are pretensioned in the various mountings (positions L, T_1 and T_2 in Fig. 5.4 and positions 17 and 19 in Fig. 1.7), the less the total springing responds when the vehicle is moving over a bumpy road; the vehicle 'copies' the road. Furthermore, the engine begins to vibrate on its mountings (especially on front-wheel drive vehicles) and 'front end shake' occurs. The 'ride comfort' also deteriorates (see Sections 5.1.3 and 5.1.2). There is also an unfavourably harder reciprocal springing when the vehicle is moving along a pot-holed road (Fig. 1.11).

5.6   Shock absorbers

The shock absorber contributes in equal degree to both driving safety and ride comfort. It is designed to prevent the wheels springing, i.e. to ensure good road-holding and eliminate oscillation in the body. Together with tyres and disc wheels, shock absorbers are one of the parts of the chassis that are most frequently exchanged for models of the owner's choice. The owner believes that the handling characteristics of the vehicle can be improved. This can apply, although associated with the risk of premature wear to the stops, if the dampers also have to limit spring travel (see Section 5.6.8). If exchanging this part causes a change to the driving, steering or braking characteristics of the vehicle, and therefore represents a danger to other road-users, in Germany the vehicle type approval and therefore also the insurance protection would automatically lapse.

The correct tyre can be recognized from the size marking and the ECE index (Fig. 2.11), just as a worn profile, the depth of which is no longer permissible, is clearly visible. The shock absorber, in contrast, is located inside the chassis, the type marking is rolled in, but mostly covered by dirt and barely legible. Furthermore, with the variety of dampers available on the market, it is likely only to be possible to find out whether the type fitted has been approved by the manufacturer or is serviceable for the vehicle by referring to manuals.

The fact that a visual inspection only indicates failure where dampers are leaky, and that inspections are rarely carried out when they are in the installed condition is likely to be one of the reasons why there are more cars on our roads with defective shock absorbers than ones with inadequate tyres.

5.6.1 Types of fitting

The top of the shock absorber is fixed to the body or the frame and the bottom to a suspension link or the axle itself. When the wheels rebound and compress, the rebound stage and the compression stage usually come into play; in both cases vibration is dampened (see Section 5.2).

The ways in which shock absorbers can be fitted on independent wheel suspensions are shown in Figs 1.1, 1.8, 1.9, 1.9a, 1.23a, 1.28, 1.28a, 1.34, 1.43a and 1.45a and on rigid axles in Figs 1.10, 1.13a, 1.22, 1.26, 1.27 and 1.43a.

The shock absorber should be arranged vertically; if it is at the angle ξ_D to the rigid axle (Fig. 5.15), the ratio i_D is included in the calculation of the damping related to the wheel on parallel springing.

$$i_D = 1/\cos\xi_D \tag{5.22}$$

The larger ξ_D becomes, the smaller the force at the wheel and the lower the path in the damper; the ratio i_D is therefore squared in the damping calculation. In the case of reciprocal springing, the distance b_D also plays a role on rigid axles; the ratio is:

$$i_{D,\varphi} = \frac{b_r}{b_D \cos \xi_D} \tag{5.23}$$

The further inside are the dampers, the less their effective distance b_D in comparison with the tread width b_r. The ratio $i_{D,\varphi}$ for reciprocal springing increases, leading to reduced body roll damping, the effect of which is unfavourable, particularly on high bodies.

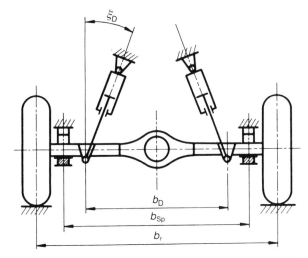

Fig. 5.15 If the dampers are fixed to a rigid axle at an angle, the angle ζ_D increases with compression with the advantage of a more unfavourable damping in the loaded condition. Moreover, the further in the dampers are positioned, the less they prevent the body roll movement.

The deviation of the damping position from the vertical is a disadvantage – in the rear and side view – even on individual wheel suspension and compound crank axles (Fig. 1.1), except that here Equation 5.22 is valid both for parallel and for reciprocal springing.

5.6.2 Twin-tube shock absorbers, non-pressurized

5.6.2.1 Design of the damper   Figure 5.16 shows the design principle. The damper consists of the working chamber A, the piston 1 fixed to the inner end of the piston rod 6, the bottom valve 4 and the rod guide 8 (Figs 5.17 to 5.19); this also takes the seal 5 and – together with the piston 1 – supports any bending moments that occur. The reservoir C, also known as the equalization chamber, which is around half filled with oil, is located between cylinder 2 and outer tube 3. The remaining volume is used for taking both the oil volume, which expands when it warms (temperatures up to +120°C are possible and briefly up to +200°C where viton seals are used), and the oil volume which is evacuated by the entry of the piston rod.

The level of the oil column in the equalization chamber must be at half full to avoid air being sucked into the working chamber through the bottom valve in the case of extreme driving conditions. This could occur if the piston rod extends fully at extremely cold temperatures (−40°C).

The inclined position of the shock absorber in the vehicle, which leads to the oil level in the equalization chamber C falling on one side, must also be considered. There is therefore a limit to the amount by which the angle ξ_D

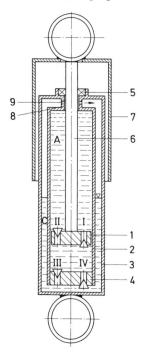

Fig. 5.16 Diagram of the twin-tube principle to explain the function.1 piston; 2 cylinder tube; 3 outer tube; 4 bottom valve; 5 piston rod seal; 6 piston rod; 7 protective sleeve; 8 piston rod guide; 9 return holes.

can deviate from the vertical (Fig. 5.15); a maximum of 45° may be reached in the full compressed condition.

5.6.2.2 Function   When the wheels compress, the damper shortens, piston 1 moves down and part of the oil flows out of the lower working chamber through the valve II into the upper one A (Fig. 5.16). The volume corresponding to the immersed piston rod volume is thereby pushed into the equalization chamber C through the valve IV in the valve body 4. This produces the main forces necessary for the compression damping and only if this does not suffice can the valve II on the piston valve become effective. As Fig. 5.18 shows, the valve II actually consists only of the conical wire spring-loaded covering plate 9. When the axle rebounds, there is overpressure between the piston 1 and piston guide 8. As this happens, the main oil volume is pushed to the settable valve I, which causes the rebound damping. The minor fluid volume is squeezed through the gap between the guide and piston rod, indicated as S_1 in Fig. 5.17, and the corner channels E and G. If the rod extends, this leads to a lack of oil in the working chamber A. The missing volume is sucked from the chamber C (Fig. 5.16) and flows through the valve III, which is also only a simple return valve. The oil pulsing back and forth between the working and equalization chamber is cooled on the outer tube 3.

5.6.2.3 Air vent and volume equalization   Twin-tube shock absorbers have

Fig. 5.17 Guide and seal set used by Sachs Boge in series production of twin-tube dampers. The finished damper is closed by rolling the outer tube 3 around the edge U of the piston rod guide 8.

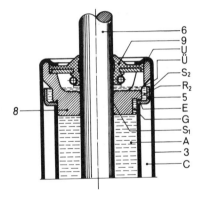

to be air vented, because air bubbles can form in the working chamber – unavoidable in this type of damper. This happens when

- the damper is stored or transported horizontal prior to installation
- the oil column in the working chamber falls when the vehicle has been standing for a long time
- the shock absorber cools at the end of a journey, the oil in the working chamber contracts and air is sucked through the piston rod and rod guide.

Without special measures, an air pocket would arise and, particularly during cold weather, unpleasant knocking, known as 'morning sickness', could occur. Designers must ensure that the oil reaching to the top of the working chamber cannot flow back into the equalization chamber when the vehicle is standing and, furthermore, that fluid fills the space that has been freed as the oil has contracted. Sachs Boge solves this problem with the angular ring 5, shown in Fig. 5.17, and several channels E and G, disposed at a right angles and pressed to the outside of the rod guide. Ring 5 creates the reservoir R_2 from which the oil can flow back via the two channels as it cools.

Another advantage is that the air that has been captured inside the working chamber can escape better. Channels E and G are used for evacuation in such cases; the air cushion quickly dissipates through the channels as a result of wheel movements. The angular ring also prevents the oil jets, which shoot from the channel E as the piston rises, from hitting the outer tube 3 directly and foaming up.

As the piston lifts, over-pressure arises above the piston, which also pushes some oil out upwards through the gap S_1 (between piston rod and guide) and the corner channels E and G. This small amount lubricates, amongst other things, the rod, collects in the reservoir R_2 and flows through the ring gap S_2 (formed by the angle ring 5 and the outer tube 3) back into the equalization chamber C. It is then cooled in the tube 3 by the wind blast of the moving vehicle. Ring gap S_1 as well as the size and number of the transverse channels G nevertheless represent a constant by-pass and their cross-sections must be considered when designing the orifices in the piston.

When subjected to compressive forces, the piston rod moves in, displacing

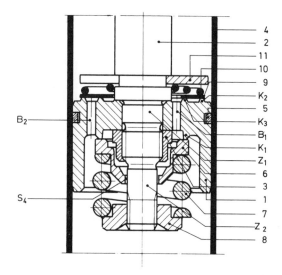

Fig. 5.18   Valve combination used by Sachs Boge on twin-tube dampers.

a certain volume and thereby creating over-pressure in the working chamber A, i.e. in the compression stage oil is also pushed through the gap S_1 and the channels E and G and cools down on the outer tube 3 as it flows back.

5.6.2.4 Rebound valve   The rebound valve in twin-tube shock absorbers is generally a combination of constant orifices and bores closed by spring-loaded valve discs (Fig. 5.18). Piston 1 is fixed by the nut 3 to the lower end of the piston rod 2. Sealing to the side of the cylinder tube 4 is provided by the piston ring 5 and the mid-centring of the piston by the spigot Z_1. The actual valve consists of the valve disc 6, which is pushed by the coil spring 7 against the sealing edge K_1. The valve spring force is regulated with the nut 8. The nozzle clearance area S_4 between the tappet Z_2 of the piston rod and the hole in the valve disc 6 produces the actual constant flow – also known as by-pass or advanced opening cross-section. When the piston moves upwards, the oil flows through the hole B_1, to pass this by-pass and, with increasing speed, (after raising the spring loaded valve disc) the actual valve.

The level of rebound damping is determined:

- at low piston speed by the constant orifice; this includes the length and size of the nozzle clearance area S_4 (the valve 6 is still closed here), the clearance between piston rod 6 and hole (in the guide 8, Fig. 5.17) and the air vent channels E and G
- at medium speeds by the aperture of the valve disc 6, i.e. by the hardness and initial tension of the spring 7
- at high piston speeds and with the valve wide open, by the number and cross-section of the holes B_{10}.

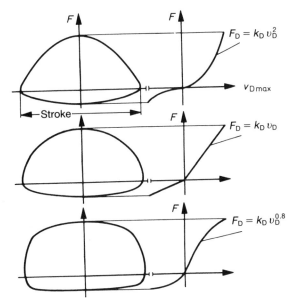

$$F_D = k_D v_D^2$$

$$v_{D\,max}$$

$$F_D = k_D v_D$$

$$F_D = k_D v_D^{0.8}$$

Fig. 5.18a The damping curve can be progressive (top), linear (centre) or degressive (bottom). The curve shape and diagram shape are directly related. The smallest area and therefore the lowest mean damping is in the diagram of the progressive curve, while the largest area is that of the degressive damping. The shape of the damping curve can be expressed in an equation by the exponent n:

$$F_D = k_D v_D^n$$

By combining these options, any valve curve, from digressive via linear through to progressive curves (Fig. 5.18a) can be set.

In the compression direction a small volume of oil flows through the nozzle clearance area S_4 on the tappet Z_2, although the main volume moves through the outer channel B_2 when the valve plate 9 is raised. This thin disc, which serves only as a check valve is held in the centre and normally seals the edges K_2 and K_3. The load is applied by the soft conical spring 10, which is in contact with the washer 11 and which can be clamped from the side. This also serves as a stop and prevents the valve from opening too far (overloading) at high piston speeds.

5.6.2.5 Compression stage valve Parts 9 to 11 shown in Fig. 5.18, which sit on top of the piston, are simply a check valve, as described at the start of Section 5.6.2.2; the compression damping forces are primarily produced by the compression valve in the bottom of the damper (part 4 in Fig. 5.16). Figure 5.19 shows a section through the configuration fitted by Sachs Boge in shock absorber types T27 and T32. The actual valve body 1 has the holes B_1, through which oil is sucked when the piston moves upwards as the wheel

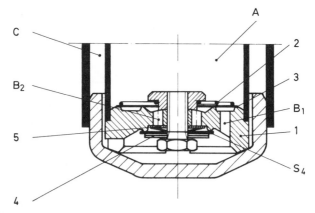

Fig. 5.19   Bottom valve of the Sachs Boge T27 and T37 twin-tube dampers.

rebounds and the volume of the extending piston rod must be replaced. The covering disc 3 loaded by the conical spring 2 lifts.

The piston rod has a diameter of 11 mm on passenger car and light van dampers; the small cross-section area of only 95 mm^2 must provide the fluid displacement which then produces the compression damping (in comparison 478 mm^2 is available for the rebound stage, corresponding to a 27 mm piston diameter minus the rod cross-section surface).

When the piston rod enters, the compression stage valve is charged by the displacing oil. This valve consists of the set of spring washers 4, the upper washer of which has the grooves S_4 as a constant orifice. The required setting can be achieved by means of the diameter of the hole B_2, the number and thickness of the spring washers and the size of the by-pass grooves S_4.

However, the constant by-pass has the disadvantage that when the vehicle is standing, the oil in the working chamber A, which is at a higher level, can flow back into the equalization chamber C. If the vehicle moves off again, after it has travelled a certain distance this equalizes out, although it may be linked to a certain unpleasant knocking noise, known as 'morning sickness'. Until the air bubble at the top of the working chamber escapes, when the wheels rebound, the oil is drawn suddenly against the piston guide. To avoid the noises this causes, Sachs Boge has added the anti-communication valve 5. This is upstream of the spring washers 4, covers the holes B_2 and therefore prevents the oil from flowing back.

The compression damping curve arises through the interplay of the bottom valve with the nozzle orifice area S_4 shown in Fig. 5.18 and the check valve 9 on the piston. There are also the air vent channels E and G shown in Fig. 5.17 and the nozzle clearance S_1 between piston rod and guide.

To ensure that this is intensively lubricated, and largely to prevent the oil from foaming and to improve cooling by means of a degree of oil circulation, there should always be a higher internal pressure in the upper part of the working chamber (i.e. between piston 1 and rod guide 8, Fig. 5.16) than between the piston and bottom valve 4. This latter part must provide more resistance to the displaced oil than the check valve II of the oil volume that flows through the piston.

5.6.3 Twin-tube shock absorbers, pressurized

The most economical form of damper design is the one that operates on the non-pressurized twin-tube system. Where certain vehicles or chassis conditions make it appear sensible or necessary to use a gas-pressure damper, the low-pressure twin-tube shock absorber is a good solution. The additional costs remain reasonable. Because compression damping continues to be provided via the bottom valve, gas pressure of around 4 bar is sufficient. This means that the piston rod extension force F_{Pr}, described in Section 5.6.4.1, remains low. This makes it possible to use these absorbers without problems on McPherson struts, with correspondingly thicker piston rods.

The basic design, the length and dimensions of the non-pressurized and pressurized designs are the same, so it does not matter which variety is used on a vehicle (e.g. for sports models), as no change to the vehicle is necessary.

The advantages of the low-pressure twin-tube design are:

- more sensitive valve response at small amplitudes
- ride comfort increases
- damping properties under extreme conditions (e.g. on pot-holed roads) are better
- hydraulic hissing noises are reduced
- shorter lengths and less friction than monotube gas pressure dampers
- the shock absorbers remain functional even after loss of gas.

Unlike the non-pressurized designs described in the previous section, the gas chamber C in the upper part of the reservoir chamber is filled with gas (Fig. 5.19a). To ensure that the working chamber A is still filled with oil even after the vehicle has been standing for a long time and the oil has cooled (after warming), Sachs Boge provides the inner tube valve 4 as a block to the gas between the piston rod seal 9 and the guide 8. This is fixed to the top of the angle ring 11 and seals to the outer edge 7 of the piston rod guide. However, the sealing lip allows the oil leakage volume pushed through the clearance S_1 (Fig. 5.17) to flow back into the equalization chamber C without allowing the gas to flow out in the opposite direction. The reservoir R_2 shown in Fig. 5.17 can therefore be omitted.

Figures 5.40 to 5.42 show similar solutions where the thicker piston rod is the reason for the slightly modified design of the gas block.

5.6.4 Monotube dampers, pressurized

5.6.4.1 Design and function The design, used almost exclusively today, with a separator piston (position 1) can easily be explained on the basis of the schematic diagram in Fig. 5.20. At the top is the evacuation chamber 3, which

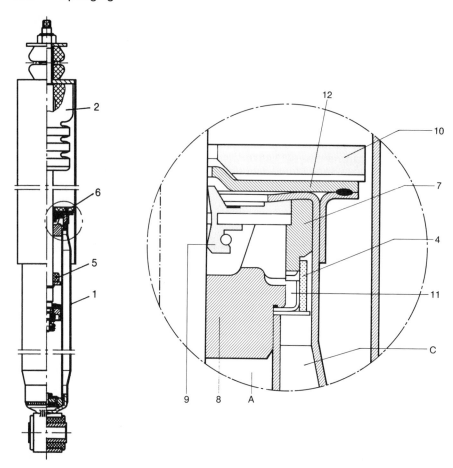

Fig. 5.19a Low-pressure twin-tube damper by Sachs Boge with special seal 9, gas lock 4 and rubber supplementary spring 2. The spring comes into contact with the plastic disc 10 when the wheel compresses; the cap 12 provides the plain contact surface and prevents the seal being damaged. The piston rod carries the rebound stop 5, which comes into contact with the guide when the wheel rebounds (see Fig. 5.32). The damper has a pin at the top and a 32 mm wide eye-type joint at the bottom. The outer tube is widened in the centre to achieve a larger equalization chamber C.

(as in the twin-tube system) must absorb the volume equalization by the oil warming and the volume displaced by the piston rod. Gas and oil are separated by the piston 1, which seals off the actual working chamber 2.

The damper piston 5 usually has a diameter of 36, 45 or 46 mm and is fixed to the piston rod 8. It carries the valves 6 and 7. As shown, the piston rod can extend upwards or downwards (Fig. 5.20a); the separator piston 1 makes it possible to install the shock absorber in any position. If the damper cylinder

Fig. 5.20 Diagram of the pressurized monotube principle with separator piston (position 1).

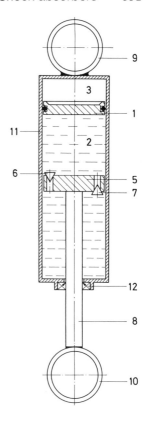

is fixed to the body or frame, the cylinder weight forms part of the sprung
mass and only the light piston rod contributes to the unsprung mass: this is a
reason for preferring the installation position shown in Fig. 5.20.

When the wheel rebounds, the oil flows through the rebound stage valve 6,
shown in Fig. 5.22 from the bottom to the top part of the operating chamber.
The gas pressure in the gas chamber 3 forces the separator piston to follow,
equalizing out the reduction in volume (caused by the piston rod extending).
If the wheel compresses, the compression valve 7 is charged (Fig. 5.23) as the
dividing piston moves upwards through the oncoming piston rod volume. The
entire piston surface is available for compression damping; this is then significantly more effective than on the twin-tube system, and the valve 7 produces
high forces at lower fluid pressure – without loss of comfort – an advantage
on vehicles with heavy rigid axles. The road-holding can be improved here by
means of responsive and correspondingly high compression damping.

The gas pressure at ambient temperature (20°C) is at least 25 bar. This
value is required to counteract the compression damping forces. If these
exceed the opposed force exercised by the gas pressure on the separator piston, the oil column will rip off at the compression valve. Therefore, for a 36

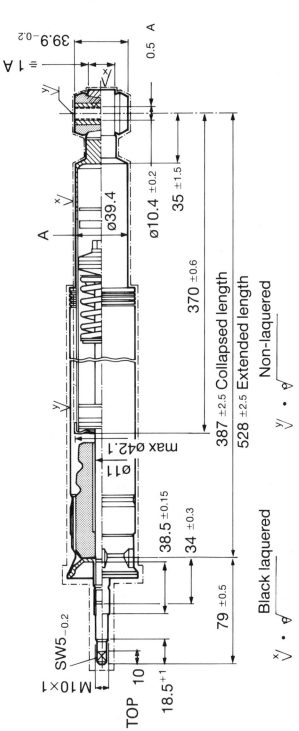

Fig. 5.20a After an original drawing by Bilstein of the front axle damper of the Mercedes Benz C class with a stroke of $s_D = 141$ mm, the fixed length $L_{fix} = 246$ mm, pin-type joint at the top (with a rolled-in spacer) and eye-type joint at the bottom; the piston rod comes out at the top. The supplementary spring shown on the left is surrounded by a short, stable tube and comes into contact with this and the support disc located above the piston rod guide when the spring compresses. The tube also carries the actual plastic sleeve, which reaches up to the damper centre.

The mechanical hydraulic compression stop sited above the piston is only fitted in this form on the six-cylinder models. The progressive coil spring, which forms part of this with its varying pitch, helps to reduce pitch and roll movement on the body and its top is carried by the piston rod by means of a washer. When the wheel rebounds, the washer comes into contact with the piston rod guide (Fig. 5.21).

mm piston diameter, 2.8 kN are needed, and for a 46 mm diameter piston, 4.6 kN.

A disadvantage of the high gas pressure is the piston rod extensive force, which amounts to

$$F_{Pr} = 190 \text{ N to } 250 \text{ N}$$

If a vehicle has soft springing (e.g. $c_f = 15$ N mm^{-1}), where gas pressure dampers are retrofitted, this can raise the body by

$$s_2 = F_{Pr}/c_f = 250/15 = 17 \text{ mm}$$

if the dampers are positioned close to the wheels. When the vehicle is running, they warm up and, at an oil temperature of 100°C, extension force and body lift increase by up to:

$$F_{Pr} \sim 450 \text{ N} \quad \text{and} \quad s_2 \sim 30 \text{ mm}$$

If gas pressure dampers are fitted as standard, this influence has already been taken into consideration by the vehicle manufacturer. If the owner subsequently changes over from twin-tube to pressurized monotube dampers it is recommended that appropriately shortened springs be fitted.

5.6.4.2 Piston rod and rod guide   Figure 5.21 shows a section through the seal package with a piston rod guide above the seal and therefore only slightly lubricated.

Unlike twin-tube dampers, a detachable piston rod guide (position 1), held by the wire snap ring 2, is used to plug the damper. The guide can be pushed down to the second snap ring 3 and the ring 2 can then be laid into the free groove in tube 4. When the load is removed, the oil pushes the guide back against ring 2.

The O-ring 5 seals the rod guide to the outside and the mono-lip seal 6 to the piston rod. The flange of this seal sits inside the guide 1 with its neck in the 'perbunane' disc 7. Internal pressure and clamping load of the closure disc 8, which is secured to the guide, ensure that the sealing neck is also pressed against the piston rod 9. The more the oil warms up when the vehicle is moving, the more the inner pressure increases and the more tightly the seal is pressed on. If a compression stop is fitted into the damper, it comes into contact with disc 8 when the wheel rebounds.

The fluid seal on the pressurized monotube damper is more dependent on the surface condition of the piston rod than on the gasket 6. The rod is therefore manufactured with particular precision. In passenger cars and light commercial vehicles monotube dampers made by Bilstein Ltd, the rods have 11 mm diameter and are made of the heat-treatable steel Ck 45 QT in accordance with the international standards ISO R683 and EN 83-70 (and the German standard DIN 1720). The strength properties are:

$$R_m = 750 \text{ N mm}^{-2} \text{ to } 900 \text{ N mm}^{-2}, R_e \geqslant 530 \text{ N mm}^{-2} \text{ and } \epsilon \geqslant 6\%$$

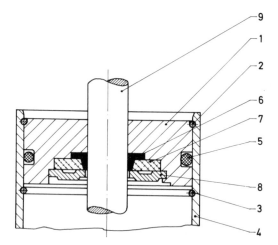

Fig. 5.21 Seal package developed by Bilstein, which keeps the temperature range of −40°C to +200°C demanded by the automobile industry. The outer piston rod guide 1 has a hard-coated hole and is made of an aluminium wrought alloy (e.g. AlMgSi 1 F 28, to the German standard DIN 1725). The piston rod 9 has the diameter $d = 11^{-0.02}$ and the hole has the tolerance range

$$d = 11 \; {}^{+0.07}_{+0.05}$$

which corresponds approximately to the ISO fit D7/h7 with a play between 0.05 mm and 0.09 mm.

The surface is raised by induction hardening to a Rockwell hardness of 58+2 HRC and is then ground to achieve a roughness depth of $R_t = 0.8$ μm to 1 μm. A hard chrome layer over 20 μm thick, subsequently applied, raises the surface hardness to 70 ± 2 HRC and the subsequent super finish treatment reduces the roughness depth to the value $R_t = 0.2$ μm needed for the seal.

5.6.4.3 Pistons and valves Due to the equalization chamber being above the working chamber, the monotube damper is longer than the one operating in the twin-tube system. To minimize this disadvantage, the separator piston 1 (Fig. 5.20) is hollowed out in the centre and a flatter working piston fitted (flatter than the one in the twin-tube system). Flat plate valves are also used.

When the piston rod extends, the oil flows past the compression valve at the top through diagonal holes to the rebound stage valve (Fig. 5.22). Both thickness and number of valve plates, as well as the support disc diameter $d_{0,D}$ and the amount of the constant orifices A_O, are critical for the level of the damping forces. The constant by-pass is created by a bottom valve plate on the compression valve (Fig. 5.23) which is smaller in diameter and does not completely cover the inclined holes. Unlike in the twin-tube system, when the piston enters, its larger diameter valve plates are charged by the entire oil

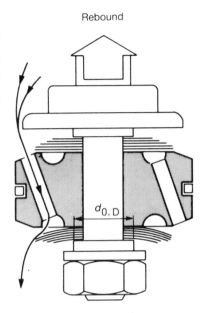

Fig. 5.22 Space-saving compression stage valve with spring plates and a supporting washer found on almost all monotube dampers. If, as shown, the piston rod moves out and upwards, the valve sits at the bottom of the piston. The piston ring shown in the illustration is used to achieve a constant flow between this and the cylinder walls.

Rebound

$d_{0, D}$

column; this causes much more intensive damping and prevents the wheels from springing – without reducing the ride comfort.

In all monotube dampers, the characteristic of the damping curve is determined exclusively by the valves on the piston and the holes. If these just have constant orifices (Fig. 5.24) there is a highly progressive curve shape with

Compression

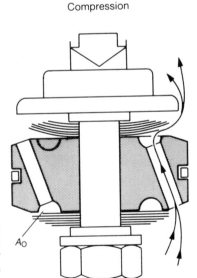

A_0

Fig. 5.23 If the piston rod moves upwards and out on monotube dampers, the spring plate valve for the compression stage is under the piston.

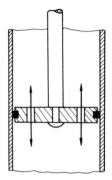

Fig. 5.24 Unshielded holes in the piston correspond to a constant flow, also known as the advanced opening cross-section or by-pass. On the monotube system they give the highly progressive damping curve shown in the following illustration. The compression and rebound forces are the same size and have very high terminal values.

high forces (Figs 5.24a and 5.18a top) on both the compression and the extension side; this also applies when there is a by-pass between the piston and cylinder tube, i.e. if the piston ring were missing. Pre-tensioned valve plates over large holes (Fig. 5.25) cause the curve to take on a degressive shape with the additional advantage of being able to set different forces on extension and compression sides (Fig. 5.25a). At higher piston speeds these only increase a little. The linear curve shown in Fig. 5.18a is achieved either through low pre-tensioned valve plates or by using a combination of constant orifices and spring-loaded valve discs (Fig. 5.18).

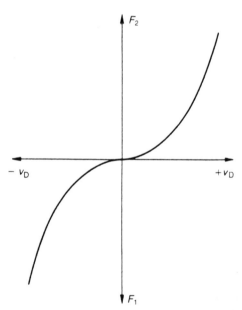

Fig. 5.24a Highly progressive damping curve achieved by holes in the piston or a gap between the piston and cylinder wall.

Fig.   5.25 Spring-loaded   valves over large holes give a degressive damping curve. The forces in the compression and rebound side can be set to different levels. The piston ring 3 prevents an additional by-pass.

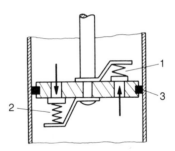

5.6.4.4 Advantages and disadvantages   The pressurized monotube damper has a series of advantages over the non-pressurized twin-tube damper:

- good cooling due to the cylinder tube 11 (Fig. 5.20) with direct air contact
- a larger piston diameter is possible with the same tube diameter (e.g. 36 mm instead of 27 mm), reducing the operating pressures
- the compression stage valve 7 sits on the piston 5 and is charged by the entire oil column
- the oil level in the oil column does not fall as it cools, so no 'morning sickness' occurs (see Section 5.6.2.3)
- due to the pressurized oil column, the oil cannot foam, resulting in good damping of even small high-frequency vibrations
- where there is a separator piston, the installation position is not restricted.

The disadvantages are that the high degree of manufacturing precision and the essential gas seal lead to higher costs. Furthermore, the greater space requirement can amount to over 100 mm in the stroke length.

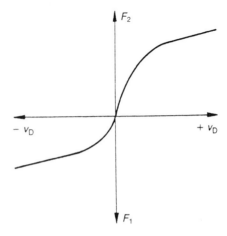

Fig. 5.25a Degressive curve with different force levels on the compression and rebound side, achieved by spring-loaded valves (see also Figs 5.18a and 5.42).

5.6.5 Monotube dampers, non-pressurized

Non-pressurized monotube dampers generally have a piston of only 20 or 22 mm diameter, an 8 to 9 mm thick piston rod and therefore absorb correspondingly lower forces. They are used as:

- engine vibration dampers
- driver seat dampers
- steering dampers (see Section 4.5).

The first two designs are installed vertically and it is only necessary to fit a compression valve (Fig. 5.19) instead of the separator piston (Fig. 5.20). As in the twin-tube system, this ensures the necessary back-pressure when the piston rod enters. The equalization chamber is above the working chamber and is around half-filled with oil and air; the two media could mix if there were no separator part, which is common on engine dampers.

Steering dampers must not have any extension force at the piston rod, otherwise the steering would be assisted in the compression direction and pulled to one side. The dampers are fitted in a lying position, so only non-pressurized monotube dampers (where the oil and air are separated) can be used.

Figure 5.25b shows a standard design, on which the flexible hose 1 performs this function; it is fixed by rolling the outer tube 3. Part 3 is bevelled off on both sides and presses the hose into pointed grooves of the cylinder tube to provide a good seal. At the same time this measure prevents displacement when the vehicle is moving. When the piston rod 17 moves in, the oil flows through the two apertures 4 of the valve in the valve body 5 and lifts the valve plate, which is loaded by the spring 7; this produces part of the compression damping.

The area between the protective tube 3 and the hose 1 acts as an equalization chamber. The hose 1 flexes when oil flows through the hole 9. As in the case of all monotube dampers, the damping valve unit (consisting of the rebound stage and the actual compression valve) is situated on the piston 10 (Figs 5.22 and 5.23). The piston ring 11 seals this off to the cylinder tube 2. The piston rod guide 12, seal 13 and support disc 14 sit between the two rolled-in grooves; the longitudinal hole in the guide acts as a pressure equalization. The eye-type joints 15 and 16 provide the installation. The advantage of this design is the short length; increasing the stroke only makes it necessary to extend the tube 2 and the equalization hose 1 with the protective tube 3. The longer tube 3 can then be a disadvantage. If it should not prove possible to house this, an alternative design with a separator sleeve could be used (Fig. 5.25c), which has the same functional parts but an in-line, welded-on, equalization chamber 8 with its inside diameter increased to 26 mm. The coil spring 19 in flat rolled steel supported on the top 18 flexes under the pressure of the oil displaced when the piston rod 17 enters. The opposed force of the spring 19 is measured such that a light pressure is applied to the oil column, but no extension force occurs. The seal between air and oil is provided by the cup seal 21, which is inserted into the guiding part 20.

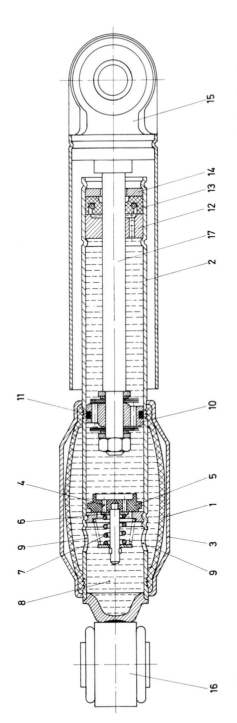

Fig. 5.25b Section through the Stabilus steering damper used on passenger cars and light vans, with its equalization chamber consisting of the elastic tube 1 and the upper part 8 above the working chamber. The piston 10 has 20 mm and the rod 8 mm diameter.

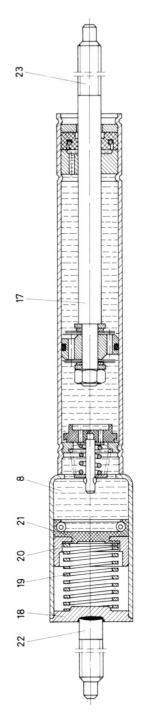

Fig. 5.25c Stabilus compact steering damper with pin-type joints on both sides (position 22 and 23), butt-welded equalization chamber 8 and spring-loaded cup seal 21.

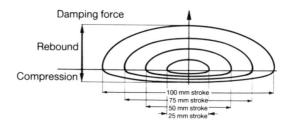

Fig. 5.26 The damping forces on the production test stand can be measured at $n = 100$ min^{-1} with increasing strokes to determine the curve.

5.6.6 Damping diagrams and characteristics

The spring force is a function of the wheel travel, whilst the damping force depends on the speed at which the two fixing points are pulled apart or pushed together. A damper, which is subject to a constant force F_D, flexes at a constant speed over the whole stroke, whilst a spring flexes immediately, but only up to a certain travel s_1, the length of which depends on the quotients of force and spring rate $c_{f \text{ or } r}$:

$$F_{Sp} = c_{f \text{ or } r} s_1 \text{ and } F_D = k_D v_D^n \text{ (see Fig. 5.18a)}$$

The spring therefore stores work and usually releases it at a moment that is not conducive to driving safety, whilst the damper destroys it by converting it into heat. The more highly the damper is stressed, the more it warms. In diagrams, the damping force F_D appears as a function of the piston speed v_D in m s^{-1}.

Figure 5.26 shows diagrams recorded on a standard test rig. At a constant rev speed ($n_D = 100$ min^{-1}), the stroke is changed step by step, but it is also possible to keep the stroke fixed and to vary the engine and therefore the test rig speed (Fig. 5.27). To record the damping curve, in both cases the maximum forces of each stroke are taken and, as shown in Fig. 5.27, entered upwards and downwards on the y-axis as a function of a maximum piston speed. The equation for calculating the individual values is:

$$v_{D,max} = \frac{\pi s_D n_D}{60} \text{ [m s}^{-1}] \tag{5.24}$$

The value $n_D = 100$ min^{-1} and $s_D = 100$ mm gives the following speed:

$$v_{D,max} = \frac{\pi \times 0.1 \times 100}{60} = 0.524 \text{ m s}^{-1}$$

Figure 5.28 shows the curve of the rear axle damping of a front-wheel drive vehicle. Damping curve and diagram shape are closely related. A progressive curve (Fig. 5.18a, top and 5.24a) has a cornered diagram with a relatively

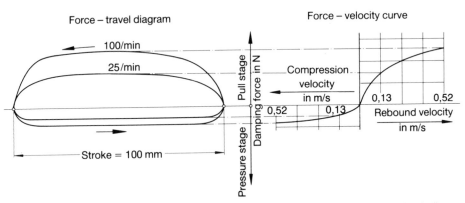

Force – travel diagram

100/min

25/min

Stroke = 100 mm

Force – velocity curve

Compression velocity in m/s

0,52   0,13

0,13   0,52

Rebound velocity in m/s

Pull stage

Pressure stage

Damping force in N

Fig. 5.27 The maximum compression and rebound forces are taken from the individual diagrams to create the damping curve formerly known as the force speed curve.

small surface, i.e. the actual mean damping, which is important for the springing behaviour, is low. The degressive curves shown in Figs 5.18a (bottom), 5.25a and 5.29 have a rounded shape and so a high mean damping.

It would be correct, but too time-consuming with conventional methods, to determine the size of the diagram's area in order to plot the resulting mean

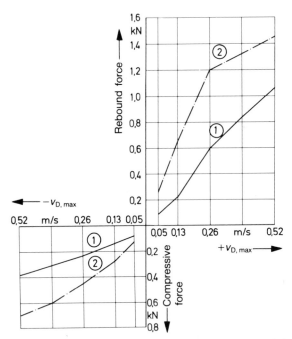

Fig. 5.28 Rear axle damping curve; 1 is the standard setting and 2 that for the heavy duty version.

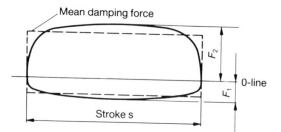

Fig. 5.29 The maximum piston speed $v_{D,max}$ and the greatest force F_2 in the rebound and F_1 in the compression direction are included in simplified form in the determination of the wheel and body damping; both are easily measurable. The actual form of the diagram, in this instance that of degressive damping (Fig. 5.18a, bottom) is ignored.

damping over the corresponding mean piston velocity, or to oppose the mean damping force to the mean piston speed $v_{D,med}$ by calculations.

$$v_{D,med} = v_{D,max}/1.62 \qquad\qquad (5.25)$$

5.6.7 Damper attachments

5.6.7.1 Requirements The damper attachments are used for fixing the damper to the frame, suspension subframe or body at the top, and to the axle housing itself or a suspension control arm at the bottom. Certain requirements must be fulfilled:

• maintenance-free and inexpensive to manufacture
• angular flexibility (to absorb the movements in fixing points) with only low counter moment, so as not to subject the piston rod to bending stress
• noise insulation (to prevent the transfer of road noise)
• precisely defined flexibility towards the damping forces; any unwanted loss of travel in the rubber components influences damping preciseness and road harshness.

On the vehicle side it must be ensured that the upper and lower fixing points align with one another in the design position (i.e. when there are three people each weighing 68 kg in the vehicle), only in this way can distortion, when the vehicle is running, and premature shock absorber wear be avoided.

5.6.7.2 Eye-type joints The requirements are best met by rubber joints. Figure 5.32 shows, on the top and bottom of the damper, the type of suspension most used: the eye-type joint, sometimes also known as a ring joint. The most common size in passenger cars is 32 mm wide, 35 mm to 36 mm diameter and has a 10 mm or 12 mm fixing hole with a +0.15 mm tolerance (Fig. 5.30). If compression stops are housed in the shock absorber or if spring forces are also concentrated in the mountings, 40 to 60 mm wide joints may be necessary (Fig. 5.36).

Fig. 5.30 The eye-type joint has 35 mm to 36 mm outside diameter, a bore of $10^{+0.15}$ mm or $12^{+0.15}$ mm and is 32 mm wide. The maximum approved distortion angles are $\alpha/2 = \pm15°$ and the cardan angles $\beta/2 = \pm4°$.

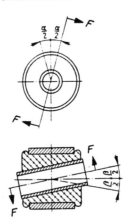

The joint itself consists of a rubber bush that is in high radial pre-tension between the outermost ring and the pressed-in inner tube. The rubber part has beads at both sides as a measure to stop it sliding out when the vehicle is moving. The size mostly used and shown in the illustration allows twisting angles up to $\alpha/2 = \pm15°$ and cardan deviations of up to $\beta/2 = \pm4°$. Greater twist angles would increase the bending moment in the piston rod and therefore need different configurations (Fig. 5.20a).

5.6.7.3 Pin-type joints   If the same angle movement occurs in all planes at the upper or lower suspension when the vehicle moves, the design solution is to use a pin-type joint (Figs 5.31 and 5.19a). This allows deviations up to $\pm6°$ in all directions and consists of two rubber snubbers, one above and one below the fixing point; the snubbers can be separated or manufactured in one piece as a 'knob snubber'. The pin usually has a cold-formed 10 mm diameter

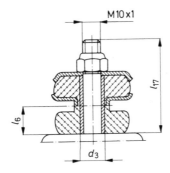

Fig. 5.31 On a pin-type joint, the preload on the rubber parts should be ensured by a spacer tube. Usually this has a wall thickness of 2 mm and 14 mm outside diameter. To avoid contact in the location hole, the upper snubber can be centred by a washer. A self-locking nut is frequently used for clamping the parts together (illustration: Sachs Boge).

and an M 10 × 1 thread at the end. The rubber parts are pre-tensioned via a dished washer and (as shown in the figures) using a self-locking nut or two lock nuts. The distance between the lower edge of washer and the damper, which is important for the function, can be achieved using a loose spacer tube (usually of 2 mm wall thickness, i.e. 14 mm outside diameter) or by means of a rolled-in tube, as shown in Fig. 5.20a.

From a design perspective, it must be ensured that even at its greatest compression and twist, the side of the pin or the spacer does not come into contact with the bodywork or axle; this would lead to unpleasant noises and increased bending stress. As shown in Fig. 5.31 on the upper snubber, contact can be avoided by the use of a washer, the outer collar of which surrounds the rubber part and grips into the hole with an edge that is turned downwards. In the case of the lower snubber, the same effect is achieved by a vulcanized collar. The fixing point itself can also be designed as a 'shim'.

5.6.8 Stops and supplementary springs

Installation of any stoppers means both the damper and the suspension strut increase in length and there must be enough space in the vehicle to allow this.

5.6.8.1 Rebound stop   Figure 5.28 shows the maximum rebound force 1.45 kN at $v_{D, max} = 0.52$ m s^{-1}. However, piston speeds of 3 m s^{-1} can occur, which lead to higher forces. If these forces can no longer be absorbed hydraulically in the shock absorber valves, rebound stops come into action (Fig. 5.4a). On passenger cars and light commercial vehicles, the most economic solution is to locate the elastic limitation of the rebound travel or the 'hydraulic stop' (Figs 5.20a and 5.42) in the damper.

The other advantage is that the slight springing effect of the top and bottom damper mountings can be additionally used to damp the rebounding wheel, and so a relatively flat, more easily manufactured bumper 5 made of rubber, polyurethane or viton, polyamide or a similar plastic is completely adequate (Figs 5.32 and 5.19a). All that is needed to fit this is a groove turned into the piston rod in which the collar on the stop disc 4 is rolled or a lock washer inserted.

In the twin-tube system, when the piston rod is extended, the snubber 5 comes into contact with the piston rod guide 6 which is smooth at the bottom (Fig. 5.32), or into contact with a disc 8 protecting the set of gaskets on monotube dampers (Fig. 5.21). Figure 5.33 shows the shapes and the progressive springing curve of the 4 to 12 mm high snubbers.

The durability of the elastic compression stop is determined by the shape and material used. It must be able to withstand oil temperatures between $-40°C$ and $+140°C$ without detrimental changes of elasticity and, in the case of sudden loads, neither scuffing nor fissures may occur. Parts coming off would get into the valves and cause the damping to fail or lock. Endurance tests carried out jointly by the respective vehicle and shock absorber manufacturers, ensure that this type of damage does not occur. For this reason, and to ensure wheel rebound travel is maintained, where dampers with snubbers are used, only those authorized by the manufacturer should be fitted.

Fig. 5.32 Sachs Boge T27 twin-tube damper with a compression stop carried by the piston rod 2. The rebound stop 5 is supported on the disc 4 rolled into a groove. The upper eye-type joint and the outer and protective tube are also dimensioned and toleranced.

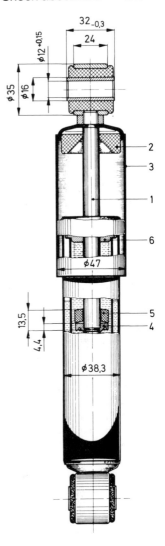

The same applies to spring dampers which, as an assembly unit, contain the compression stop and the supplementary spring as shown in Fig. 5.36.

5.6.8.2 Compression stops Compressions stops act close to the end of the wheel travel and are designed to limit compression travel without generating noise. The stop parts are housed in the top of the protective tube (Fig. 5.32), which represents a low-cost solution and today creates no difficulties, either from a technical point of view or in respect of the service life. As explained in Section 5.6.8.1, the damper mountings are designed in such a way that they can transfer relatively large forces and usually only a slight reinforcement is

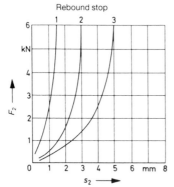

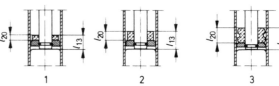

Fig. 5.33 Sachs Boge rebound stop in a twin-tube damper with 26 mm and 30 mm piston diameter (types S_{26} and S_{30}); shown here are body shapes and compression travel s_2 as a function of the tensile force F_2 up to 6 kN. The heights l20 are at position 1: 4 mm, position 2: 9 mm and position 3: 12 mm; snubbers up to 18 mm high are used.

necessary if additional forces occur through compression stops or supplementary springs.

The jounce bumper 2 shown in Fig. 5.32 is carried by the piston rod 1; when the wheels bottom out, it comes into contact with a cap surrounding the outer tube and is supported – fully compressed – on the steel protective tube 3. In the case of an incorrect shape or non wearproof rubber or plastic mixture, dust can occur that gets into the piston rod seal and renders it ineffective (Fig. 5.17). The consequence would be escaping oil, a reduction in the damping effect and destruction of the (not always oil-proof) bumper. Figure 5.34 shows the progressive springing curves of three compression stops of different length and the shape of those shown in part 2 of Fig. 5.32.

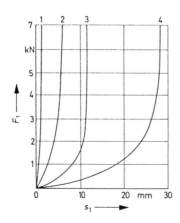

Fig. 5.34 Compression travel s_1 on the Sachs Boge compression stops for the T27 and T32 twin-tube dampers at forces up to F1 = 7 kN. Configurations 1, 2 and 3 are l12 = 8 mm, 15 mm or 23 mm high in their unladen condition and are the same shape as part 2 in Fig. 5.32. The supplementary spring (position 4) is 44 mm high.

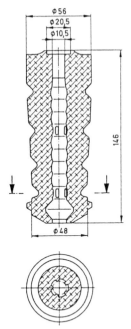

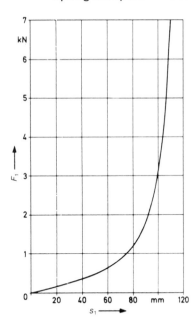

Fig. 5.35 Supplementary spring manufactured by Elastogram in a polyurethane elastomer on the rear spring dampers of the VW Golf. Material properties and shape make the highly progressive springing curve possible. At 146 mm overall height, it can be compressed by 110 mm and accept an impact load of over 700 kg or a force of $F_1 \geqslant 7$ kN.

5.6.8.3 Supplementary springs   Flat compression stops barely allow any reasonably shaped springing curve. Reduced impacts or the desire for a soft cushioning necessitate installation of a supplementary spring made of polyurethane elastomer or a hollow bumper (Figs 5.4a and 5.9). Figure 5.34 contains at position 4 a springing curve of a 44 mm high supplementary spring suitable for twin-tube dampers and Fig. 5.35 shows a design used for strut dampers. As shown in Fig. 5.36, this is carried by the piston rod and comes into contact with a cap or disc when it compresses.

5.7   Spring dampers

The spring damper, a device carried over from the motor cycle, is used by more and more passenger car manufacturers, not only on independent wheel suspensions, but also on rigid and compound crank axles. This force centre, formerly described as a suspension strut, does not carry the wheel-like McPherson struts, but comprises all parts of a wheel suspension that are necessary for springing and damping:

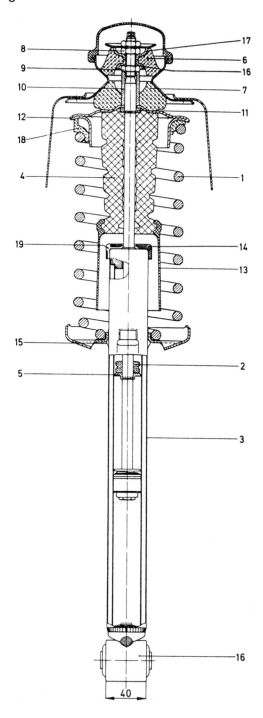

coil spring 1, rebound stop 2, supplementary spring 4 (Fig. 5.36) and, as the supporting element, the shock absorber.

The coil spring can be retrofitted and supported with rubber insulators on the body or pre-assembled into the unit, in which case two bolts are used to fix the entire assembly. Fitted spring dampers can be seen in Figs 1.8, 1.28, 1.37, 1.38, 1.41, 1.44, 1.45, 1.57 and 1.59.

5.8   McPherson struts and strut dampers

5.8.1 McPherson strut designs

The McPherson strut also carries and controls the wheel; the piston rod, which is strengthened from 11 mm to 18 mm up to 25 mm diameter on passenger cars (and up to 28 mm on light commercial vehicles), can absorb longitudinal and lateral forces and replaces the upper suspension link, including its three mountings. The designs, which are known today as McPherson struts, are divided into those

- on which the steering knuckle is solidly fixed to the outer tube (Fig. 5.37), and those
- with a bolted on steering knuckle (Figs 5.39, 5.40, 1.4 and 1.39)

and in terms of the damper part into

- wet suspension struts on which the damper part is directly mounted into the carrier tube (Figs 5.39 and 5.40) and

---

Fig. 5.36 Sachs Boge rear spring damper on the VW Golf and Vento with coil spring 1 and rebound stop 2 visible in the cross-section. This is carried by the 11 mm thick piston rod and is located 107 mm above the (27 mm diameter) piston so that it has an adequate minimum bearing span in the fully extended condition; the stop ring 5 is rolled into a groove of the piston rod.

The upper fixing is a pin-type joint which transfers the springing and impact forces to the body via the large noise-insulating rubber snubbers 6 and 7. The two parts are drawn together by the hexagonal nuts 8 and 9; the tube 10 and the bushes 16 and 17 ensure that a precise preload is achieved. The lower washer 11 comes into contact with a wire snap ring (which sits in a half-round groove) and both the spacer tube 10 and the upper spring seat 12 come into contact with the washer. The spring seat supports the coil spring 1 via the elastic ring 18 and also the polyurethane supplementary spring 4, which has a round bead at the bottom to take the plastic protective tube 13.

If the axle compresses, part 4 comes into contact with the cap 14. This ensures the piston rod seal is not damaged. The cap has a groove (position 19) through which the air in the supplementary spring can escape when it is compressed. The lower spring seat is supported at three points (position 15), which protrude from the outer tube and the outside diameters of which must have a tolerance of ±0.5 mm.

To ensure the rubber part only flexes a little under vertical forces, the eye-type joint 16 was made 40 mm wide. The attachment to the axle shown in Fig. 1.41 is made with an M 10 bolt.

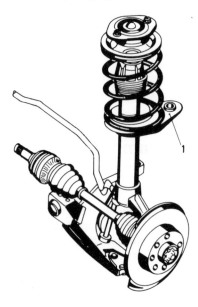

Fig. 5.37 Driven McPherson front axle of a Opel/Vauxhall model. The outer tube is press fitted to the steering knuckle, whilst the steering arm 1 sits relatively high up.

- cartridge designs in which the damper part is inserted into the carrier tube and screwed together (Fig. 5.38).

A decision in favour of one of the solutions is mainly a question of the manufacturer's preferences, although whether the outer tube needs to be included for transferring steering forces, i.e. whether the steering arms sit on it, is also a consideration (Figs 5.37, 1.40, 3.79, 4.0 and 4.31).

Wet suspension struts are better at conducting heat away from the damper and, where they are detachably linked to the steering knuckle, offer the advantage that they do not need to be able to be dismantled and that, if the damping fails, the actual damping part can be easily exchanged. This design also makes it possible to close the strut by means of indentations in the outer tube (Fig. 5.40), rolling it (edge 6 in Fig. 5.39) or welding it to the sealing cap.

If, as shown in Fig. 5.37, the steering knuckle is press-fitted to the suspension strut, a screwed closure cap is necessary for exchanging the damper cartridge.

5.8.2 Twin-tube McPherson struts, non-pressurized

The suspension strut shown in Fig. 5.39 operates on the twin-tube principle; it operates in the same way as the non-pressurized twin-tube damper (see Section 5.6.2). To have a sufficient minimum bearing span l–o (Fig. 1.6) in the fully rebounded condition, the rebound stop 13 has been set high. This measure, together with the PTFE-coated guide bush 11, reduces friction.

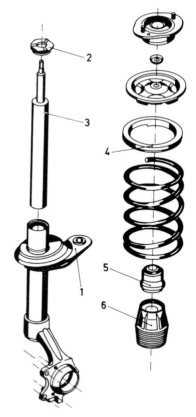

Fig. 5.38 If the damping on the Opel/Vauxhall suspension strut fails, the bolted closure cap must be undone and the shock absorber cartridge 3 changed. The elastic ring 4, located above the coil spring, the supplementary spring 5 and the dust bellow 6 can be seen clearly.

5.8.3 Twin-tube McPherson struts, pressurized

The development of the pressurized McPherson strut has met with significant difficulties for many years. Direct transfer of the monotube principle, as used in the shock absorber, is not possible because of the high extension force. There are solutions that keep the rod small and transfer the wheel control to the cylinder tube, but these are expensive and involve high levels of damper friction.

The pressurized twin-tube system is a good compromise. Here, the oil is only under a pressure of 0.6 to 1.0 MPa (depending on the manufacturer) and the extension force of the 18 to 28 mm thick piston rod is therefore limited.

Figure 5.40 shows a section through a McPherson strut. The spring seat 22 and the lower bracket for fixing to the steering knuckle are welded to the outer tube 2. The piston rod 1 is solid but can be hollow to reduce the weight; the piston has valve plates on both sides, depending on the desired damping curve, or a twin-tube damper valve operating only in the extension direction (see Sections 5.6.2.4 and 5.6.4.3). This can be an advantage where degressive valve curves are requested.

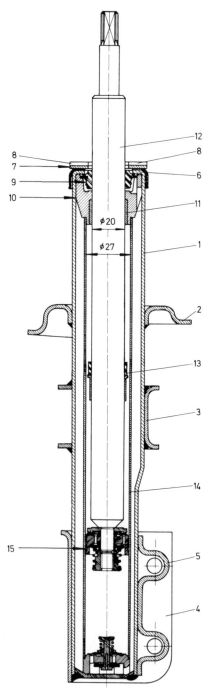

Fig. 5.39 McPherson strut of the Fiat Panda manufactured by Monroe: the spring seat 2 for taking the coil spring, the tab 3 (for fixing the steering arm) and the bracket parts 4 and 5 to which the steering knuckles are bolted to the outer tube 1. The stop disc 7 is supported on the rolled edge 6 of the outer tube, and its two transverse grooves 8 ensure that the supplementary spring cannot create overpressure in the interior; this would press dirt and deposits into the seal 9. The bush 11 is pressed into the sintering iron rod guide 10 from the bottom and its surface conditioned to reduce friction (to the piston rod 12). The rod is 20 mm diameter and, in the mid-range, carries the rebound stop 13; when the wheel is fully extended, the minimum bearing span (centre bush 11 to centre piston) is 120 mm.

The rod 12 is drawn in at the bottom to provide space for the rebound stage and check valve (see Fig. 5.18). The low-friction ring 15 provides the seal between the piston, which is 27 mm diameter, and the cylinder tube 14.

Fig. 5.40 Low-pressure twin-tube McPherson strut by Sachs Boge, drawn with the piston rod 1 fully in. The lower end 23 is drawn in and threaded to mount the compression stage valve; the upper tappet 24 gripping into the upper strut mount on the wheel house has two surfaces for retention.

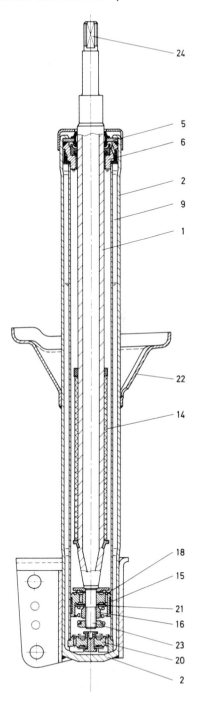

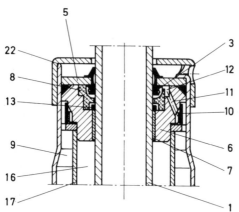

Fig. 5.41 Rod guide and seal unit of the Sachs Boge low-pressure twin-tube McPherson strut.

The studs of the hollow piston rod are made in a special cold-forming process; sensibly, the upper one is given a hexagonal socket (or two flat surfaces) for holding during assembly and the lower tappet must be oil and gastight. The rebound stop is made of plastic, is tight to the rod and transmits the vertical forces via the tube 14 to a zone of the rod that is not subject to bending. To keep friction low, the seal between the piston and cylinder wall is the broad PTFE ring 15. The extension stage valve 16 is similar to that shown in Fig. 5.18. The forces in the pressure stage are applied jointly by the valves 18 and 20 (see Section 5.6.2.5).

The constant orifice on the piston, also known as a by-pass or advanced opening cross-section, is created by punched holes in the lower valve plate 21 and a similar by-pass plate for the compressive stroke is used on the compres-

Fig. 5.42 A pressurized twin-tube McPherson strut by Sachs Boge with a hydraulic-mechanical rebound stop and electrically adjustable three-stage damping. The coil spring 11 fixed to the inner diameter of the cylinder tube underneath the piston rod guide carries a shim at the bottom, which comes into contact with the ring 12 when the wheel rebounds and therefore increases the rebound damping at the end of the stroke.

To vary the damping, the two solenoid valves positioned on the outer tube 1 grip into the ring channel 5, which is linked with the working chamber in the cylinder tube 3 via the hole 4. If the valve body 7 opens the inlet hole 6, the oil from the ring area meets the set of spring discs 9 via the advanced opening 8 and flows through this into the equalization chamber 10. The damping curve results from the sum of oil flow resistances at the piston (rebound stage) or in the bottom valve (compression stage) and with the solenoid valve opened parallel. The hardest setting is obtained if both valves remain closed and there are two softer settings when either of the two valves is opened. The electronic control module needed to control these valves was developed by the company VDO.

The degressive force–speed curve indicates the setting range (see also Figs 5.18a and 5.25a).

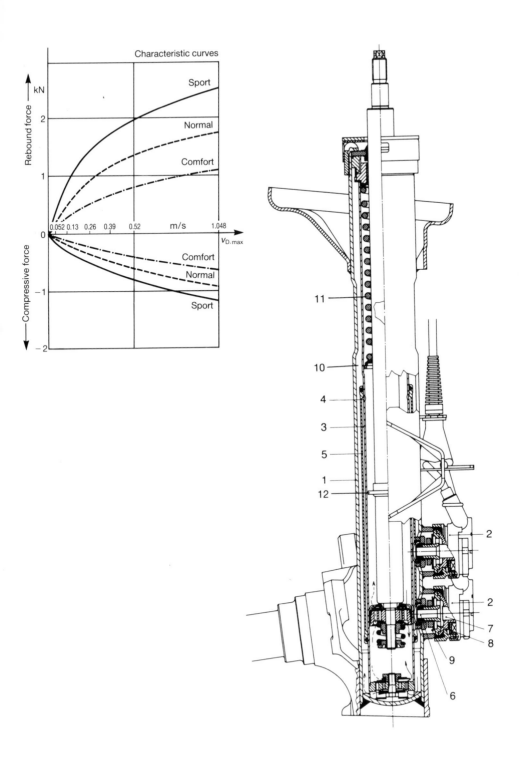

Characteristic curves

Rebound force (kN)

Sport
Normal
Comfort

0.052 0.13 0.26 0.39 0.52   m/s   1.048

$v_{D.max}$

Comfort
Normal
Sport

Compressive force

sion valve 20. In order not to influence the efficiency of this constant opening on the damping curve too much, the clearance area between guide bush 7 and piston rod 1 is sealed in a controlled manner using the PTFE ring 13 (Fig. 5.41). In the non-operative condition (as shown) it is at the bottom, but during operation, i.e. when there is pressure in the working chamber 16, it comes into contact with the spacer 8. This has transverse grooves of a precisely fixed cross-section which provide the necessary ventilation.

As described in Section 5.6.2.3, when the oil cools after a journey, an air bubble can form in the top of the pressurized twin-tube damper. On the strut damper, the pressure in the oil column in the equalization chamber 9, together with the inner tube valve 10, should significantly delay this. However, if at very low temperatures pressure is reduced and oil concentrated, the ventilation facility becomes important again.

The internal pressure in the upper part 16 of the working chamber increases on both rebound and compression damping. The residual oil volume flowing through the clearance between rod 1 and bush 7 collects in the high ring channel 12 and is passed through the inclined holes 11 into the lower channel, which is formed by an angle ring and the tube valve 10. This latter part lifts and allows the oil to flow back into the equalization chamber 9. The chamber is around half full of oil and is pressurized by gas. The tube 10 acts as a lock and prevents ingress of gas in the reverse direction on the rod seal 3 (see also Fig. 5.42). Foaming of the oil and forming of air bubbles in the valve, known as cavitation, is prevented by the inner pressure of $p = 0.6$ MPa to 1.0 MPa. If, for some reason, the gas should escape, the damping function remains largely intact due to the existent bottom valve 20: a designed-in safety feature. The suspension strut can be closed by welding, or as shown in Fig. 5.41, by several beads, which press the closure plate against the guide unit 6, and press this in turn against the cylinder tube 17, which then presses the valve body 20 shown in Fig. 5.40 against the bottom of the cold-sunk outer tube 2. The gasket 3, with the dust lip that protrudes upwards forms a unit with the closure plate 5, which is covered from the top by the cap 22. The supplementary spring comes into contact with this cap at full jounce.

5.8.4 Damper struts

Damper struts only carry the wheel without transferring vertical springing forces; there is no spring seat. However, rebound stops and supplementary springs are arranged as in suspension struts (Figs 1.25 and 1.63).

5.9 Variable damping

The dampers and McPherson struts described in the previous sections have a fixed curve over the entire operating range that depends only on piston velocity. It is determined by the vehicle manufacturer for a given vehicle type (Fig. 5.28). Various special designs take account of the wish to be able to vary this. If an increase in damping force is required at the end of the damper stroke, hydraulic or mechanical/hydraulic rebound stops provide a useful solution (positions 11 and 12 in Figs 5.42 and 5.20a). The demand for a travel-dependent influence over longer stroke ranges can be met with longitudinal grooves in the cylinder tube. Far more effective is the use of electrical dampers and suspension struts that are adjustable from the outside. This can be done in the mounted condition and whilst the vehicle is travelling. Figure 5.42 shows a suspension strut design offered by BMW. The illustration contains a description of how it works.

6

Chassis and vehicle overall

6.1 Vehicle and body centre of gravity

6.1.1 Centre of gravity and handling properties

Depending on the problem posed and the topic, the following are important variables in vehicle engineering:

- vehicle centre of gravity S
- body centre of gravity B
- axle centres of gravity U_f or U_r.

The distance of the centres of gravity S and B from the front or rear axle and their height above ground are necessary for

- calculating the climbing ability
- designing brake systems and springing
- vibration considerations
- driving stability investigations
- determining mass moment of inertia etc.

Low centres of gravity are always desirable, as they are associated with fewer driving dynamic problems and increased vehicle performance during cornering and braking, but in practice the design options are relatively restricted.

The position of a vehicle centre of gravity S and the body centre of gravity B is highly dependent on the load; when people get into the vehicle or luggage is loaded in the boot or onto the roof, the centre of gravity changes *vis-à-vis* the empty condition, both in the longitudinal direction (*x*-axis) and upwards (*z*-direction). The body lowers when it is loaded, i.e. its centre of gravity B drops. The centre of gravity of the people and, in particular, that of the luggage carried on the roof, is higher than that of the body so the end result is usually a higher overall centre of gravity S (distance h_S, Fig. 6.10).

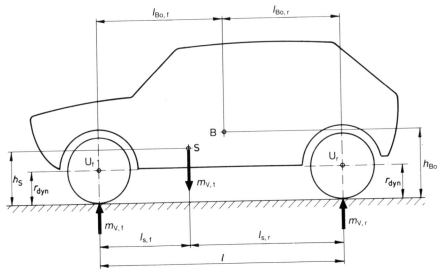

Fig. 6.1 Designation of the paths for determining the centres of gravity S of the over-all vehicle and B of the body. The centres of gravity U_f and U_r of the front and rear axles can be regarded as being in the centres of the wheels.

6.1.2 Calculating the vehicle centre of gravity

Calculating the position of the centre of gravity is likely to be possible only with great difficulty and considerable effort. It is much simpler to determine the position experimentally by weighing. For this, both the empty vehicle should be observed and when it is occupied by two or four people, approximately 176 cm tall and weighing around 68 kg.

6.1.2.1 Centre of gravity distance to front and rear axle   Figure 6.1 contains the paths and angles necessary for calculating the centres of gravity and Fig. 3.1 the position of the coordinate system. When the vehicle is weighed, it must be standing on a completely horizontal plane and with each axle on a weighbridge. So as not to distort the weighbridge, it must be possible to turn the wheels freely. The weighed front axle load $m_{V,f}$ and the rear axle load $m_{V,r}$ give the total weight $m_{V,t}$ of the vehicle:

$$m_{V,t} = m_{V,f} + m_{V,r} \; [\text{kg}] \tag{6.1}$$

The balance of moments around $m_{V,f}$ or $m_{V,r}$, in conjunction with the wheel-base l in the longitudinal direction, gives the centre of gravity distances l_f to the front and l_r to the rear axle:

$$l_f = \frac{m_{V,r}}{m_{V,t}} \, l; \qquad l_r = \frac{m_{V,f}}{m_{V,t}} \, l = l - l_f \tag{6.2}$$

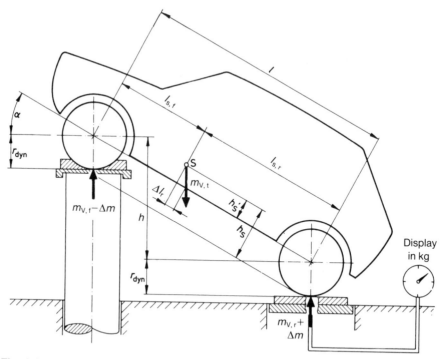

Fig. 6.2 Vehicle on a weighbridge with forces and paths for deriving the equation for vehicle centre of gravity height h_S included.

If the lateral distance of the centre of gravity (y-direction) from the vehicle centre-line is required the wheel loads must be weighed to be able to calculate first of all the lateral offset of the centres of the front and rear axles from the centre-line via similar equations made up from the rear view, and then similarly for the vehicle centre of gravity from the top view (see Equations 5.14 and 6.24).

6.1.2.2 Centre of gravity height   To calculate h_S, first the front and then the rear axle must be lifted as high as possible (by the amount h) with an elevating mechanism (autohoist, jack, crane), with the other axle standing in the centre of a weighbridge (Fig. 6.2). The following would need to be ensured.

- The vehicle must be prevented from falling off by inserting wedges from the outside on the axle to be raised. The brake must be released and the gear box must be in neutral. It must be possible to turn the wheels on the platform easily; the platform would otherwise distort and the result be imprecise.
- The wheels are held still on the centre of the platform, the vehicle forward movement must be even when the vehicle is raised.

- The vehicle should be in the on-road condition, i.e. full tank, tools, spare wheel, etc. (as per the kerb weight, see Section 5.3.1).
- Both axles must be prevented from compressing or rebounding before the vehicle is raised. The locking device must be of an adjustable variety so that the amount by which the body sinks when there are two or four people and luggage in the vehicle can be taken into consideration.
- To eliminate tyre springing during the measurement, it is recommended that the tyre pressure on both axles be increased to 3.0 to 3.5 bar.

Mathematical observation of the measurement is as follows (Fig. 6.2):

$$h/l = \sin \alpha$$

The angle α is known; but $h_S = h'_S + r_{dyn}$ is sought whereby

$$h'_S = \Delta l_r / \tan \alpha$$

To be able to determine Δl_r, the equation of moments produced around the centre of the front axle is set up:

$$m_g (l_f + \Delta l_r) \cos \alpha = (m_{V,r} + \Delta m) \, l \cos \alpha$$

$\cos \alpha$ is eliminated so:

$$\Delta l_r = \frac{(m_{V,r} + \Delta m) \, l}{m_{V,t}} - l_f, \quad \text{whereas} \quad l_f = \frac{m_{V,r}}{m_{V,t}} \, l, \text{ therefore}$$

$$\Delta l_r = \frac{\Delta m}{m_{V,t}} \, l, \quad \text{hence} \quad h'_S = \frac{l \, \Delta m}{m_{V,t} \tan \alpha} \quad \text{and} \tag{6.3}$$

$$h_S = \frac{l}{m_{V,t}} \frac{\Delta m}{\tan \alpha} + r_{dyn}$$

In Equation 6.3 the angle α can be expressed through the easily measurable vehicle stroke height h and so the equation can be simplified:

$$h_S = \frac{l}{m_{V,t}} \frac{\Delta m}{h} (l^2 - h^2)^{1/2} + r_{dyn} \tag{6.4}$$

With $\Delta m/h$ or $\Delta m/\tan \alpha$ there is a constant in the equation. When it is weighed, in each instance, only the changes caused by the vehicle lifting on one side, namely Δm and the raised dimension h, need to be determined.

The other values such as wheelbase l, vehicle weight $m_{V,t}$ and the dynamic rolling radius r_{dyn} remain the same. The centre of gravity height is required for calculating various vehicle conditions, i.e. for the travelling vehicle, so the dynamic rolling radius r_{dyn} of the tyre must be added to h'_S and not the somewhat lower static rolling radius that only applies to the standing vehicle. In

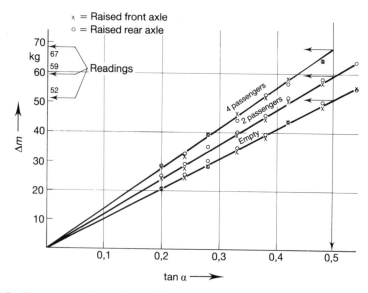

Fig. 6.3 The measured axle load differences Δ_m are entered separately for the front and rear axles as a function of tan α, depending on which axle was on the weighbridge. A mean straight line, which must go through the zero point, can be used for the most precise possible determination of the quotient Δ_m/tan α.

accordance with Equation 2.2 r_{dyn} must be calculated from the rolling circumference C_R.

To determine the centre of gravity height, it is recommended that the measurements be carried out from $\alpha = 10°$, increasing the angle by steps of 2.5° up to $\alpha = 25°$. Front and rear axles should be raised in turn with 2.5° corresponding to a lift of around 100 mm. The axle load difference Δm of the two weights can then be entered separately for front and rear axles as a function of tan α. The result is divergent measurement points (Fig. 6.3) through which a mean should be drawn starting from the zero point. To obtain Δm/tan α, the quotient must be produced at any point on this straight line (in this case at tan $\alpha = 0.5$). In the example and with an empty vehicle this would be:

Δm/tan $\alpha = 52/0.5 = 104$ kg

Table 6.4 shows a form with the numerical values of Fig. 6.3 entered and the calculated centres of gravity heights.

$h_{S,0} = 528$ mm, $h_{S,2} = 530$ mm and $h_{S,4} = 537$ mm

6.1.2.3 Ratio i_{ul} for the empty condition   The known centre of gravity height $h_{S,0}$ of the empty vehicle can be compared with the empty height h_{ul} of the unladen vehicle. For the passenger car, this would be $h_{ul} = 1380$ mm, so that the ratio would be:

$$i_{ul} = h_{S,0}/h_{ul} = 0.377$$

If the centre of gravity height $h_{S,0}$ of a 4 to 5-seater passenger car is not known, it can be judged using i_{ul}:

$$h_{S,0} = (0.38 \pm 0.02)\, h_{ul} \tag{6.4a}$$

6.1.2.4 Influence of loading   The value $h_{S,0}$ applies to the kerb weight; when the vehicle is laden, the centre of gravity generally moves upwards, i.e. the path h_S increases, unlike the vehicle height, which reduces. The amount by which the centre of gravity of the vehicle as a whole rises when there are two, four or five people in it, is a question of the spring rate on the front and rear axles, the seat heights and the weights and sizes of the occupants (Figs 5.7 to 5.10). The following can be an approximate figure for the centre of gravity height $h_{S,pl}$ (index pl = partly laden):

$$h_{S,pl} = h_{S,0} + \Delta h_S \tag{6.4b}$$
Two people $\Delta h_{S,2} \sim +12$ mm
Four people $\Delta h_{S,4} = -8$ mm to $+29$ mm

A fifth person on the rear seat or load in the boot causes the body to go down, so the overall centre of gravity sinks.

6.1.3 Axle weights and axle centres of gravity

If, instead of the height of the centre of gravity h_S of the vehicle as a whole, the height h_{Bo} of the body centre of gravity is required, it can be determined by assuming that the centre of gravity of the unsprung mass $m_{u,f}$ (front) $m_{u,r}$ (rear) is approximately at the centre of the wheel, i.e. at the distance of the dynamic rolling radius r_{dyn} to the ground (Figs 6.1 and 6.5). Furthermore, their weight should be known, determined by weighing or calculated by approximation.

$$m_{u,f\,or\,r} = \frac{i_{m,f\,or\,r}\, m_{V,f\,or\,r}}{1 + i_{m,f\,or\,r}} \tag{6.4c}$$

The following approximate values can be included in the equation:

Front axle	$i_{m,f} \sim 0.12$
Non-driven rear axle	$i_{m,r} \sim 0.13$
Driven rear independent wheel suspension	$i_{m,r} \sim 0.14$
Driven rear rigid axle	$i_{m,r} \sim 0.22$

Table 6.4 Measuring sheet with values of a passenger car entered with additional information on size and weight of the people in the vehicle during the measurement. Source: Technical laboratory, Polytechnic of Cologne.

Vehicle	Passenger car					
Year of manufacture						
l	2570 (mm)					
r_{dyn}	296 (mm)					
Tyres	195/65R14					
State when measured	M	*Passengers in front*			*Passengers in rear*	
Empty	0	Weight (kg)	Size (cm)		Weight (kg)	Size (cm)
2 passengers	2	72.5	183		60	170
4 passengers	4	68.2	180		71	178
		$\Sigma 140.7$			$\Sigma 131$	

M	h (mm)	$\sin\alpha = \dfrac{h}{l}$	α (°)	$\tan\alpha$	$m_{V,f}$ (kg)	$\Delta m_{V,f}$ (kg)	$m_{V,r}$ (kg)	$\Delta m_{V,r}$ (kg)
	1200	0.47	27.92	0.53	683.2	54.5	577.4	54.2
	1100	0.43	25.42	0.48	678.0	49.3	573.6	50.4
	1000	0.39	22.96	0.42	672.6	43.9	567.4	44.2
0	900	0.35	20.56	0.38	667.0	38.3	562.8	39.6
	800	0.31	18.06	0.33	662.3	33.6	558.8	35.6
	700	0.27	15.66	0.28	657.6	28.9	553.5	30.3
	600	0.23	13.30	0.24	653.2	24.5	549.3	26.1
	500	0.20	11.54	0.20	648.9	20.2	544.1	20.9
	0	0	0	0	628.7	0	523.2	0
	1200	0.47	27.92	0.53	—	—	654.9	64.3
	1100	0.43	25.42	0.48	760.9	57.3	648.5	57.9
	1000	0.39	22.96	0.42	755.1	51.5	642.4	51.8
	900	0.35	20.56	0.38	748.1	44.5	636.4	45.8
2	800	0.31	18.06	0.33	742.7	39.1	630.8	40.2
	700	0.27	15.66	0.28	742.9	39.3	625.8	35.2
	600	0.23	13.30	0.24	731.6	28.0	620.0	29.4
	500	0.20	11.54	0.2	727.7	24.1	615.3	24.7
	0	0	0	0	703.6	0	590.6	0
	1100	0.43	25.42	0.48	—	—	759.6	63.8
	1000	0.39	22.96	0.42	785,3	58.2	752.2	56.4
	900	0.35	20.56	0.38	772.9	45.8	739.3	43.5
	800	0.31	18.06	0.33	772.9	45.8	739.3	43.5
4	700	0.27	15.66	0.28	765.7	38.6	734.6	38.8
	600	0.23	13.30	0.24	759.3	32.3	728.8	33.0
	500	0.20	11.54	0.20	754.6	27.5	724.0	28.2
	400	0.16	8.98	0.16	749.0	21.9	—	—
	0	0	0	0	727.1	0	695.8	0

$m_{V,t}$ (kg)											
	0	1151.9	$I_{S,f} = \dfrac{m_{V,r}}{m_{V,t}} l$	0	1164.1	$I_{S,r} = l - I_{S,f}$	0	1398.9	$\dfrac{l}{m_{V,t}}$	0	2.23
	2	1294.2		2	1169.6		2	1393.4		2	1.98
	4	1422.9	(mm)	4	1253.3	(mm)	4	1309.7	(mm/kg)	4	1.80
	0	52	$h_S = \dfrac{l}{m_{V,t}} \cdot \dfrac{\Delta m}{\tan\alpha} + r_{dyn}$	0	$2.23 \cdot \dfrac{52}{0.5} + 296$	$h_{S0} = 528$ mm					
Δm by $\tan\alpha = 0.5$	2	59	$\left(\text{mm} = \dfrac{\text{mm}}{\text{kg}} \cdot \text{kg} + \text{mm}\right)$	2	$1.98 \cdot \dfrac{59}{0.5} + 296$	$h_{S2} = 530$ mm					
	4	67		4	$1.8 \cdot \dfrac{67}{0.5} + 296$	$h_{S4} = 537$ mm					

A passenger car which has a front axle load $m_{V,f} = 609$ kg in the unladen condition can be used as an example:

$$m_{u,f} = \frac{i_{m,f} \, m_{V,f}}{1 + i_{m,f}} = \frac{0.12 \times 609}{1 + 0.12} = 65.3 \text{ kg}$$

Section 5.2 contains further details.

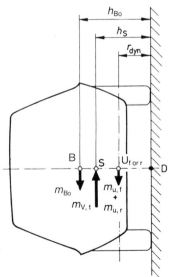

Fig. 6.5 Vehicle shown tipped to derive the equations of moments for the height h_{Bo} of the body centre of gravity.

6.1.4 Body weight and body centre of gravity

Taking into consideration both axles, the body weight is:

$$m_{Bo} = m_{V,t} - (m_{u,f} + m_{u,r})$$ (6.5)

and the distances of the centres of gravity to the axle centres shown in Fig. 6.1 become:

$$l_{Bo,f} = \frac{m_{Bo,r}}{m_{Bo}} l; \quad l_{Bo,r} = \frac{m_{Bo,f}}{m_{Bo}} l = l - l_{Bo,f}$$ (6.6)

$m_{Bo,f}$ and $m_{Bo,r}$ are the proportions of the body weight over the front or rear axle:

$$m_{Bo,f} = m_{V,f} - m_{u,f} \text{ and}$$ (6.6a)

$$m_{Bo,r} = m_{V,r} - m_{u,r}$$ (6.6b)

The height h_{Bo} of the body centre of gravity B is easy to calculate by observing the vehicle when it is tipped forwards (Fig. 6.5) using an equation of moments, assuming that the individual weights act as forces at their respective distance on the ground:

$$h_{Bo} = \frac{m_{V,t} h_S - (m_{u,f} + m_{u,r}) \, r_{dyn}}{m_{Bo}}$$ (6.7)

Depending on the loading condition and the weight of the unsprung mass, the body centre of gravity h_{Bo} is 20 to 40 mm higher than that of the vehicle as a whole h_S.

6.2   Mass moments of inertia

From mechanics it is known that when a body is accelerated in a straight line the inertia force

$$F_c = m \, a_x = \text{mass} \times \text{acceleration [N]}$$

In comparison to this, in the case of accelerated rotational movement, the acceleration moment is influenced by the rotation mass J.

The rotation mass – equivalent to the mass moment of inertia J [kg m^2] and also known as second degree mass moment – is a measure of inertia on rotating bodies. In vehicles, three important rotational movements occur in the various vehicle conditions, to which the variables of the mass moments of inertia J are related.

- The vehicle moment of inertia $J_{V,z}$ around the vertical axis (z-axis, Fig. 3.1) is required for driving stability studies or even for reconstructing road traffic accidents.
- The body moment of inertia $J_{Bo,x}$ around the vehicle's longitudinal axis (x-axis) is essential for generally studying body movement (roll behaviour) during fast lane changes in the driving direction.
- The body moment of inertia $J_{Bo,y}$ around the transverse axis (y-axis) is the determining variable for calculating pitch vibration behaviour.

In addition to this, in general, the inertia moments of power units (engine–gearbox unit) and individual rotationally symmetrical elements, such as steering wheels, tyred wheels, etc. are of importance.

The position of its centre of gravity and the variables of the moment of inertia are usually determined with the basic design of a vehicle (drive, wheelbase, dimensions and weight).

In addition to the type of drive, the vehicle's moment of inertia $J_{V,z}$ around the vertical axis is the determining factor for its cornering performance. Manoeuvrability increases as the inertia moment decreases, whilst driving stability when the vehicle is moving in a straight line and on S bends decreases by the same amount.

$J_{V,z}$ comprises the mass $m_{V,T}$ of the vehicle as a whole and the radius of gyration $i_{V,z}$ squared:

$$J_{V,z} = m_{V,t}\, i_{z,V,z}^2 \ [\text{kg m}^2] \tag{6.8}$$

For the vehicle, the size of the path $i_{V,z}$ depends on the length and width of the body, the position of the engine and its weight, the distribution of the occupants in the interior and the height of the luggage load. Series tests with saloons have shown that the radius of gyration is mainly a function of the load status and only varies within narrow limits from vehicle to vehicle. Table 6.6 shows the average values. Only the vehicle weight $m_{V,t}$ in the occupancy or load condition to be investigated is necessary for determining the approxi-

Table 6.6 The approximate radius of gyration $i_{Bo \ or \ V}$ (valid for medium-sized saloon cars) for the inertia moment J_{Bo} of the body or J_V of the vehicle as a whole, shown as a function of the loading condition and the pivot axis (Fig. 3.1).

Load	Inertia radius in metres		
	Car body only		Whole vehicle
	x-axis	around the y-axis	z-axis
Empty	0.65	1.21	1.20
2 passengers in front	0.64	1.13	1.15
4 passengers	0.60	1.10	1.14
4 passengers and luggage	0.56	1.13	1.18
Formula sign	$i_{Bo,x}$	$i_{Bo,y}$	$i_{V,z}$

mate moment of inertia $J_{V,z}$ (see Section 5.3.6). The values shown in Table 6.6 relate to medium-sized saloons. If the vehicle has a five, six or eight-cylinder engine, a difference value must be added:

$$\Delta i \approx 0.05 \text{ m to } 0.1 \text{ m}$$

If vehicle length L_t and wheelbase l are included in the following equation, an accuracy of at least 98% can be achieved; only a correction factor needs to be added:

$$J_{V,z} = 0.1269 \, m_{V,t} \, L_t \, l \, [\text{kg m}^2] \tag{6.9}$$

Nevertheless, this equation only applies to the usual vehicle loading. Higher loads in the boot (or a roof load) must be considered separately:

$$J^*_{V,z} = 0.1269 \, m_{V,t} \, L_t + \Delta m \, l^2_{\text{ml}} \, [\text{kg m}^2] \tag{6.10}$$

l_{lm} is the distance of the loading mass Δm to the vehicle centre of gravity.

The moment of inertia J_{Bo} of the body is not so easy to calculate. In this instance, the weights $m_{u,f}$ and $m_{u,r}$ of the unsprung masses must be known and their distances to the respective coordinate axis drawn through the vehicle centre of gravity (see Equations 6.4c and 6.6 and Fig. 3.1); it is easier, in this case, to use Table 6.6:

$$J_{Bo,y} = m_{Bo} \, i^2_{Bo,y} \, [\text{kg m}^2] \tag{6.11}$$

A front-wheel drive passenger car with two occupants can be used as an example for the pitch vibration calculation (around the y-axis):

Axle load front, partly laden (index pl) $m_{V,f,pl} = 609$ kg
Axle load rear, partly laden (index pl) $m_{V,r,pl} = 393$ kg

The weight of the axle mass is:

front $m_{u,f} = 67$ kg and rear $m_{u,r} = 59$ kg

The radius of gyration is $i_{Bo,y} = 1.13$ m. Equation 6.5 gives:

$$m_{Bo,pl} = m_{V,f,pl} + m_{V,r,pl} - (m_{u,f} + m_{u,r}) = 609 + 393 - (67 + 59)$$
$$m_{Bo,pl} = 876 \text{ kg}$$

In accordance with Equation 6.11 the mass moment of inertia of the body is then:

$$J_{Bo,y} = m_{Bo,pl} \, i^2_{Bo,y} = 876 \times 1.13^2, \ J_{Bo,y} = 1119 \text{ kg m}^2$$

The same applies to body roll movements around the x-axis. The values in the table should also be used here:

$$J_{Bo,x} = m_{Bo} \, i^2_{Bo,x} \, [\text{kg m}^2] \qquad (6.12)$$

6.3 Braking behaviour

6.3.1 Braking

When the driver brakes, the equivalent braking force acts as a reaction force at the centre of gravity of the vehicle as a whole (Fig. 6.1).

$$F_{B,t} = \mu_{x,w} \, F_{V,t} \qquad (6.13)$$

i.e. the coefficient of friction $\mu_{x,w}$ times the weight force $F_{V,t}$ of the vehicle as a whole, whereas $\mu_{x,w}$ can be equated to the deceleration a_x in m s^{-2}, divided by gravity:

$$\mu_{x,w} = a_x/g \qquad (6.13a)$$

At an international level, the formula z is used for the braking function.

$$z \simeq \mu_{x,w} \text{ and as a percentage: } z = \mu_{x,w} \, 100(\%) \qquad (6.13b)$$

i.e. during braking $z = 80\%$ (corresponding to $a_x = 7.85$ m s^{-2}) the coefficient of friction $\mu_{x,w} = 0.8$ is necessary (Fig. 2.21).

The braking force $F_{B,t}$ acting at the vehicle's centre of gravity causes longitudinal forces $F_{B,f}$ and $F_{B,r}$ at the centres of wheel contact of the front and rear axles, and an increase in axle load $+\Delta F_{V,0}$ at the front and a reduction $-\Delta F_{V,0}$ at the back when the vehicle is observed as a rigid body. In accordance with Fig. 6.7 the equations would then be

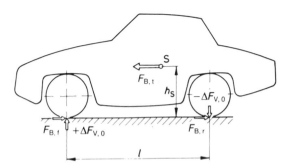

Fig. 6.7 A braking force $F_{B,t}$ acting at the centre of gravity S of the vehicle causes the axle load transfer $\pm\Delta F_{V,0}$ and the braking forces $F_{B,f}$ on front and $F_{B,r}$ on rear axle. If the aerodynamic and rolling resistances are ignored, the forces can be easily calculated.

$$\kappa = h_S/l \tag{6.13c}$$

$$\Delta F_{V,0} = \mu_{x,w} F_{V,t} \kappa [kN] \tag{6.14}$$

$$F_{V,f,dyn} = F_{V,f} + \Delta F_{V,0} \text{ and } F_{V,r,dyn} = F_{V,r} - \Delta F_{V,0} \tag{6.15}$$

The lower the centre of gravity and the longer the wheelbase, the less is the (undesirable) load transfer $\Delta F_{V,0}$. The braking force related to one axle is then

front $F_{B,f} = \mu_{x,w} F_{V,f,dyn}$ and $\tag{6.16}$

rear $F_{B,r} = F_{B,t} - F_{B,f} = \mu_{x,w} F_{V,r,dyn}$ $\tag{6.17}$

Half the braking forces per axle multiplied by the dynamic rolling radius r_{dyn}, gives the braking moments M_b at the wheels (see Equation 6.25a), which are:

front $M_{b,f} = 0.5 F_{B,f} r_{dyn}$ $\tag{6.18}$

rear $M_{b,r} = 0.5 F_{B,r} r_{dyn}$ $\tag{6.19}$

The larger is r_{dyn}, the higher is the moment to be generated by the brake. This is one reason for using tyres with an $r_{dyn} \leqslant 300$ mm on a medium size passenger car ($r_{dyn} = C_R/2\pi$, see Equation 2.2).
    The sizes of $F_{B,f}$ and $F_{B,r}$ depend both on the vehicle and its loading condition and on the road, i.e. the coefficient of friction $\mu_{x,w}$ possible on it (see Section 2.7). A front-wheel drive vehicle and the calculation of the braking forces for two possible cases with an unfavourable loading can be used as an example to indicate the range of the braking force distribution

$$\Phi_f = F_{B,f}/F_{B,t} \text{ (times 100 as a \%)} \tag{6.20}$$

$$\Phi_r = (1 - \Phi_f) = F_{B,r}/F_{B,t} \tag{6.20a}$$

with an unchanged centre of gravity height h_S. As described in Section 6.1.2.4, however, h_S alters based on the load and the pitch angle (see Section 6.3.3.5).

6.3.1.1 Braking on dry concrete with only two people in the vehicle

$F_{V,f} = 6.9$ kN; $F_{V,r} = 4.2$ kN; $l = 2.49$ m
$\mu_{x,w} = 0.9$; $h_S = 0.58$ m

$$\Delta F_{V,0} = 0.9 \times 11.15 \times \frac{0.58}{2.49} = 2.34 \text{ kN}$$

$F_{B,f} = 0.9 \,(6.95 + 2.34) = 8.36$ kN
$F_{B,r} = 0.9 \,(4.2 - 2.34) = 1.67$ kN

6.3.1.2 Braking on ice with a fully laden vehicle

$F_{V,f} = 7.1$ kN; $F_{V,r} = 7.0$ kN; $\mu_{x,w} = 0.15$; l and h_S as previously
$\Delta F_{V,0} = 0.49$ kN, $F_{B,f} = 1.14$ kN and $F_{B,r} = 0.98$ kN

In the first case, the front axle must accept as a percentage share:

$$\Phi_f \times 100 = \frac{8.36}{11.15} \times 100 = 75\%$$

and accordingly the rear axle 25%. In the second example, the braking force distribution is 54% and 46%. In the usual distribution of 75% to 80% on the front and 20% to 25% on the rear, the axle could lock in the first case (because its brake applies too high a moment); on ice it would be the front axle.

6.3.2 Braking stability

If both wheels of an axle lock, i.e. if they slide on the road, there is not just reduced friction in the longitudinal direction (Fig. 2.21), but also lower friction in the lateral direction. If the rear axle locks, as shown in Fig. 6.8, relatively large braking forces $F_{b,f}$ are possible on the rolling wheels of the front axle (in front of the braking force $F_{B,t}$ and the vehicle centre of gravity S), i.e. the condition is unstable.

Lateral forces or irregularities in the road acting on the body can cause the vehicle which, to this point, has been travelling in a straight line to leave its direction of travel. A reinforcing yawing moment occurs (Fig. 6.9), which seeks to turn the vehicle sideways to its previous direction. There is a danger of lateral roll over. However, if the front axle locks, the vehicle has control on the rear wheels (which are still rolling in this case) (Fig. 6.10). The braking forces $F_{b,r}$ are behind the centre of gravity S. The condition is stable.

The position is different if the braking moments on the wheels of one axle are of different sizes. The brakes pull to one side due to different lining coefficients of friction or unequal coefficients of friction on the left and right wheels (μ split, see Section 1.7.1).

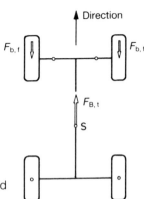

Fig. 6.8 Locking rear wheels lead to an unstable driving condition.

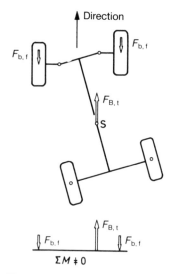

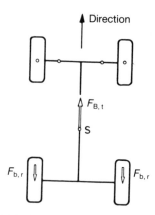

Fig. 6.10 When the front wheels lock, the vehicle condition remains stable although the vehicle can no longer be steered.

Fig. 6.9 When the rear wheels lock, a reinforcing yawing moment occurs even when the vehicle only slightly leaves the direction of travel.

Figure 6.11 shows a higher braking force $F_{b,f,l}$ on the left front wheel (than on the right one). The difference force of the two wheels $\Delta F_{b,f} = F_{b,f,l} - F_{b,f,rs}$, with the lever of half the tread width, gives the yawing moment $M_z = \Delta F_{b,f}\, 0.5\, b_f$ which introduces rotation to the left into the vehicle. In addition, there is also the steering moment $M_{S,b}$, which causes the steering to turn in the same direction.

Where the brake is on the outside (at the wheel), the size of this moment depends on the length of the wheel offset $+r_S$, and is:

$$+M_{S,b} = \Delta F_{b,f}\, r_S \cos \sigma \tag{6.21}$$

In the case of negative $-r_S$, there is counter steering (Fig. 6.12), and if $r_S = 0$, only the yawing moment ($-M_z$, Fig. 6.13) occurs. This also applies to centre axle steering (Fig. 3.87a).

A differential braking moment is less noticeable on the rear axle. Firstly, the braking forces $F_{b,r}$ are smaller and secondly there is a stable condition. The different sized forces $F_{b,r,l}$ and $F_{B,r,rS}$ are behind the centre of gravity S (Fig. 6.14).

6.3.3 Calculating the pitch angle

The pitch angle, i.e. the angle Φ_B by which the body turns around the lateral axis when the brakes are applied, can be calculated as a function of the braking force $F_{B,t}$ (Fig. 6.15, see also Section 5.4.3).

Fig. 6.11 As the static calculation below indicates, unequal braking forces $F_{b,f,l}$ and $F_{b,f,rs}$ at the centres of tyre contact of the front wheels cause the vehicle to rotate around the vertical axis. In the case of a positive wheel offset at ground (positive scrub radius), there is also a steering input in the same direction of rotation.

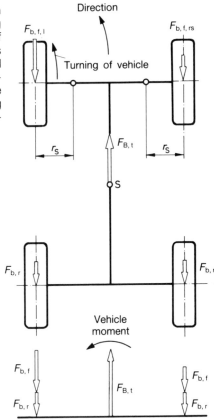

6.3.3.1 Data of the calculation example   A passenger car with the following data can be used to clarify the relationships:

Axle load front	$F_{V,f} = 6.95$ kN
Axle load rear	$F_{V,r} = 4.20$ kN
Axle weight force front	$F_{u,f} = 0.80$ kN
Axle weight force rear	$F_{u,r} = 0.70$ kN

Spring rate based on only one axle: front $c_f = 11.5$ N m^{-1}
rear $c_r = 14$ N m^{-1}

Dynamic rolling radius of tyre $r_{dyn} = 0.288$ m
Braking $z = 0.8$ i.e. $\mu_{x,W} = 0.8$ (see Equation 6.13b)
Wheelbase $l = 2.50$ m
Centre of gravity height $h_s = 0.58$ m

Details relating to the numerical values can be found in Sections 5.3.6.1, 6.1.3, 2.2.5.4 and 6.12.

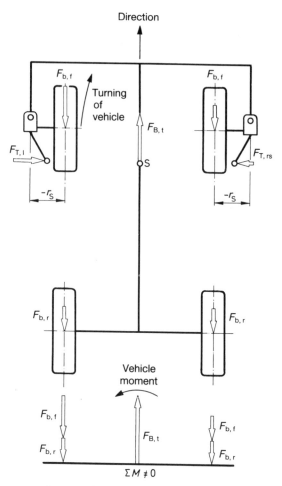

Fig. 6.12 In the case of a negative wheel offset (or elastokinematic toe-in alteration, Figs 3.79 and 3.64), the steering is turned by the front wheel, which must transfer the greater braking force $F_{b,f}$ (the left wheel in the illustration) opposite to the direction in which the vehicle is turned by the outer yawing moment. The static calculation shown indicates this. This leads to an equalization which, even in the case of different braking forces at the front, largely prevents deviation from the direction of travel.

6.3.3.2 Opposed springing forces When the body is observed as a rigid mass, the spring opposed forces $\Delta F_{Sp,f}$ and $\Delta F_{Sp,r}$ (related to one axle's side, front and rear, Fig. 6.7) correspond to half the axle load transfer $\pm\Delta F_{V,0}$ and, irrespective of whether the brakes are outside of the wheel or inside on the differential, the forces can be calculated easily on the basis of Equation 6.14.

$$\pm F_{V,0} = 0.8 \times 11.15 \times (0.58/2.50) = 2.07 \text{ kN}$$

Fig. 6.13 If, where the brake is on the inside, the longitudinal force lever $r_a = 0$ or, where the brake is on the outside, the wheel offset at ground (positive scrub radius) is 0, then unequal braking forces $F_{b,f}$ on the front wheels have practically no effect on the steering. The steering rod forces F_T are almost zero.

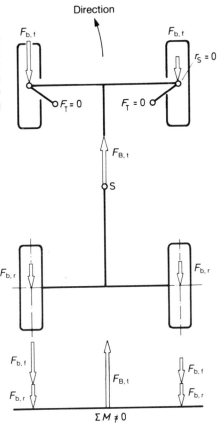

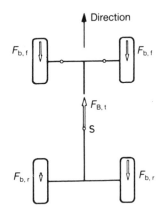

Fig. 6.14 A rear wheel brake, which is unevenly pulled, hardly has any effect on the steerability of a vehicle.

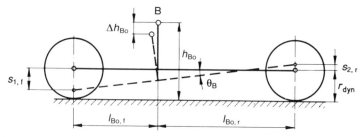

Fig. 6.15 If the body goes down more at the front than it rebounds at the rear, the body centre of gravity B moves down by Δh_{Bo}. The braking force $F_{B,Bo}$ would then act at the height $(h_{Bo} - \Delta h_{Bo})$ at point B. The pitch angle θ_B is also shown.

The vehicle goes down at the front and rebounds at the back. The spring rates are quoted in N mm^{-1} and also relate to one wheel. That is 1 N mm^{-1} = 1 kN m^{-1}, so we can assume $\Delta F_{V,0}$ divided by two. In accordance with Equation 5.10, with linear springing the following theoretical values would result:

Compression travel front $s_{1,f} = \Delta F_{V,0}/(2\ c_f) = 2.07/23 = 0.09$ m
Rebound travel rear $s_{2,r} = \Delta F_{V,0}/(2\ c_r) = 2.07/28 = 0.074$ m

6.3.3.3 Pitch angle with linear springing The pitch angle θ_B is (see also Equation 5.15):

$$\theta_B = \frac{s_{1,f} + s_{2,r}}{l} \tag{6.22}$$

and (times $360°/2\pi$)

$$\theta_B = 57.3 \times \frac{s_{1,f} + s_{2,r}}{l} \tag{6.23}$$

In this example the result is then:

$$\theta_B = 57.3 \times \frac{0.09 + 0.074}{2.50} = 3.76° = 3°46'$$

6.3.3.4 Pitch angle with progressive springing In order to determine the travel on the front and rear axles, the spring characteristics must be known. Travel is entered here in mm and wheel load in kg. The required values should therefore be calculated from the axle load.

Wheel load, front, normal $m_{1,f} = \dfrac{F_{V,f}}{2g} = \dfrac{6950}{2 \times 9.81} = 354$ kg

Wheel load, front, max.	$m_{1,max} = \dfrac{F_{V,f} + \Delta F_{V,0}}{2g} = \dfrac{6950 + 2070}{2 \times 9.81} = 460 \text{ kg}$
Wheel load, rear, normal	$m_{1,r} = \dfrac{F_{V,r}}{2g} = \dfrac{4200}{2 \times 9.81} = 214 \text{ kg}$
Wheel load, rear, min.	$m_{1,min} = \dfrac{F_{V,r} - \Delta F_{V,0}}{2g} = \dfrac{4200 - 2070}{2 \times 9.81} = 108 \text{ kg}$

In spite of the harder springing, the highly progressive curve shown in Figs 5.8 and 5.10 can be used as an example. The spring travel is:

front at 354 kg = 112 mm, and at 460 kg = 134 mm
rear at 214 kg = 110 mm, and at 108 kg = 44 mm

The vehicle would therefore:

front: go down by $s_{1,f} = 134 - 112 = 22$ mm and
rear : rebound by $s_{z,r} = 110 - 44 = 66$ mm

This would then give a pitch angle of only:

$\Theta_B = 2.02° \approx 2°1'$

6.3.3.5 Change of the centre of gravity height   Point B rises or falls when the brakes are applied based on how far the vehicle rebounds front and rear and how far the body centre of gravity is away from the axle centres. With the path entered in Fig. 6.15 and the weight forces plus Equations 5.14 to 5.14b, the result is then the change of height:

$$\Delta h_{Bo} = -s_{1,f} \frac{F_{Bo,f}}{F_{Bo}} + s_{2,r} \frac{F_{Bo,r}}{F_{Bo}} \tag{6.24}$$

When the springs compress, the body goes down and so $s_{1,f}$ becomes negative. The values in accordance with Equations 6.5 to 6.6b are:

$$F_{Bo,f} = F_{V,f} - F_{u,f} \text{ and} \tag{6.24a}$$

$$F_{Bo,r} = F_{V,r} - F_{u,r} \tag{6.24b}$$

$$F_{Bo} = F_{Bo,f} + F_{Bo,r} \tag{6.24c}$$

With the numerical values of the calculation example (see Section 6.3.3.1) and with linear springing, the result is then:

$F_{Bo,f} = 6.95 - 0.80 = 6.15$ kN; $F_{Bo,r} = 4.20 - 0.70 = 3.5$ kN
$F_{Bo} = 6.15 + 3.5 = 9.65$ kN, $s_{1,f} = 0.09$ m and $s_{2,r} = 0.074$ m

$$\Delta h_{Bo} = -0.09 \times \frac{6.15}{9.65} + 0.074 \times \frac{3.5}{9.65} = -0.03 \text{ m}; \quad \Delta h_{Bo} = -0.03 \text{ m}$$

The static centre of gravity height of the body, calculated using Equations 6.7 and 6.24a to c, is $h_{Bo} = 0.625$ m and that which occurs when the brakes are applied is:

$$h_{Bo} - \Delta h_{Bo} = h'_{Bo} = 0.595 \text{ m} \tag{6.24d}$$

The centre of gravity therefore goes down 4.8%. The resulting height h'_s of the vehicle centre of gravity can be calculated from the value $h'_{Bo} = 0.595$ m using the transformed Equation 6.7. This can be more easily done if the axle weight forces $F_{u,f}$ and $F_{u,r}$ are ignored. The error involved is less than 0.5%:

$$\Delta h_s = -s_{1,f} \frac{F_{V,f}}{F_{V,t}} + s_{2,r} \frac{F_{V,r}}{F_{V,t}} \tag{6.25}$$

The axle weight forces (unsprung masses, see Section 6.1.3) must be known if the pitch poles are to be included in the equation.

6.3.4 Influence of pitch poles

6.3.4.1 Pre-conditions for calculations
Pitch poles are only effective when the brakes are outside the wheel. The entire calculation has to be done differently because not only do the braking force portions of the body $F_{B,Bo,f}$ and $F_{B,Bo,r}$ act at the pitch poles O_f (front) and O_r (rear), but so do the vertical forces $\Delta F_{Bo,f}$ and $\Delta F_{Bo,r}$ acting against brake dive (Fig. 3.119).

Figures 3.83 and 3.87 show the static situation and, with Equations 6.13 to 6.17, the braking force related to one wheel can be calculated:

$$F_{b,f \text{ or } r} = F_{B,f \text{ or } r}/2 \tag{6.25a}$$

6.3.4.2 Forces on the pitch poles of both axles
Figure 3.120 shows how the forces are calculated on one axle side with double wishbones and Fig. 6.6 shows the calculation based on the entire axle and using the pitch poles required in this instance. To be able to calculate the forces $\Delta F_{Bo,f}$ supporting the body vertically when the brakes are applied, the equation of moments must be formed with the pivot at the centre of tyre contact. Paths e and c define the point O_f (present on both sides) in the illustration:

$$\Delta F_{Bo,f} = \frac{F_{B,Bo,f}\, e + F_{B,u,f} r_{dyn}}{c} = \Delta F_{V,f,2} \tag{6.26}$$

The axle load difference $\Delta F_{V,f}$ is the same size as $\Delta F_{Bo,f}$ and opposes compression when the brakes are applied. The forces that also appear in Equations 6.26 and 6.29 can be determined using Equations 6.6a and 6.6b:

$$F_{B,Bo,f \text{ or } r} = \mu_{x,w} F_{Bo,f \text{ or } r} = \mu_{x,w}\, m_{Bo,f \text{ or } r}\, g \tag{6.27}$$

$$F_{B,u,f \text{ or } r} = \mu_{x,w} F_{u,f \text{ or } f} = \mu_{x,w}\, m_{u,f \text{ or } r}\, g \tag{6.28}$$

To calculate $\Delta F_{Bo,f}$ (and also $\Delta F_{Bo,r}$) only the braking forces $F_{B,f}$ and $F_{B,r}$, which occur in the centres of tyre contact and relate to the axle as a whole, need to be divided up into the proportion affecting the wheel suspension $F_{B,u,f}$ and $F_{B,Bo,f}$, which is critical to the body. The same applies to the rear axle. The index r appears in this instance.

6.3.4.3 Numerical values With the values of Section 6.3.3.1 the result for the front axle is then:

$$F_{B,f} = \mu_{x,w} F_{V,f} = 0.8 \times 6.95 = 5.56 \text{ kN}$$
$$F_{B,u,f} = \mu_{x,w} F_{u,f} = 0.8 \times 0.8 = 0.64 \text{ kN}$$
$$F_{B,Bo,f} = F_{B,f} - F_{B,u,f} = 4.92 \text{ kN}$$

and for the rear axle:

$$F_{B,r} = \mu_{x,w} F_{V,r} = 0.8 \times 4.20 = 3.36 \text{ kN}$$
$$F_{B,u,r} = \mu_{x,w} F_{u,r} = 0.8 \times 0.70 = 0.56 \text{ kN}$$
$$F_{B,Bo,r} = F_{B,r} - F_{B,u,r} = 2.8 \text{ kN}$$

The following dimensions should apply to the pitch poles (Figs 6.16 and 6.17):

front $c = 1.0$ m, $e = 0.15$ m
rear $d = 0.5$ m, $g = 0.25$ m so that the result is:

front $$\Delta F_{Bo,f} = \frac{F_{B,Bo,f} e + F_{B,u,f} r_{dyn}}{c}$$ (see 6.26)

$$\Delta F_{Bo,f} = \frac{(4.92 \times 0.15) + (0.644 \times 0.288)}{1.0} = 0.922 \text{ kN} = \Delta F_{V,f,2}$$

$$\Delta F_{Bo,r} = \frac{F_{B,Bo,r} g + F_{B,u,r} r_{dyn}}{d}$$

rear $$\Delta F_{Bo,r} = \frac{(2.8 \times 0.25) + (0.56 \times 0.288)}{0.5} = 1.723 \text{ kN} = \Delta F_{V,r,2}$$ (6.29)

The pitch poles should be as close as possible to the wheel and as high as possible. Because of the more favourable position of the rear poles, $\Delta F_{Bo,r} > \Delta F_{Bo,f}$, i.e. also $\Delta F_{V,r,2} > \Delta F_{V,f,2}$.

6.3.4.4 Spring opposing forces The spring opposing forces $\Delta F_{V,1}$ that result from the axle load transfer $\Delta F_{V,0}$ (see Equation 6.14) and the weight force differences $\Delta F_{V,2}$ with existing pitch poles are critical to the pitch angle desired Φ_B:

front: $\Delta F_{V,f,1} = \Delta F_{V,0} - \Delta F_{V,f,2}$ (6.30)

rear: $\Delta F_{V,r,1} = \Delta F_{V,0} - \Delta F_{V,r,2}$ (6.31)

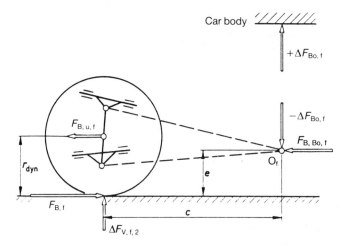

Fig. 6.16 Paths and forces when the body is supported around the front pitch axis. The higher O_f (path e) and the closer to the wheels (path c), the larger the difference in force $\Delta F_{Bo,f}$ supporting the body and the smaller the pitch angle θ_B.

Where $\Delta F_{V,0} = 2.07$ kN (see Section 6.3.3.2) related to the entire axle, the values are:

$\Delta F_{V,f,1} = 1.148$ kN
$\Delta F_{V,r,1} = 0.347$ kN

and thus therefore relating to only one wheel

$F_{z,W,f,1} = 0.574$ kN
$F_{z,W,r,1} = 0.174$ kN.

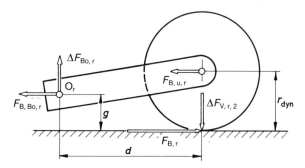

Fig. 6.17 Paths and forces when the body is supported at the rear pitch poles O_r which are relatively close to the wheels. Fig. 3.123 shows the forces, based only one axle side, and Fig. 3.122 shows further kinematic aspects.

6.3.4.5 Pitch angles   At the spring rates $c_f = 11.5$ N mm^{-1} or $C_r = 14$ N mm^{-1} based on one axle side and in accordance with Equation 5.10, the paths needed for determining Φ_B (see also Section 5.4.3) are:

front $s_{1,f} = 0.05$ m and rear $s_{2,r} = 0.012$ m

With the linear springing assumed, the angle decreases. Without a pole it was 3°46' and now in accordance with Equation 6.23 it is:

$$\Theta_B = 57.3 \times \frac{0.05 + 0.012}{2.50} = 1.42° = 1°25'$$

Section 6.3.3.4 contains the calculation of Φ_B with progressive springing.

6.3.4.6 Pitch poles on one axle only   If, for example, only the rear axle has pitch poles $s_{2,r}$ must be determined using $\Delta F_{V,r,1}$, whereas $\Delta_{FV,0}$ alone is critical on the front axle. At the value $\Delta F_{V,0} = 2.07$ kN in Section 6.3.3.2 the compression travel was $s_{1,f} = 0.09$ m.

6.3.5 Anti-dive control and brake reaction support angle

Automobile manufacturers frequently quote the anti-dive control k_ϵ as a percentage in publications. It can easily be calculated on the basis of Fig. 6.15 and 6.16.

front $k_{\epsilon,f} = \Delta F_{V,f,2} / \Delta F_{V,0}$ (6.32)
$k_{\epsilon,f} = 0.922/2.07 = 0.45$
$k_{\epsilon,f} = 45\%$

rear $k_{\epsilon,r} = \Delta F_{V,r,2} / \Delta F_{V,0}$ (6.32a)
$k_{\epsilon,r} = 1.723/2.07 = 0.83$
$k_{\epsilon,r} = 83\%$

The brake reaction support angle ϵ entered in Fig. 3.124 in the examples (Figs 6.16 and 6.17) is:

front: $\tan \epsilon_f = e/c$ (6.33)
$\tan \epsilon_f = 0.15/1.0 = 0.15$; $\epsilon_f = 8°30'$
rear: $\tan \epsilon_r = g/d$ (6.33a)
$\tan \epsilon_r = 0.25/0.5 = 0.5$; $\epsilon_r = 25°32'$

On production passenger cars, ϵ_f is usually below 10° and ϵ_r between 30° and 40°.

6.4 Traction behaviour

6.4.1 Drive-off

The relationships when the vehicle moves off and accelerates are somewhat different to those when the brakes are applied. As shown in Fig. 3.87, the tractive force F_A must be shifted to the centre of the rolling wheel if the differential is fixed to the body or the engine (i.e. separately from the wheel suspension) and the drive-off moment is concentrated in its suspension (Fig. 3.85). This applies on all front independent wheel suspensions and is equivalent to one pitch pole O_f coming into effect. As shown in Fig. 3.119a, in such cases, the squat can be reduced by angling the two double wishbones in the same direction. The same applies to the rear axle in terms of the take-off dive (Fig. 3.124). The diagonal springing angle κ is then positive.

The picture is different when the differential is in the axle housing on a driven rigid axle (Fig. 1.27). The drive pinion connected to the prop shaft is vertical to the axle shaft connected to the wheels (Fig. 1.12), i.e. torque in and output form a 90° angle. The result is that the tractive force F_A, which occurs at the centres of wheel contact, is supported exclusively in the suspension parts of the axle. Where there are pitch poles, the body is pushed upwards into these points O_r and the tail only dives a little, as shown in Fig. 3.123 using the example of the opposed braking forces. The same effect is achieved with trailing link pairs at an angle to one another (Fig. 3.125); there are pitch poles here too.

Figure 6.18 shows the forces generated during acceleration. Those acting on the body are:

- the aerodynamic force F_L, which can be ignored at speeds below 25 km h^{-1}, and the
- excess force $F_{i,x}$, which is equal to the inertia in the x-direction.

The rolling resistance forces (see Section 2.6)

$$F_{R,t} = F_{R,f} + F_{R,r} = k_R (F_{V,f} + F_{B,r}) = k_R F_{V,t} \qquad (6.34)$$

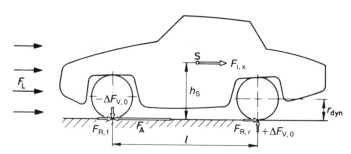

Fig. 6.18 Forces occurring in the vehicle centre of gravity S and at the centres of tyre contact when a front-wheel drive vehicle accelerates.

Fig. 6.19 When standard vehicles are in the direct (mostly fifth) gear, no pair of gears of the manual gearbox is engaged. However, the lower gears require two pairs of gears to transfer the engine moment.

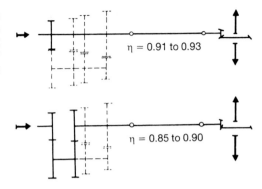

$\eta = 0.91$ to 0.93

$\eta = 0.85$ to 0.90

and the opposed tractive forces F_A act at the wheels. The additional forces necessary to accelerate the turning masses, can be determined using the rotating mass factor.

The equations for calculating the drive-off are:

$$F_{i,x} = F_A - (F_L + F_{R,t})\ [\text{N}] \tag{6.35}$$

$$F_A = \frac{M_{M,max}\ \eta\ i_D\ i_G}{r}\ [\text{N}] \tag{6.36}$$

The following terms are used in the equation:

$M_{M,max}$	the maximum engine torque in N m
η_δ	the total efficiency (Figs 6.19 and 6.20)
i_D	the ratio of the final drive (differentials)
i_G	the ratio of the gear engaged
r	the static rolling radius r_{stat} must be inserted in m at speeds below 25 km h^{-1}, and above 60 km h^{-1} the dynamic rolling radius $r_{dyn} = C_{R,dyn}/2\pi$ ($C_{R,dyn}$ = rolling circumference in m, see Equation 2.1d and 2.2)

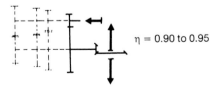

$\eta = 0.90$ to 0.95

Fig. 6.20 If, on a front-wheel drive or rear-engine vehicle, the engine is longitudinal, on a manual gearbox one pair of gears is always engaged to transfer the drive moment, regardless of what gear has been selected and whether the vehicle has a four, five or six-speed box. On transverse engines, the degree of efficiency can be better than $\eta = 0.9$.

A compact front-wheel drive passenger car with a 1.3 l transverse engine can be used as an example. When the acceleration in first gear from around 5 km h^{-1} is observed, the necessary data are:

$M_{M,max} = 94$ N m, $i_D = 3.94$, $i_{G,1} = 3.55$, $\eta_T = 0.90$

tyres 155 R 13 78 S, $r_{stat} = 0.263$ m

$$F_A = \frac{94 \times 0.90 \times 3.94 \times 3.55}{0.263} = 4499 \text{ N} = 4.5 \text{ kN}$$

Because the weight is taken off the front axle as the vehicle moves off (see Equation 6.37), r becomes 10 to 15 mm greater than r_{stat} and the driving force around 5% smaller.

The vehicle has a curb weight of $m_{V,ul} = 875$ kg. With two people each weighing 68 kg in the vehicle the actual weight would be:

$m_{pl} = 1011$ kg and $F_{V,t} = m_{pl} \times g = 9918$ N $= 9.92$ kN

The forces in the longitudinal direction at k_R from Fig. 2.20 are:

$$F_{R,t} = k_R F_{V,t} = 0.012 \times 9.92 = 0.12 \text{ kN}, \; F_L = 0$$
$$\text{and} \quad F_{i,x} = F_A - F_{R,t} = 4.5 - 0.12 = 4.38 \text{ kN}$$

The axle load transfer $\Delta F_{V,0}$ is determined using Equation 6.14; the vehicle data are:

$l = 2.52$ m, $h_{ul} = 1.4$ m and $\mu_{x,W} = 1.05$ (see Fig. 2.21)

The height of the centre of gravity $h_{S,0}$ can be obtained, using Equation 6.5, from the unladen height h_{ul} of the vehicle:

$h_{S,0} \approx 0.38 \times 1.4 \approx 0.532$ m

When there are two people in the vehicle the centre of gravity rises by 10 to 15 mm (see Section 6.1.2.4); therefore $h_{S,2} = 0.546$ m is assumed.

$\Delta F_{V,0} = \mu_{x,W} F_{V,t} (h_{S,2}/l) = 1.05 \times 9.92 \times (0.546/2.52)$
$\Delta F_{V,0} = 2.26$ kN

As shown in Fig. 1.21, when there are two people in the vehicle, approximately 60% of the weight is carried on the front axle:

$F_{V,f} = 0.6 F_{V,t} = 5.95$ kN

Unlike when the brakes are applied, when the vehicle accelerates the weight is taken off the front axle by $\Delta F_{V,0}$:

$$F_{V,f,dyn} = F_{V,f} - \Delta F_{V,0} = 5.95 - 2.26 = 3.69 \text{ kN} \tag{6.37}$$

The coefficient of friction required is then:

$$\mu_{x,W} = F_{i,x}/F_{V,f,dyn} \tag{6.37a}$$
$$\mu_{x,W} = 4.38/3.69 = 1.19$$

When the vehicle accelerates fast from slow speeds, the driven front wheels would spin due to the load alleviation. This disadvantage is particularly evident in the range of maximum engine torque. The coefficient of friction needed $\mu_{x,W} = 1.19$ is too high. Taking into consideration the load alleviated and therefore larger tyre radius r, $\mu_{x,W}$ would drop to around 1.13 but not solve the problem.

On rear wheel drive vehicles, Equation 6.35 is exactly the same. It is simply a matter of shifting force F_A shown in Fig. 6.18 to this axle and adding the axle load shift to $F_{V,r}$. The result would be:

$$F_{V,r,dyn} = F_{V,r} + \Delta F_{V,0} \tag{6.38}$$

$$\Delta F_{V,r,1} = \Delta F_{V,0} - \Delta F_{V,r,2} \tag{6.31}$$

If the driven rigid axle of the vehicle under investigation has pitch poles (Figs 1.27, 3.125 and 1.12), $\Delta F_{V,0}$ and $\Delta F_{V,r,2}$ must first be calculated to obtain $\Delta F_{V,r,1}$ (Equations 6.14 and 6.31).

Equation 6.23 is again used for calculating the pitch angle Φ_A and Equations 6.32a and 6.33 can be used for determining the take-off drive control $k_{\kappa,r}$ and the drive-off reaction support angle κ_r as these values here are of the same size as $k_{\epsilon,r}$ and ϵ_r, i.e. are produced when the brakes are applied (Fig. 3.124). Only in the case of independent wheel suspensions and rigid axles with a separate differential (De Dion axles) does the actual angle κ need to be taken into consideration.

6.4.2 Climbing ability

The climbing ability q is quoted as a percentage and relates to the vertical height h_z reached at the end of a path s_x measured on the horizontal:

$$q = h_z/s_x \, 100(\%) \tag{6.39}$$

$$\tan \alpha = h_z/s_x \tag{6.40}$$

The inclination the vehicle can theoretically climb in first gear (e.g. in the range of the greatest engine torque) can be calculated using the excess force $F_{i,x}$ and the total weight $F_{V,t}$ of the vehicle. In the previous example, in Equation 6.35 only the rolling resistance would be somewhat smaller. The force $F_{R,t}$ must be multiplied by $\cos \alpha \sim 0.9$. $F_{i,x}$ would increase from 4.38 kN

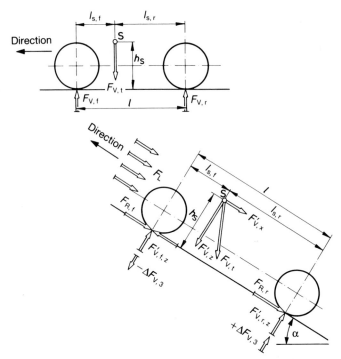

Fig. 6.21 Paths and forces necessary for calculating the skid point, shown on a front-wheel drive vehicle that is travelling up an incline at a constant speed.

to 4.39 kN, a negligibly small difference of only 0.2%. The climbing ability of the example vehicle is:

$$\sin \alpha = F_{i,x}/F_{V,t} = 4.38/9.92 = 0.44$$
$$\sin \alpha = 26.1°, \ \tan \alpha = 0.49 \ \text{and} \ q = 49\%$$

On inclines, an axle load transfer of $\pm \Delta F_{V,3}$ occurs, i.e. a reduction of F_{V},f on the driven front axle. In accordance with Fig. 6.21 it is:

$$\Delta F_{V,3} = F'_{V,x} \, h_S/l = F_{V,t} \sin \alpha \ (h_S/l) \ [\text{kN}] \tag{6.41}$$

and $\mu_{x,w} = F_{i,x}/(F_{V,f} - \Delta F_{V,3})$ \hfill (6.42)

The values of the example inserted will give:

$$\Delta F_{V,3} = 0.92 \times 0.44 \times \frac{0.546}{2.52} \qquad \Delta F_{V,3} = 0.946 \ \text{kN}$$

and $\mu_{x,w} = \dfrac{4.38}{5.952 - 0.946} = 0.87$

To travel a 49% incline evenly the example vehicle only needs a coefficient of friction of $\mu_{x,w} = 0.87$ in the range of the greatest engine torque, a value to be found on dry concrete.

6.4.3 Skid points

More powerful engines would be able to climb theoretically steeper inclines with either front or rear-wheel drive, if the grip of the road surface were to permit it. To have realistic values the skid points should therefore be determined, i.e. the inclination (as a percentage) on the road surface of which the driven wheels do not yet quite slip; $\mu_{x,w} = 0.8$ would be the correct coefficient of friction as an initial value.

Using Fig. 6.21 the equations can be derived that are necessary for calculating $\alpha = f(\mu_{x,w})$. In this the x'-direction is in the climbing plane and the z-direction is vertical to it. Breaking down the total weight force $F_{V,t}$ at the centre of gravity S gives

$$F'_{V,z} = F_{V,t} \cos \alpha \quad \text{and} \quad F'_{V,x} = F_{V,t} \sin \alpha$$

and $F'_{V,z}$ causes the axle loads

$$F'_{V,f,z} = F_{V,f} \alpha \quad \text{and} \quad F'_{V,r,z} = F_{V,r} \cos \alpha$$

As can be seen at the top of the illustration, $F_{V,f}$ or $F_{V,r}$ are the axle loads weighed on the vehicle standing on the flat. The component $F'_{V,x}$, known as the vehicle load downhill, is the same as the excess force $F_{i,x}$ previously calculated. This causes a load reduction on the front axle by $-\Delta F_{V,3}$ (Equation 6.41) and an increase in axle load on the rear axle of $+\Delta F_{V,3}$. The value h_S/l, which appears in the equation, shows that the longer the wheelbase l and the lower the centre of gravity S, the smaller is the axle load transfer (which is unfavourable on front-wheel drive). The condition that the sum of all forces in the x-direction equals 0, would be met if:

$$F_A = F'_{V,x} + F_{R,f} + F_{R,r} + F_L$$

$F_{R,f}$ and $F_{R,r}$ together give:

$$F_{R,t} = k_R F'_{V,z}$$

The solution, based on a driven front axle, in accordance with Equation 6.42 is:

$$\mu_{x,w} = \frac{F'_{V,x} + k_R F'_{V,z} + F_L}{F'_{V,f,z} - \Delta F_{V,3}}$$

$$\mu_{x,w} = \frac{F_{V,t} \sin \alpha + k_R F_{V,t} \cos \alpha + F_L}{F_{V,f} \cos \alpha - F_{V,t} (h_S/l) \sin \alpha}$$

Numerators and denominators divided by $F_{V,t}$ give:

$$\mu_{x,w} = \frac{\sin \alpha + k_R \cos \alpha + F_L/F_{V,t}}{(F_{V,f}/F_{V,t}) \cos \alpha - (h_S/l) \sin \alpha}$$

The speeds achievable on steep inclines do not exceed 25 km h^{-1} so F_L can be ignored. However, on flatter inclines this counter force must be included in the equation. To simplify matters, numerators and denominators are divided by cos α.

$$\text{Front wheel drive} \quad \mu_{x,w} = \frac{\tan \alpha + k_R + F_L/(F_{V,t} \cos \alpha)}{F_{V,f}/F_{V,t} - (h_S/l)(h_S/l) \tan \alpha} \qquad (6.43)$$

The coefficient of friction needed to travel a given incline can be determined using this equation. To obtain the climbing capacity, i.e. tan α, as the result, it is necessary to transform the equation:

$$\text{Front wheel drive} \quad \tan \alpha = \frac{\mu_{x,w}(F_{V,f}/F_{V,t}) - k_R - F_L/(F_{V,t} \cos \alpha)}{1 + \mu_{x,w}(h_S/l)} \qquad (6.44)$$

tan $\alpha \times 100$ = climbing capacity q as a % (see Equations 6.39 and 6.40).

If F_L needs to be considered, α needs to be estimated provisionally for it to be possible to insert cos α. It may be necessary to correct this later in such cases. However, the numerical value is relatively small.

The formula clearly indicates that the higher is the front axle load $F_{V,f}$ and the smaller the value h_S/l, the greater the angle α becomes. The picture is completely reversed on a rear-wheel drive vehicle (Fig. 1.21), in this instance the equation is:

$$\text{Rear wheel drive} \quad \tan \alpha = \frac{\mu_{x,w}(F_{V,r}/F_{V,t}) - k_R - F_L/(F_{V,t} \cos \alpha)}{1 - \mu_{x,w}(h_S/l)} \qquad (6.45)$$

On this type of drive h_S/l and rear axle load should be large. If the coefficient of friction necessary for a given incline is required, the following formula applies:

$$\text{Rear wheel drive} \quad \mu_{x,w} = \frac{\tan \alpha + k_R + F_L/(F_{V,t} \cos \alpha)}{F_{V,r}/F_{V,t} + (h_S/l) \tan \alpha} \qquad (6.46)$$

To produce the diagram of the driving and climbing performance, half the payload in the total weight force should be considered, whilst to determine the skid point the different loading conditions must be assumed. These do not just lead to a change in $F_{V,t}$ but also in the axle load distribution, which is included in the equation as $F_{V,f}/F_{V,t}$ or $F_{V,r}/F_{V,t}$. The three most important loading conditions are (see Section 5.3.6 and Fig. 1.21):

- two people each weighing 68 kg in the front
- four people each weighing 68 kg
- full payload.

The payload F_{ml} (see Section 5.3.3) must be distributed so that, in accordance with Equation 5.1, the permissible rear axle load $F_{V,r,pe}$ is achieved. Therefore, the front axle is usually not fully loaded. The wheelbase l and the changing centre of gravity heights:

$h_{S,2}$, $h_{S,4}$ and $h_{S,pe}$ (see Section 6.1.2.4).

must also be known. Figure 6.22 shows a diagram of the (calculated) skid points on three different coefficients of friction

$\mu_{x,W} = 0.8$ (dry), $\mu_{x,W} = 0.5$ (wet) and $\mu_{x,W} = 0.15$ (ice)

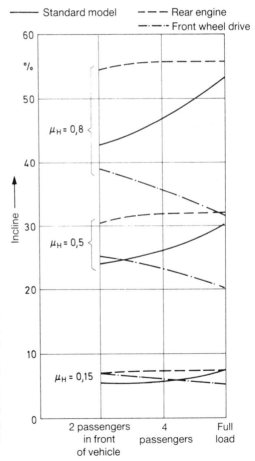

Fig. 6.22 Skid points as a function of three different coefficients of friction $\mu_{x,W} = 0.8$, 0.5 and 0.15 and the loading condition and type of drive. Fully loaded rear-wheel drive vehicles can negotiate the largest inclines, whilst front-wheel drive vehicles have the best climbing capacity at low loads, i.e. with only two passengers in the front and a relatively empty fuel tank, particularly on ice.

as a function of loading condition and type of drive. The mean percentage axle load distribution from Fig. 1.21 and $h_S/l = 0.23$ is considered in this calculation, whilst F_L is ignored. As a series of investigations showed, on standard passenger cars the climbing capacity at $\mu_{x,W} = 0.8$ with two people in the vehicle averages 45%, increasing to around 52% when the vehicle is fully laden. Climbing capacities over 60% quoted by manufacturers are not realistic. The wheels would spin because of the lack of friction (see the calculation in Section 6.4.1).

Publications should therefore not base their values on the (purely theoretical) engine performance, but rather on the climbing capacity as a function of the coefficient of friction $\mu_{x,W} = 0.8$ produced by the road. This applies even more so to a front-wheel drive vehicle. The picture for four-wheel drive vehicles is rather different. Here, the engine torque and the ratio in the manual gear box and differential are the deciding factors along with the higher rolling resistance on uneven roads k_R (see Fig. 2.20 and Equation 6.36). These are:

$$\tan \alpha = \mu_{x,W} - k_R \tag{6.47}$$

In the case of $\mu_{x,W} = 0.8$ and $k_R = 1.5 \times 0.012$ an incline of 38% could be climbed.

Subject index

Cars and their components have a vocabulary all of their own, with several different terms often being used to mean the same thing. With around 2 000 keywords, this index covers the most common terms; cross-references make it easy to find the word you want.

- A figure number (in italics, e.g. *2.6*, *3.23b*) means the keyword will be found in the figure or the caption underneath. Tables are numbered consecutively with figures, and are also italicised.
- An equation number (e.g. e2.6, e3.23b) means it appears as a symbol in the equation concerned, or is explained immediately before or after the equation in the text.

A page number (e.g. 26, 323) means the word is located in the text of that page.

A useful hint when looking for words:

All keywords relating to 'tyres' will be found under that heading, and can be found there (or in the table of contents). Words with multiple meanings, such as joint, motion, force, bearing, load, suspension, and angle are classified by type or design (e.g. supporting ball joint, guide joint). Any relevant keywords will then be found under the base word concerned.

Index of car manufacturers*

Index of suppliers*

* Suppliers to the car industry, as mentioned in the text.